Skates Made of Bone

Skates Made of Bone
A History

B. A. Thurber

McFarland & Company, Inc., Publishers
Jefferson, North Carolina

This book has undergone peer review.

ISBN (print) 978-1-4766-7390-5
ISBN (ebook) 978-1-4766-3737-2

Library of Congress and British Library
cataloguing data are available

Library of Congress Control Number 2020000983

© 2020 B. A. Thurber. All rights reserved

*No part of this book may be reproduced or transmitted in any form
or by any means, electronic or mechanical, including photocopying
or recording, or by any information storage and retrieval system,
without permission in writing from the publisher.*

On the cover: *clockwise from top left* woodcut from Alessandro Guagnini,
"Arma Magni Ducis Moschouiæ," in *Sarmatiae Europeæ descriptio,
quae Regnum Poloniæ, Lituaniam, Samogitiam, Russiam, Masouiam, Prussiam,
Pomeraniam, Liuoniam, & Moschouiæ, Tartariaeque partem complectitur*
(Krakow: Matthia Wirzbietae, 1578), folio 16v; drawing from J. Romilly Allen,
"The Primitive Bone Skate," *The Reliquary and Illustrated Archaeologist* 2 (1896),
32; Ullr on his magic bone, image from *Olaus Magnus, Historia de gentibus
septentrionalibus* (Rome: J. M. de Viottis, 1555), 122; Late Bronze Age horse
metatarsus skates from Ivanovice na Hané, Czech Republic, *Archeologické
rozhledy* 43, no. 1 (2011): fig. 5.1–2, courtesy of David Parma

Printed in the United States of America

*McFarland & Company, Inc., Publishers
Box 611, Jefferson, North Carolina 28640
www.mcfarlandpub.com*

Table of Contents

Acknowledgments   vii
Preface   1

1. Skating Before Skates   5
2. How to Skate on Bones   10
   - 2.1. Sources and Approaches   10
   - 2.2. Selecting Bones for Skating   11
   - 2.3. Making Skates   14
   - 2.4. Attaching the Bones   20
   - 2.5. The Pole   24
   - 2.6. Skating!   25
   - 2.7. How Fast Did They Go?   31
   - 2.8. Wear and Discard   33
3. The Study of Bone Skates   36
   - 3.1. Skaters and Scholars   36
   - 3.2. Identifying Bone Skates in Written Records   39
   - 3.3. Identifying Bone Skates in the Archaeological Record   43
4. How Ice Skating Came to Be   51
   - 4.1. An Origin Story   51
   - 4.2. The Steppes As a Homeland   54
   - 4.3. Skates, Skis and Horses   56
   - 4.4. Skating Across Europe   62
5. Tools or Toys?   66
   - 5.1. The Question of Use   66
   - 5.2. Bone Type   70
   - 5.3. Complexity   74
   - 5.4. A Note on the Earliest Skate Candidates   83
6. Skating and Skiing in Medieval Scandinavian Literature   85
   - 6.1. Skates and Skis   85
   - 6.2. Skaters and Skiers   87
   - 6.3. Skating and Skiing   90
   - 6.4. *Skríða* As a Generic Verb of Motion   93
   - 6.5. The Similarity of Bone Skates and Skis   98

7. Skating on Bones in the Middle Ages — 100
  7.1. The Scandinavian Expansion 100
  7.2. Bone Skates as Scandinavian Artifacts in Great Britain 107
  7.3. Bone Skates on the Continent 114
  7.4. Directions for Future Research 117

8. The End of the Bone Age — 119
  8.1. The Emergence of Metal-Bladed Skates 119
  8.2. The Spread of the New Style 127
  8.3. Where to Go from Here 132

Appendix: Modern Descriptions — 135
  A.1. Germany and Poland 135
  A.2. Central Europe 138
  A.3. Great Britain 141
  A.4. The Northeast 142
  A.5. Scandinavia 142

Chapter Notes — 147

Bibliography — 169

Index — 183

Acknowledgments

This book grew out of my interest in ice skating. I spent a decade competing in figure skating, and as an undergraduate, I was the president of the MIT Figure Skating Club. When I had to write a senior thesis, of course I decided to do it on ice skating. Therefore, my acknowledgments must start with Anne McCants, who advised that thesis, and Harry Merrick and Josh Sosin, my other thesis committee members.

When I was in graduate school at Cornell, I joined the Old Norse reading group. This group eventually began to hold an annual conference, the Fiske Conference on Medieval Icelandic Studies (colloquially known as Norsestock). This provided an outlet for my interest in bone skates and where they appear (and don't appear) in the Icelandic sagas. Section 2.7 and part of section 7.1 are based on papers presented at this conference in 2007 and 2006, respectively.

More recently, I benefited from conversations with students and colleagues at Shimer College. The students who were working on theses always seemed to find it amusing that I was still working on my undergraduate thesis. Particular accolades are due to Colleen McCarroll, Shimer's librarian, who appreciated my "interesting" interlibrary loan requests and was actually able to fulfill many of them, Maria Pondo, who helped with Polish-language resources, and Harold Stone, who helped with some of the Latin. Parts of this book grew out of publications and presentations developed during my time at Shimer. Section 7.2 is based on a paper presented at the International Congress on Medieval Studies in 2016. Chapter 6 is a revised and expanded version of B. A. Thurber, "The similarity of bone skates and skis," *Viking and Medieval Scandinavia* 9 (2013): 199–217, doi:10.1484/J.VMS.1.103882, which was based on a paper presented at the International Congress on Medieval Studies in 2012.

Outside of Shimer but within the Chicago area, I used library resources at the Chicago Public Library, the Evanston Public Library, the Field Museum, the Mitchell Museum of the American Indian, the Newberry Library, Northwestern University, North Park University, and the University of Chicago. The butchers at Publican Quality Meats provided the radius I used in some of my skating experiments. Chris Hyland let me show bone skates to his ice dance class at Robert Crown Community Center in Evanston.

On the internet, I benefited from email exchanges with Alice Choyke, Rune Edberg, Hans Christian Küchelmann, Frits Locher, and William R. Short. The feedback of two anonymous reviewers and conversations with Gary Mitchem at McFarland improved the manuscript immensely. Alice Choyke and Frits Locher helped connect me with people who could provide illustrations, and the latter provided a hard-to-find source. Renie Heije of the Archaeology Department in the Hague, Åslaug Midtdal of the Holmenkollen Ski Museum in Norway, Gabriella Németh, David Parma, Olivier Putelat, Marie Jonasson Schmidt of the Lödöse Museum in Sweden, Beáta Tugya, and Kees Zandvliet of the Amsterdam Museum graciously gave me permission to reproduce images.

Preface

> "As anthropologists for the most part begin their studies with the Stone Age, so may we begin with the Bone Age."—G. Herbert Fowler, *On the Outside Edge* (1897)

People have been ice skating for thousands of years. The first ice skates were not exactly skates, not in the modern sense. They were made from the leg bones of large animals, usually horses and cattle. Their only similarity to modern skates is that they were used on ice. Allen[1] calls bone skates "only a kind of sledge" rather than proper skates, and Balfour[2] suggests calling them "runner-skates" because they lack the edges that enable skaters on metal-bladed skates to push with their feet. Instead, skaters push themselves along with poles, which is why Herteig[3] calls bone skates "en mellomting mellom skøyter og ski" (a cross between skates and skis),[4] which helps explain why Hall[5] suggests that they should perhaps be called "ski-skates" instead.

The most likely inventors of bone skates are pastoral nomads living in the Eurasian steppes during the Neolithic or the Bronze Age. These are the people who domesticated horses and spoke Proto-Indo-European. They had the opportunity to learn about wooden skis during their travels and may have adapted skiing to the resources available to them in their homeland, namely, the bones of the horses they rode. Like skis, skates could have been used for travel, hunting, and recreation. When people from the steppes migrated to Europe, they brought bone skates with them. Skating eventually became a popular way to have fun and flourished in regions with suitable climates during the Middle Ages.

In addition to being fun to use, bone skates are useful in the study of settlement, migration, and cultural interaction. This book shows how they fit into patterns of human migration and interactions between cultures. Because most of the evidence for bone skates is from medieval sites in northern Europe, much of this book is about speakers of Germanic languages, especially Scandinavians. They were neither the first nor the only skaters, but they were enthusiastic enough about skating to take bone skates with when they moved to new areas and to mention skating in their literature. Bone skates were so popular among these groups that Robert Munro[6] attributed their invention to "the early Teutonic races who inhabited the shores of the Baltic."

Munro's paper appeared at the end of the nineteenth century, an important time in the study of bone skates. George Herbert Fowler explains what was known about bone skates at this time in his little book on the history of ice skating, *On the Outside Edge*. This includes the fact that Fowler and other scholars were still living in the Bone Age: bone skates remained in use in some rural areas of Europe until well into the twentieth century.[7] As Fowler was writing his book, other antiquaries were sharing stories of skating on bones in their youth, and people in rural areas were skating on bones. Bone skates remained in use for over 800 years after the invention of metal-bladed skates no later than about 1225 CE.

Research on bone skates has come a long way since Fowler's time. In the last century, thousands of bone skates have been found at sites all over northern and central Europe. Arthur MacGregor, one of the heroes of bone skate studies, has pointed out that it is time to take bone skates seriously:

> Between their initial introduction and their final displacement by steel skates (a process which began on the continent in the 14th or 15th century but which was, on the evidence cited above, only completed within living memory), a great deal is known about them from historical and archaeological as well as ethnographic sources. Far from being shrouded in obscurity, therefore, bone skates may be counted among our better-known antiquities and should henceforth be treated with the certitude merited by several millennia of widespread use.[8]

While strides have been made in the forty-some years since MacGregor's article, there is still a need for a large-scale study that puts bone skates into a broad context. To date, the literature on bone skates mainly consists of articles by archaeologists describing how bone skates were made and used. These articles are generally broad surveys of the evidence or detailed analyses of sets of artifacts from a certain geographic area.[9] Reports of excavations often mention skate finds in the context of particular sites, but do not discuss these finds or their implications in detail. As a result, there is a great deal of information about bone skates available and ripe for analysis. Much of it has been brought together in the database assembled by Hans Christian Küchelmann and Petar Zidarov as they worked on their bone skates project and kept updated since then.[10]

This book fits bone skates into the broader narrative of European history and shows what they have the potential to contribute. They provide clues to local climate conditions and information about interactions among groups of people. Mentions of skates in literature and ethnographic accounts of their use provide insight into how people thought about skates and skating and how skating affected their lives. The information about the past that bone skates can yield has not yet been fully mined. This book provides a glimpse of how much potential they have.

One thing that quickly emerges from this analysis is that skating on bones was fun. In fact, many of the medieval and modern descriptions of skating on bones focus on how much fun it was. This is supported by the archaeological evidence: Edberg and Karlsson[11] argue that the hundreds of bone skates found at Birka and Sigtuna, two medieval sites in Sweden, were preferentially used by children for recreation. Because skating is so much fun, bone skates have made their way into the cultural imagination. They are often mentioned in popular books about the Vikings, and they have been featured in books about figure skating since 1772, when Robert Jones opened his *Treatise on Skating* with William fitz Stephen's description of young men on bone skates in twelfth-century London.[12] More recently, bone skates have appeared in popular fiction. In Middle-earth, the Lossoth "can run on the ice with bones on their feet."[13] In *Bracelet of Bones*, the main character, Solveig, makes a pair of bone skates to sell in Kiev and is disappointed when someone who lives in the region points out that skates fetch better prices in the north than the south.[14] In *How to Steal a Dragon's Sword*, the ninth book in the *How to Train Your Dragon* series, the characters "could race down [icy tunnels] at quite a speed on their bone skates, shooting through tunnels that looped in and out of each other like a web spun by a spider that had gone crazy."[15] Bone skates are an important element of the plot of Joan Lennon's *Ice Road*, book three of *The Wickit Chronicles*, which revolves around traveling across a frozen marsh.[16]

It seems like everybody knows something about bone skates, but no book has been devoted to them until now. The present volume is the first book-length study of bone skates and their place in the history and culture of Europe. Like Fowler's book, it makes "a contribution" to the "*Geschichte des Schlittschuhlaufens: Ein Beitrag zur Kenntnis der höheren Psychophysik*" (*History of Ice Skating: A Contribution to the Understanding of Higher Psychophysics*), which, over a century ago, Fowler thought would be published soon by one Schwingenbein Schlangenbogen,[17] distinguished Professor of Useless Arts at the University of Whoknowswhere.

The primary purpose of this book is to describe the spread of bone skates through Europe from their invention alongside skis in the Eurasian steppes until they meet skis again in southern Scandinavia. It does not address the question of whether bone skates were used in North America and is limited to skates made from the leg bones of large animals; bone sled runners and skates made from ribs and jawbones are not discussed in detail, nor are metal-bladed skates, though they become important in the final chapter. Bone skates probably first appeared close to early skis, but the two technologies spread in different directions—skates directly east and skis northeast. This book begins its journey with the invention of bone skates and skis in the steppes, follows skates through Europe until they meet skis again in Scandinavia, and ends with what happened afterward.

The following chapters address various questions about bone skates. How were they made and used? What purpose did they serve? When and where were they invented? How did they spread across Europe? Not all the answers are readily available, but all these questions are addressed. The first chapter summarizes the history of bone skates and situates them in the context of how skate technology developed over time. It also summarizes some of the most important literature on bone skates.

Chapter 2 provides a practical guide to making and using bone skates, from selecting suitable bones to gliding across ice. It is based on evidence from archaeological, ethnographic, and experimental sources, including my own experience making and using bone skates. Historical sources and a mathematical model are used to estimate how fast skaters were able to go.

Chapter 3 describes the conversation about bone skates that began in the middle of the nineteenth century. The scholars who studied bone skates were interested in the North in general, and ice skating—on metal blades or bones—went with the territory. However, the study of such subjects was in its infancy, and many mistakes were made. In translating Old Norse texts, they saw skating where there was none or left themselves open to misinterpretation by not drawing a clear distinction between skating and skiing. To add to the confusion, while some scholars recalled skating on bones, others denied the very existence of bone skates, instead interpreting them as textile smoothers or other tools. Such difficulties have continued to the present, and current questions about how to identify bone skates in the archaeological record and whether particular artifacts are skates are discussed.

Chapter 4 presents the hypothesis that bone skates were invented in the Eurasian steppes between 3300 and 1500 BCE. Their most likely inventors are pastoral nomads who rode horses, drove wagons, and spoke Proto-Indo-European, the ancestor of many modern European languages, including English. There is reason to suppose that these people encountered skiers in the Altai Mountains, which means that bone skates could have been inspired by skis. I propose that bone skates were skis adapted to the conditions and resources of the steppes. Since the evidence currently available is not sufficient to propose this as anything other than a hypothesis, I describe types of evidence that would support or refute it to provide avenues for future research.

Chapter 5 addresses the question of use. Were bone skates primarily tools or toys? Where and when were they used for each purpose? Edberg and Karlsson[18] argue that bone skates were primarily toys for children and teenagers in medieval Scandinavia. This chapter evaluates their argument and extends it to other places and times based on the archaeological and literary evidence. The end result is that bone skates were probably most commonly used by young people for recreation from the Late Bronze Age onward but also had some practical uses.

Chapter 6 digs into the evidence from medieval Scandinavian literature. Although direct references to bone skates are rare, the ability to skate was clearly worth boasting about. I identify oblique and direct references to bone skates and argue that Scandinavian authors saw bone skates as small skis for use on ice. In Old Norse literature, the generic verb *skríða* (to slide) is used to refer to skiing and, probably, to skating.

Chapter 7 is based on a statistical analysis of nearly three thousand medieval bone skates. The medieval enthusiasm for skating is reflected in the archaeological record by the appearance of bone skates at many sites along the coasts of the North Sea and the Baltic Sea during. These areas were both raided and settled by Scandinavians. If it becomes possible to determine whether a given skate belonged to a Scandinavian or someone already living there, then, it will be possible to mine bone skates for information on the Scandinavian expansion. This chapter takes the first steps toward doing this. Great Britain is used as a case study.

Chapter 8 describes what happened after the Middle Ages. Today, metal-bladed skates have taken the place of bone skates. The transition from bone to metal seems to have been a long, slow process that may have begun as early as about 1225 CE, the date of the oldest metal-bladed skates found so far, and was not complete until at least 1972. This chapter attributes the end of the Bone Age of ice skating to gradual cultural change rather than any one specific cause. But bone skates never completely disappeared. They endure to this day through the work of experimentalists and re-enactors who want to understand how people lived in the Middle Ages.

Finally, the Appendix lists the modern (mainly nineteenth-century) descriptions of bone skates I referred to while putting together this book. They range from brief mentions of sightings to detailed explanations of how bone skates were used in the author's childhood. These descriptions provide important information about the modern use of bone skates and show that they were not abandoned when metal-bladed skates came long. Instead, bone skates enjoyed a long period of use even after an alternative became available. People did not stop skating on bones until sometime in the last 50 years. The most recent description notes that people were still skating on bones in 1972.

1

Skating Before Skates

One of the earliest extant descriptions of ice skating is in William fitz Stephen's twelfth-century *Description of London*. This description is probably familiar to readers of skating books; it is very popular because of its vividness and clarity. It paints a familiar picture of people having fun playing outside in winter with a casual disregard for safety. The type of fun they are having is less familiar to modern eyes. It sounds a bit like skating in that it involves special footwear and occurs on ice, but the skates and the technique are very different from those known today. It is as follows:

> Cum est congelata palus illa magna, quæ mœnia urbis aquilonalia alluit, exeunt lusum super glaciem densæ juvenum turmæ.... Sunt alii super glaciem ludere doctiores, singuli pedibus suis aptantes et sub talaribus suis alligantes ossa, tibias scilicet animalium, et palos ferro acuto supposito tenentes in manibus; quos cum aliquando glaciei illidunt, tanta rapacitate feruntur, quanta avis volans, vel pilum balistæ. Interdum autem magna procul distantia, ex condicto duo aliqui ita ab oppositis veniunt; concurritur, palos erigunt, se invicem percutiunt; vel alter vel ambo cadunt, non sine læsione corporali; cum post casum etiam vi motus feruntur ab invicem procul, et qua parte glacies caput excipit, totum radit, totum decorticat. Plerumque tibia cadentis vel brachium, si super illud ceciderit, confringitur; sed ætas avida gloriæ, juventus cupida victoriæ, ut in veris præliis fortius se habeat, ita in simulatis exercetur.[1]

> (When the great marsh that washes the north wall of the city is frozen over, swarms of young men issue forth to play games on the ice.... Others, more skilled at winter sports, put on their feet the shin-bones of animals, binding them firmly round their ankles, and, holding poles shod with iron in their hands, which they strike from time to time against the ice, they are propelled swift as a bird in flight or a bolt shot from an engine of war. Sometimes, by mutual consent, two of them run against each other in this way from a great distance, and, lifting their poles, each tilts against the other. Either one or both fall, not without some bodily injury, for, as they fall, they are carried along a great way beyond each other by the impetus of their run, and wherever the ice comes in contact with their heads, it scrapes off the skin utterly. Often a leg or an arm is broken, if the victim falls with it underneath him; but theirs is an age greedy of glory, youth yearns for victory, and exercises itself in mock combats in order to carry itself more bravely in real battles.[2])

The skates these skilled young men used are bone skates. These are the skates that were available before modern skates were invented, and the people who used them did not know how to skate in the modern sense. In their most basic form, these skates were bones from large ungulates, usually horses and cattle, like the one in the drawing shown in figure 1. Skaters simply stood on them while pushing themselves across the ice using a pole with a sharp metal point. Because they never lifted their feet from the ice, they did not have to attach the skates to their feet, even though William fitz Stephen's skaters chose to do so. It is questionable whether this can truly be called skating because it is so different from the modern hands-free technique, but it is gliding across ice, which is the basic essence

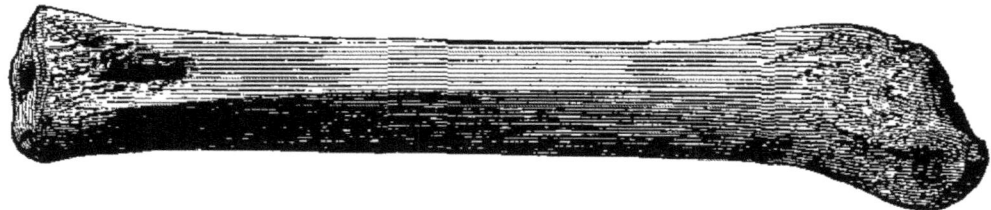

Figure 1: A simple bone skate. Note that it is upside down—the side in contact with the ice faces upward to show the wear on its surface (Robert Munro, "Notes on Ancient Bone Skates," 186).

of ice skating. These skates may have been the ancestors of modern skates or at least have provided some of the inspiration for them.

The ancestors of bone skates may have included skis; they may even have been adaptations of skis, which appeared in the Altai Mountains long before the earliest evidence for bone skates that has been collected so far, to the climate and resources of the Eurasian steppes. The oldest bone skates found to date are Bronze Age specimens from central Europe. They were generally horse bones (usually metapodia, but sometimes radii) with slight modifications. One surface was flattened to make gliding easier. Sometimes holes were drilled in them to attach them to the skater's feet or to make them easier to store and carry between skating sessions. Over the centuries, the basic bone skate did not change much, though cattle bones became more frequent.

These skates were generally shorter than skates made from the corresponding horse bones, which made them the right size for children. Edberg and Karlsson[3] argue that such small skates were probably not meant for practical adult uses at all. Instead, they were primarily toys for children. These toys seem to have been quite popular during the Middle Ages. Thousands of medieval bone skates have been found by archaeologists. A few sites, mainly in Sweden and Norway, have yielded hundreds of bone skates. Medieval skates have also been found at many sites scattered across Europe, including places where Scandinavians raided and settled during the Viking Age. Medieval Scandinavians were such enthusiastic skaters that they mentioned bone skates in their literature. They were not the only enthusiastic medieval skaters, but they were the only ones who wrote about skating.

Numerous observations and descriptions show that bone skates remained in use in Europe until well into the twentieth century. The people who wrote them include archaeologists and anthropologists both reminiscing about their own childhoods and observing people skating in rural areas. The most recent record of bone skates in use outside of reenactment contexts that I have found is the description of a pair of skates seen by Ulf Erik Hagberg in a museum in Iceland that were reportedly still in use in 1972.[4] Among people interested in the past, bone skates have been used much more recently than that. In addition to being the subject of experiments by archaeologists, they are a favorite of reenactment groups and medieval studies classes.[5] Races on bone skates have been held in Ezinge, the Netherlands, as recently as 1996.[6] Edberg and Karlsson[7] provide an account of a skating event held at the Lödöse Museum in Sweden in 2015, and Marie Jonasson Schmidt[8] has confirmed that this museum has offered ice skating on bone skates during the holiday program every February since 2016. Each year, 30–40 families have participated, and some of the children have become quite good skaters. Finally, I demonstrated the use of bone skates to a skating class at Robert Crown Center in Evanston, Illinois, in October 2017, and at the World Figure Championship in Vail, Colorado, in September 2018. Less formally, I showed off my bone skates on public skating sessions between 2016 and 2018. It is safe to say that bone skates remain in use to this day, though now the fun of skating on them is associated with the thrill of living history.

Occasionally, the claim that bone skates were used in North America is made. I have seen a few supposedly Native American bone skates for sale on eBay and other auction sites, and this claim sometimes comes up in books, especially popular histories of hockey and ice skating. For example, in her description of the Kiowa, Marriott[9] writes, "The children slid on the ice, or skated on it with skates made of buffalo rib bones, in the winter." Marriott does not provide a date for this, but her book is about the period between the arrival of the Spaniards and the twentieth century. However, skating is not mentioned in Culin's *Games of the North American Indians*, a comprehensive catalog of games played by Native Americans. The closest game Culin[10] does describe is shinny, which bears a certain resemblance to hockey and was sometimes played on ice. Bone skates would have been quite unsuitable because it is extremely difficult to stop or turn on them. Scandinavian immigrants to North America, who arrived by the thousands in the nineteenth and early twentieth centuries, are other possible users of bone skates. Unfortunately, despite searches at the Newberry Library and North Park University, I have been unable to find any evidence that they used bone skates. The only concrete evidence I have been able to uncover for the use of bone skates in North America is a set of three skates. These skates, made from bone and ivory and found among Canada's Arctic islands, are attributed to the Dorset culture and date to between 500 and 1000 CE.[11] If bone skates were ever used in North America, they seem not to have been widespread.

This book is about ice skates made from the leg bones of large animals. Skates were also made from other bones: occasionally ribs were used,[12] and there is some evidence for skates made from lower jaw bones with a wooden platform attached. Sleds for small children were also made from the lower jaws of large animals, and sleds for adults sometimes had bone runners. These runners can be difficult to distinguish from bone skates because they were made from the same bones and often exhibit the same wear patterns. Figure 2 shows some examples of the types of bone runner that are not covered in this book.

Wooden skates are also mostly beyond the scope of this book. Wooden skis are similar to bone skates and may have been related to them, but making working hands-free skates entirely out of wood would have been very difficult because it is hard for the edge of a piece of wood to grip the ice as a metal edge does. Wooden skates from Transylvania are described by Balfour[13] as "little, if at all, superior to those of bone," and such skates seem not to have been widespread. In contrast, metal-bladed skates with wooden platforms to attach to the skater's shoes were very popular for centuries; the very first metal-bladed skates, discussed in Chapter 8, were of this type. Such "wooden" skates were used by Hans Brinker, who was too poor to afford proper metal skates, in the popular children's story.[14] References to wooden skates are normally either to skates of this type or to skis. From the late seventeenth century until the nineteenth century, the word "skate" was used to refer to both skates and skis.[15] Percy[16] translates the Old Norse word *skiðum* (skis) as "skates of wood" in his popular translation of an Old Norse poem. This means that care must be taken to figure out whether older texts actually refer to skates or skis when the difference is important.

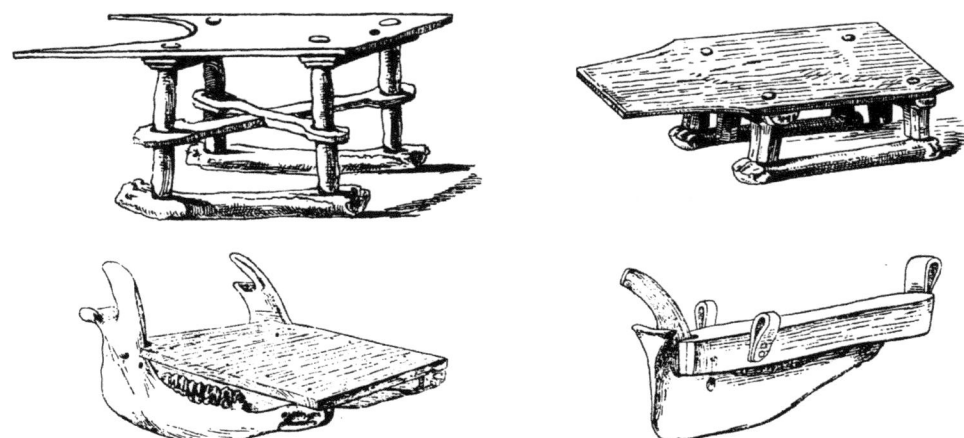

Figure 2: Other bone sliding technologies. Top row: Bavarian (left) and Bosnian (right) sleds with bone runners. Bottom row: A sled and a skate from Pomerania. The sled (left) is a cattle mandible with a wooden platform attached, and the skate is made from a sheep's jawbone mounted on wood with leather loops for fastening (upper row, left: F. von Luschan, "Mitteilungen aus dem Museum der Gesellschaft," 327, fig. 13; right: Henry Balfour, "Sledges with Bone Runners in Modern Use," 249, fig. 7a; lower row: Rudolf Virchow, "Einige Ueberlebsel in pommerschen Gebräuchen," 362, fig. 1–2).

Bone skates are between skis and metal-bladed skates functionally and perhaps also historically. They are like skis in that they are used with poles and like skates in that they are used on ice. The earliest known metal-bladed skates, which date to the first half of the thirteenth century, consist of a wooden platform for the foot with a band of metal along the bottom to glide on. This band looks a bit like a bone skate, and the first metal-bladed skates may have been used with poles, just like bone skates. The revolution in skating came with the discovery that metal blades could bite into the ice, which allowed people to push with their feet and do away with the poles. Metal-bladed skates also made stopping and turning much easier.

The discovery of foot-pushing ushered in a new era of ice skating, and over the next few centuries, metal-bladed skates evolved into their current form. After the Middle Ages, they gradually replaced bone skates, and by the end of the nineteenth century, bone skates were only used in rural areas. This led to some confusion about whether bone skates were actually skates, with some nineteenth-century ethnographers remembering using bone skates in their youth and others arguing that bone skates could not possibly have been skates. This confusion was because not everyone understood how skating on bones worked. Arthur MacGregor[17] seemingly settled the debate by making a pair of bone skates, skating on them, and comparing the resulting wear patterns with those found on archaeological specimens. The similarity made it highly likely that the latter were also skates.

Thousands of bone skates have been dutifully excavated and cataloged over the last century and a half. The best currently available tool for studying bone skates is the database assembled by Hans Christian Küchelmann and Petar Zidarov while they were working on their bone skates project.[18] They have kept the database updated, and as of May 2018, it includes over 2,600 bone skates. The majority of skates in the database date to the Middle Ages. Medieval Scandinavians seem to have left bone skates behind in Denmark, Sweden, Germany, and Russia. Many Anglo-Scandinavian skates have been found in Great Britain, most notably in York; details of these are found in MacGregor's review.[19] The database also includes many skates from other parts of Europe and other times, from the Bronze Age to the twentieth century.

This database represents an important and exciting step in the study of bone skates because it enables large-scale statistical analyses of these artifacts. Databases are becoming more common tools in archaeology; Martin's[20] study of Anglo-Saxon cruciform brooches, is an example of a similar analysis. I rely heavily on this database for the analysis presented in this book, especially the statistical work on medieval skates in Chapter 7. The analysis also includes numerous bone skates that have not yet made their way into the database. The most important additional dataset is the one assembled by Edberg and Karlsson,[21] which includes 769 skates from two sites in Sweden: 290 from Birka and 389 from Sigtuna. The 290 skates from Birka date to between the latter part of the eighth century and the end of the tenth century. The 389 skates from Sigtuna begin where the Birka skates leave off and go through the latter part of the thirteenth century. This report is significant because it is the largest and most detailed study of bone skates conducted to date. The authors have enough data to apply statistical methods, which they use to show that bone skates were toys for children and adolescents whose feet had not yet reached their adult size.

This collection of skates builds on over a century of scholarship. The foundations were laid by Robert Munro[22] and Otto Herman,[23] whose papers remain useful resources to this day. Gösta Berg contributed thoughtful analyses of bone skates, skis, and other sliding technologies used in Sweden in three different languages over a period of nearly thirty years.[24] Shortly after Berg's last paper, Arthur MacGregor brought archaeology and ethnography together with two papers on bone skates[25] that summarize the evidence from Europe and Asia and provide detail of the skates from Great Britain. His book[26] puts bone skates into context in the broader field of worked bone and antler.

Following MacGregor, many others, including Cornelia Becker[27] and Alice Coyke with various coauthors,[28] have studied bone skates at individual sites and worked on putting them into regional contexts. Becker[29] focuses on skates from Berlin-Spandau, and Choyke focuses on skates from central Europe. These and other studies of bone skates on local and regional levels provide the data for Küchelmann's database and the basis for this book. The other main resources I used in writing this book are medieval and modern descriptions of bone skates and their use, including the modern descriptions quoted in the Appendix.

This book brings all these resources together to show how bone skates can be used in analyses of human migrations, interactions between cultures, and climate history. They are much more than a footnote to the history of ice skating. Thousands have been found, but no study has brought them all together and placed them in a broader context until now. The following chapters paint a picture of the Bone Age of ice skating and suggest ways bone skates can contribute to the understanding of the past.

2

How to Skate on Bones

2.1. Sources and Approaches

This chapter describes how bone skates were made and used in the sequence skaters and skate-makers followed, from finding suitable bones to actually skating. This is a general overview of the entire process; more specific details that pertain to the arguments of later chapters are presented in those chapters. The information presented in this chapter comes from sources of three main types:

1. Archaeological sources: This category includes both skates that have been found by archaeologists and analyses of how skates were made. Two prominent examples are Arthur MacGregor's *Bone, Antler, Ivory and Horn: The Technology of Skeletal Materials since the Roman Period*, which describes the making and use of bone skates alongside similar analyses of numerous other types of bone artifact, and "Skating with Horses: Continuity and Parallelism in Prehistoric Hungary" by Alice M. Choyke and László Bartosiewicz, which describes early bone skates.
2. First-hand accounts: This category consists of medieval and modern descriptions of how bone skates were made and used by people other than modern experimenters. The main medieval sources are Olaus Magnus's *Historia de gentibus septentrionalibus (Description of the Northern Peoples)*, published in 1555, and William fitz Stephen's twelfth-century *Descriptio nobilissime ciuitatis Londonie (Description of the Most Noble City of London)*. The most helpful modern sources are the reminiscences of antiquaries preserved in the proceedings of their meetings, such as those published in the *Verhandlungen der Berliner Gesellschaft für Anthropologie, Ethnologie und Urgeschichte (Proceedings of the Berlin Society for Anthropology, Ethnology and Prehistory)*.
3. Reports of experiments: This category comprises modern reconstructions of bone skates. Numerous experiments with bone skates have been conducted by archaeologists, most notably Arthur MacGregor,[1] Hans Christian Küchelmann and Petar Zidarov,[2] and Ulf Erik Hagberg,[3] authors of popular books on medieval Scandinavians, such as William Short[4] and Magnus Magnusson,[5] and others, including G. Herbert Fowler,[6] who describes experiments with bone skates in his book on the history of figure skating, and Federico Formenti and Alberto E. Minetti,[7] who study the biomechanics of skating. The results of my own experiments with bone skates, conducted between August 2015, and February 2017, are included in this category.

This chapter is organized according to the process of creating and using skates. The first steps were selecting and shaping suitable bones. Shaping may or may not have included adding a system for attaching the bones to the skater's feet. Once on skates, a skater needed a pole or two to push with. Then, it was time to skate. The discussion of skating includes material on biomechanics based on experiments conducted by Formenti and Minetti and my own mathematical model. Finally, skaters discarded their broken or worn-out skates to prepare a new pair.

2.2. Selecting Bones for Skating

The things to look for in bones to be used as skates are flatness (the most important criterion) and length (the ideal skate is approximately as long as the skater's foot). The bones should be as flat as possible, especially on the dorsal side (the front of the bone when it is attached to the animal, which was usually in contact with the ice surface). Flatter bones ease the preparation process because they are already close to the right shape. It was preferable for the two skates in a pair to be similar in size and shape; pairs of similar skates have been found at Sigtuna in Sweden,[8] a site near Budapest[9] and Gyoma 133[10] in Hungary, and Baumgarten an der March, Austria,[11] and in the Czech Republic.[12]

The bones that best satisfy these criteria are the metapodia (metacarpi and metatarsi) of large ungulates. These are hand- and foot-bones in humans, but in these animals, they are the long bones in the lower legs, right above the hoof (see figure 3). They are sometimes called cannon bones or shin bones. Metapodia are a good choice for skates because they are very strong due to their high density and thick walls.[13] Horse bones were the most popular, with cattle bones as a common second choice. Metapodia of other animals, including deer and donkeys, have also been used for skates.[14] Radii, the next bones up in the front legs, were used less frequently, and sometimes ribs were used.[15]

In the 1970s, Ulf Erik Hagberg[16] and a group of students tested both horse and cattle bones and found that the former generally worked better because of their shape and size. Because horse metapodia comprise one bone rather than the two fused bones of cloven-hoofed mammals such as cattle, they are easy to shape into skates. Cattle metacarpi are naturally concave on the dorsal side, which makes them more difficult to shape into skates. Figures 4 and 5 show a pair of medieval skates from Lödöse, Sweden, made from horse metapodia from two different perspectives.

Bone skates seem to have been made by individual skaters rather than by professional skate-makers and were probably neither bought nor sold.[17] Therefore, when they started a new pair of skates, skaters would have selected the best bones from those available to them,[18] which may have required them to resort to second-choice bones at times. Although horse bones were the best, they may not have been as readily available as cattle bones. The availability of bones in the past, as it is today, was probably influenced by the local diet. But hippophagy is not a prerequisite for making skates from horse bones; the Sarmatians, who used bone skates made from horse metapodia during the Roman Iron Age and Migration Period, did not eat horse meat.[19] The most important factor is the availability of horses. Edberg and Karlsson[20] suggest the high percentage of horse bone skates found at Sigtuna in comparison to Birka was because horse bones were easier to obtain there.

The best modern description of the process of obtaining bones for skating is that of

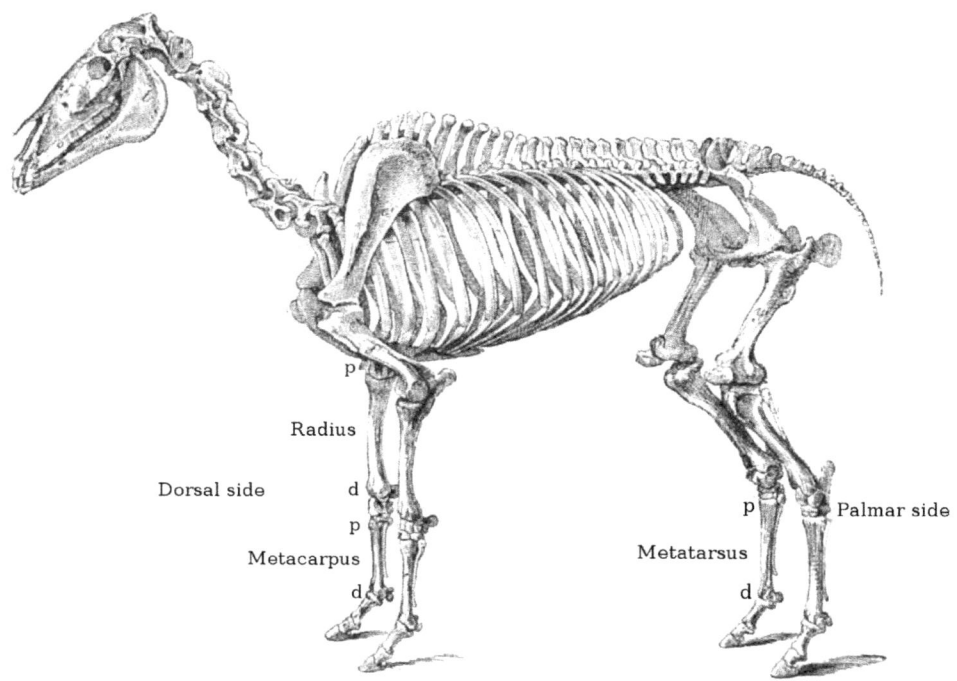

Figure 3: A horse skeleton showing the locations and orientations of the bones most commonly used for skates; "p" and "d" denote the proximal and distal ends of the bones, respectively (I. Vaughan, *Strangeways' Veterinary Anatomy*, 18, modifications by B. A. Thurber).

Brückner.[21] During the nineteenth century, he and his friends looked for horse bones that were between 10 and 13 inches long. Because Brückner describes removing skin and flesh, it seems likely that he and his friends removed the bones from the bodies of dead animals themselves. One of the Roman Iron Age skates from Gyoma 133, a site in Hungary, may have been made from aged bone.[22] This suggests that perhaps young people also scouted for bones in prehistory. Few others have described this part of the process. Even modern experimenters are mostly silent on the subject. Of the accounts that do exist, Küchelmann and Zidarov[23] were allowed to use the lions' dinner leftovers at the Sofia Zoo, and Marie Jonasson Schmidt[24] used cattle metacarpi purchased from a butcher to prepare skates for the Lödöse Museum's skating parties.

    I made my own skates from Barkworthies smoky beef shins (cattle metacarpi sold as chew toys for dogs) purchased at a pet store. One of my metacarpus skates (the one I use with my right foot) is slightly larger than the other. It is approximately 23.5 cm long and 6.5 cm wide at the front (distal) end, 3.3 cm wide in the middle, and 6.6 cm wide at the back. The smaller skate is approximately 23.4 cm long, 5.7 cm wide at the front, 3.1 cm wide in the middle, and 5.8 cm wide at the back. They are shorter than the hiking boots I used to skate in (which are approximately 27.5 cm long), but they work. I also made a single skate from a cattle radius obtained from a local butcher shop. At 28.9 cm, my radius skate is significantly longer than my metacarpus skates. Unless otherwise specified, the following discussion is based on my metacarpus skates, shown in figure 6.

2.2. Selecting Bones for Skating

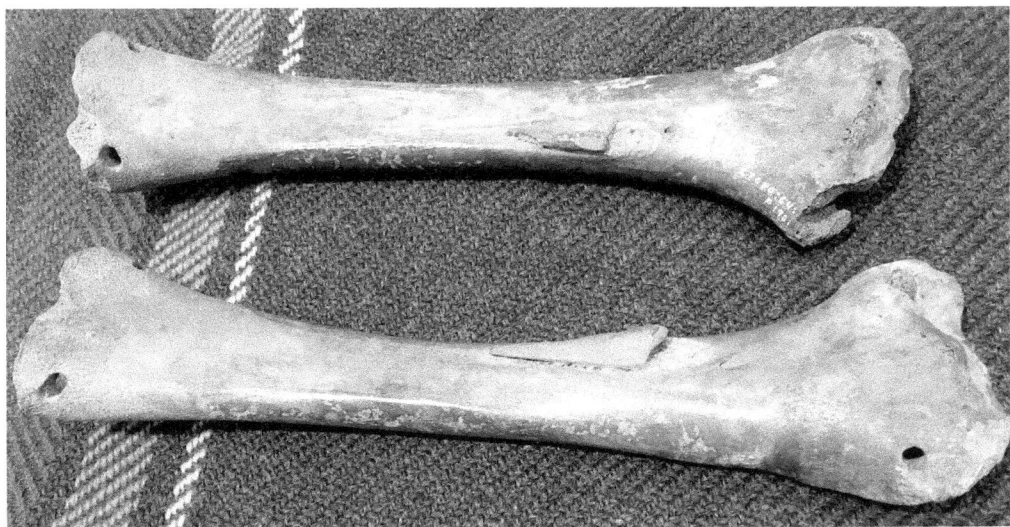

Figure 4: A pair of medieval skates from Lödöse, Sweden, made from horse metapodia with transverse holes for attachment at the ends. The side the skater stood on is up (courtesy Lödöse Museum, Sweden).

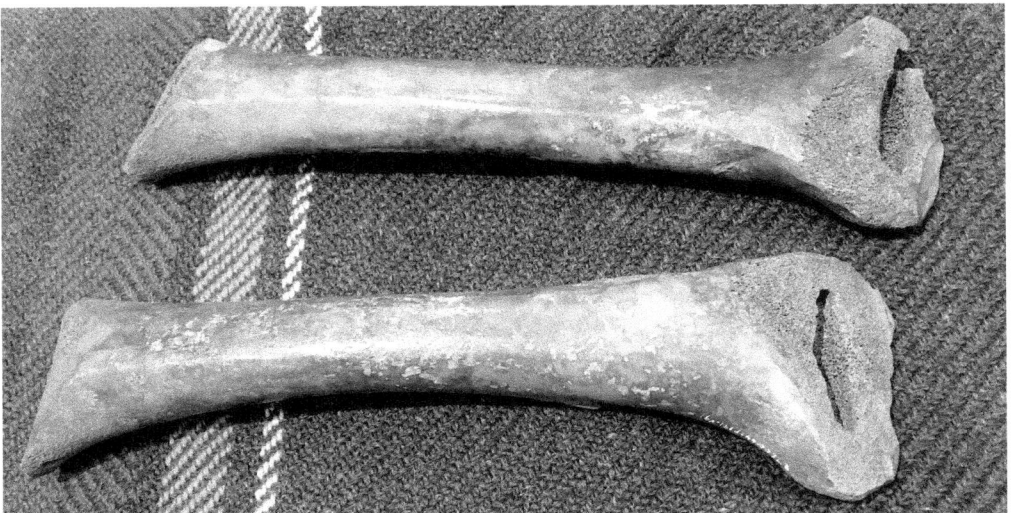

Figure 5: Another view of the medieval horse metapodium skates from Lödöse, Sweden. The side that was in contact with the ice is up (courtesy Lödöse Museum, Sweden).

The metacarpi I used for my skates are slightly larger than the cattle metacarpi commonly used in the Middle Ages because cattle are generally larger now. However, they are within the size range of the metapodia used for medieval skates. On a given animal, the metacarpi are shorter than the metatarsi, and the radii are usually longer than either. Species are not directly comparable; for example, a horse metacarpus is likely to be longer than a cattle metacarpus, but a cattle metatarsus could be longer than a horse metacarpus, depending on the particular animals. Edberg and Karlsson[25] cite measurements of 1655 Swedish metapodia that date to between 0 and 1300 CE. The cattle metacarpi are typically

14   2. How to Skate on Bones

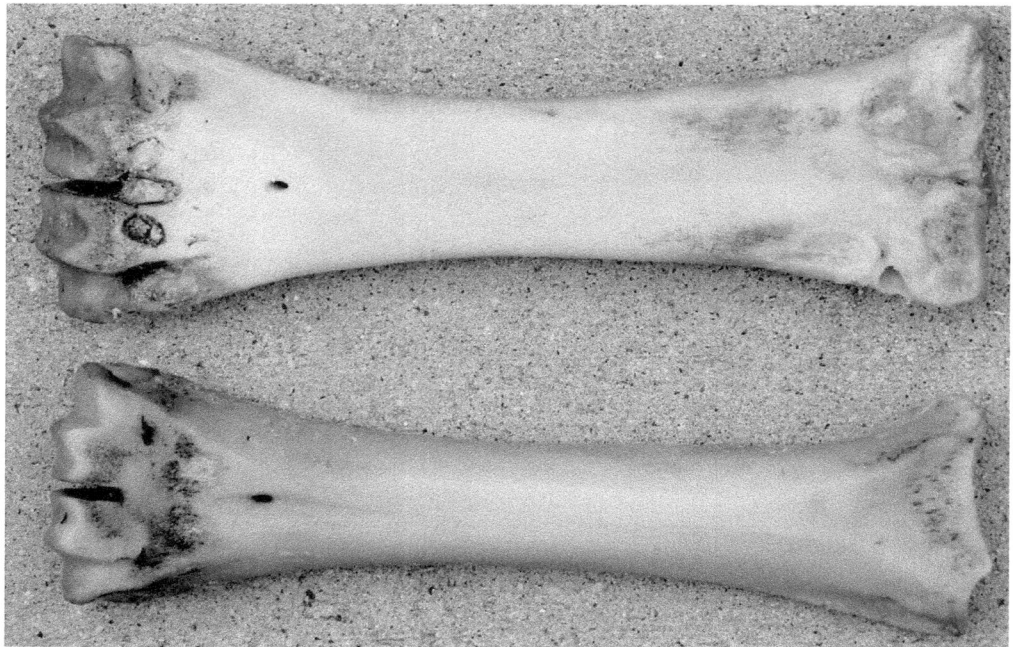

Figure 6: Author's bone skates. The top skate shows where the foot rests (the palmar side), and the bottom skate shows the surface that was in contact with the ice (the dorsal side). The front of the skates (the distal end of the bones) is to the left.

between 15.1 and 20.5 cm long, and the cattle metatarsi and horse metapodia are slightly longer (16.3–23.5 and 19.0–27.5 cm, respectively). The additional length of horse bones is one of the points in favor of using them over cattle bones.

2.3. Making Skates

Medieval sources are mostly silent on the way bones were turned into skates; Olaus Magnus is the only medieval author to mention the preparation process at all. Ethnographic observations and reports of modern users of bone skates (as opposed to experimenters) are helpful but generally less detailed. The most detailed information on how skates were made comes from the reports of modern experimenters and may not be historically accurate. Archaeological specimens are also helpful and perhaps the most important source of information. While questions about the process of making skates remain, some parts of it are quite clear.

The first step is to clean the bones. The clearest description of the process is from the nineteenth century: Brückner[26] reports that "Sie wurden mit dem Taschenmesser von Fleisch und Haut sorgfältig gereinigt" (they were carefully cleaned of flesh and skin with a pocketknife). This seems quite simple, but getting bones really clean is difficult, and modern experimenters have found ways to complicate the process. To make skates for the Lödöse Museum's activities, Schmidt[27] scraped the bones clean, heated them slowly (without boiling) to remove the marrow, and finished by cleaning them in dilute chlorine to disinfect them for use by children. Short[28] and Küchelmann and Zidarov[29] both boiled the bones to deflesh them, the latter using soda (sodium carbonate) following a practice common in tanneries. There is no historical evidence for simmering or boiling, but people could have cooked bones to make broth for dinner and then used them for skates—provided they didn't break them. Boiling during the preparation process could lead to early breakage. Short[30] thinks this explains why his bone skates cracked and flaked during the five winters he used them. More generally, MacGregor[31] warns against heating bones to be used as tools because it can change their mechanical properties, and Formenti and Minetti[32] were careful to avoid heating the horse metatarsi they used for skates in their experiments because of this concern.

Brückner's pocket knife method seems likely to be historically accurate. Olaus Magnus explains that bone skates work so well because of their "connaturalem lubricitatem" (inherent greasiness),[33] which a competitive skater would want to preserve as much as possible. In fact, according to Olaus Magnus, the skaters who won races greased their skates with pork fat because "quia gelidas aquæ guttis velut per poros glaciei in vehementi frigore surgentibus, tibiæ sic vnctæ impediri, aut constringi non possunt"[34] (when the cold drops of water rise as it were through the pores of the ice during fierce cold, the bones smeared in this way cannot be hampered or kept in check.[35])

Choyke and Bartosiewicz[36] concur with the effect of greased bones, though perhaps not with the underlying mechanism. They note that the natural greasiness of bones is good for skating and therefore, people ought to have simply used bones without any additional processing. Boiling is an unlikely step because it removes the grease. Leaving the bones in as close to their natural state as possible seems to have been considered advantageous for skating. However, the extent of this advantage is unclear—bone skates certainly work after being boiled. They also work after having had centuries to dry out, as Magnus Magnusson[37] showed when he skated on a pair from a museum collection.

My own metacarpus skates underwent a process that may have been even worse than boiling. The bones had been baked before I purchased them, presumably to sterilize them. In their study of trabecular bone from cattle vertebrae, Fantner et al.[38] found that baked bones are weaker than boiled bones, which are weaker than raw bones, and that boiled and baked bones are, respectively, more and less elastic than raw bones. Trabecular (or cancellous) bone is the soft bone that forms inside the ends of bones, in contrast to cortical (or compact) bone, which is the hard layer on the outside of bones. Cortical bone is the most relevant to skating because it is what forms the surface of skates, but some trabecular bone may appear on the gliding surface—when flattening my own skates, I filed the ends down so far that some trabecular bone appeared; skating also wears the bone down enough to bring it to the surface. Such material is apparent on archaeological finds. Because all types of bone are made up of an organic matrix and a mineral component, these conclusions can be used to understand the effects of these processing techniques conceptually. Despite having been baked, my skates work well. It remains to be seen whether they crack or flake over time. A study of the effects of boiling, greasing, and other processing techniques on the friction between bone skates and ice may help determine whether Olaus Magnus's supposition was correct. But in the absence of Olympic competition, whether a pair of skates had been boiled, baked, or simply scraped with a pocketknife to remove the flesh may not matter very much.

Once defleshed, the bones could be used as-is (especially horse metapodia, which were naturally quite flat) or worked in a variety of ways. Softening the bones may have made working them easier. Choyke[39] suggests that bones were softened, perhaps by "steaming or soaking in urine," before being worked with metal tools to avoid damage due to the brittleness of dry bone. MacGregor[40] discusses the process of softening bones using acid. Soaking bones in acid is likely to cause permanent damage, as boiling does, but soaking in cold water appears to be less damaging. I soaked my bones in cool water for approximately 48 hours before working them. I repeated this several times because my process was iterative: I soaked the bones, worked on them, tried to skate, and repeated the cycle until I was satisfied. I found that soaking the bones made them slightly easier to work. The water became a bit dirty, which may mean that my bones lost some of their remaining organic content.

Historical bone skates show no signs of having been softened before being made into skates, and modern descriptions of how skates were made, including Brückner's description and a description provided by a Romanian child from the early twentieth century,[41] do not mention any such process. The Romanian child reports simply waiting for the bones to dry and then hacking them into shape with an axe. The result of this hacking is an example of the numerous modifications that have been observed on archaeological bone skates. Küchelmann and Zidarov developed a classification system for these modifications. The most basic are the following[42]:

1. Flattening the skates. This is the most important modification because a flat gliding surface is necessary for skating. The amount of flattening varied from simply removing the protruding ends of the bones to deliberately flattening the bottom surface of the skates. It can be hard to tell how much flattening was deliberate in archaeological specimens because the process is self-reinforcing: the more a skate is used, the flatter it gets. Von Luschan[43] reports that some children recruited less fortunate schoolmates for this task "durch Geld oder Schläge" (through money or violence).

    Depending on the bone's original shape, some deliberate flattening was often in order. Brückner[44] remarks that because he was the miller's son, his job was to flatten the bottoms of his and his friends' skates on the millstone. That, and defleshing, were all they did to the bones they found before putting them to use as skates. Skates were sometimes flattened using a knife, stone tools, or an axe.[45] MacGregor[46] notes that shaping skates and other bone tools using axes seems to have been a common practice. Steve at the Dark Ages Re-Creation Company[47] and Küchelmann and Zidarov[48] found shaping skates using a hatchet and a drill quite simple. Jacobi[49] used a grinding machine and a drill to make three pairs of skates quite quickly.

    Choyke[50] describes how a horse metacarpus was typically modified to create a bone skate. The remnants of the second and fourth metacarpals were removed, and the dorsal side of the bone was ground flat. Measurements of 15 (proximal) and 18 (distal) Sarmatian skate fragments from Gyoma 133 show that the amount of grinding performed was significant. In some cases, over thirty percent of the bone's thickness was removed. Grinding was particularly extensive on the proximal (heel) ends of skates, probably because metapodia are naturally thicker at that end.[51] However, it is difficult to tell how much of

2.3. Making Skates

this flatting was deliberate and how much was the result of use, especially on skates that show significant wear.

Many skates were also flattened on the top and sides. Flattening the top presumably made it easier to stand on the skates. Some skates show signs of deliberate roughening on their upper surfaces, which may have helped keep the skater's feet from sliding off; Short[52] observed this problem when skating in medieval turnshoes. This may also explain why the skaters at the Lödöse Museum had to press on the skates to keep them on while skating in period footwear or socks.[53] A skate made from a cattle metacarpus that has been chopped on both sides is shown in figure 7, with an unchopped cattle metacarpus from the same site is shown for contrast in figure 8.

2. Adding a binding apparatus. Some skates have holes through which cords may have been passed to tie the skates to the skater's feet. These holes also provided a convenient way to carry or hang the skates,[54] which may have been their primary purpose. Brückner[55] reports drilling holes through bone

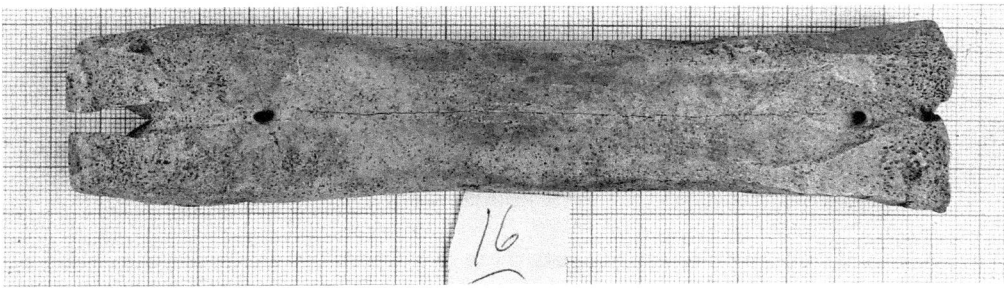

Figure 7: A medieval cattle metacarpus skate from Birka, Sweden, that has been flattened on both sides. The distal end (on the left) has also been pointed and upswept. The palmar side (where the skater's foot rested) is visible (Rune Edberg and Johnny Karlsson, *Isläggor från Birka och Sigtuna. En undersökning av ett vikingatida och medeltida fyndmaterial*, 19, fig. 3.2, courtesy Rune Edberg).

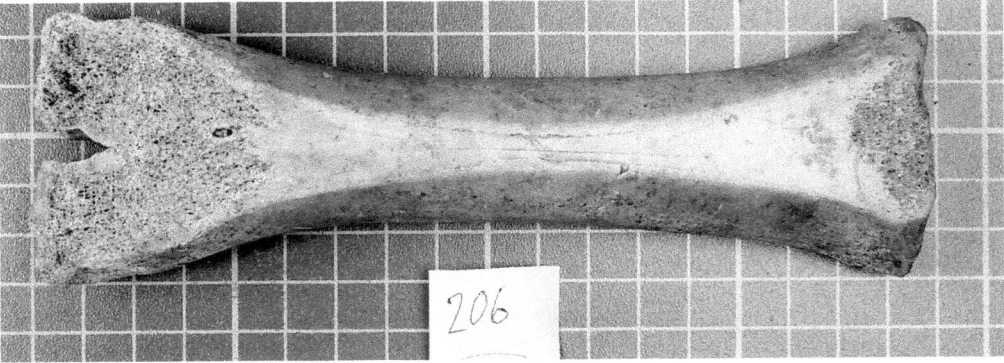

Figure 8: A medieval cattle metacarpus skate from Birka, Sweden, that has not been chopped flat on the sides. The distal end, to the left, is upswept. The dorsal side, which was in contact with the ice, is visible (Rune Edberg and Johnny Karlsson, *Isläggor från Birka och Sigtuna. En undersökning av ett vikingatida och medeltida fyndmaterial*, 19, fig. 3.3, courtesy Rune Edberg).

skates only to make them easier to carry. Roes[56] notes that the holes in skates are normally round, which she takes as evidence that cords were used for bindings. This is borne out by several skates found in Trondheim, Norway,[57] and one skate from Lödöse, Sweden,[58] that were found with the remains of leather thongs, which could have been used for any of these purposes. Figure 9 shows a pair of skates with binding holes from Lödöse, Sweden.

The holes in bone skates are usually transverse (horizontal), but some skates, like the ones in figure 9 and the skate from Sigtuna in figure 10, have paired holes. Some bone skates feature axial holes (holes drilled along the length of the bone), especially at the proximal end. These were used to hold iron loops or wooden pegs for attaching the skates to the skater's feet. Bones with vertical holes were probably sled runners rather than skates: the legs of the sled extended downward into the holes.

When there were holes, they were generally drilled. The Romanian child cited above describes the process as follows:

—hát itt ahol egy kicsit kiemelkedik, mind egy gerinc, átalfúrtam, mer osztan lássa, ebbe a két lyukba van fűzve az ispárga na s...[59]

(—and here, where it is a bit elevated like a backbone, I drilled the bone through, because you see, this is where the string is threaded through, well...[60])

However, this was not always the case. Thunig[61] describes burning holes through the bones in the local smithy as a child in Germany, and Roes[62] describes a skate with "a neat rectangular hole that must have been made for a leather strap." Some skates feature holes that have been hacked into the bone instead of drilled. Roes[63] suggests these may have been the result of hurried

Figure 9: A pair of medieval skates from Lödöse, Sweden, made from cattle metapodia, showing the dorsal side, which was in contact with the ice (top) and the palmar side, where the skater's foot rested (bottom). Note the upswept distal end visible at the left side of the top skate and the holes for bindings (courtesy Lödöse Museum, Sweden).

2.3. Making Skates

skate-making, perhaps in preparation for the type of "'urgent' competition" Küchelmann and Zidarov[64] express concern about.

There may have been a geographic component to the choice of binding apparatus, if any. Herman[65] reports that skates were used with binding holes in some parts of Hungary and without in other parts. Similarly, Edberg and Karlsson[66] suggest that there may have been "loose" and "fast" traditions in different parts of Sweden. This points to the possibility of regional skate styles, which is discussed in Chapter 7.

When I made my pair of bone skates, I removed the small part of the fifth metacarpal that remains on a cattle metacarpus using a hammer and chisel. Then, I filed the dorsal side down using a surform file. I did not get the bones completely flat. My skates arch up somewhat in the middle but work quite satisfactorily. The important thing seems to be to get the bones flat enough that standing on them on a hard, flat surface feels stable. Bones that have broad, flat surfaces at their ends are good enough even if the middle is concave.

I filed down the tops of my skates (the palmar side) to flatten them out a bit, paying special attention to the distal end, where the condyles protruded. Then, I drilled holes through both ends of the bone using a ¼" masonry drill bit. Drilling was fairly easy, except when I tried to drill through a slanted part. Then, it was difficult for the drill to get started because it slid off the bone. Küchelmann and Zidarov[67] also had this problem with their horse metacarpi and suspect that flattening the sides of the bones helped resolve it, which may be why this was sometimes done in the past.

Making the skates was hard work; it took me several hours to file them down sufficiently, but an advanced skater may have been able to make do with skates that were not as flat as mine. Selecting bones based on how flat they were (if numerous bones were available) would have decreased the preparation time. Still, it seems to me that a suitable pair of skates could be made in the half hour proposed by Küchelmann and Zidarov[68] only if the skater was not very picky about flatness or if the bones were selected very carefully indeed. Jacobi[69] reports an even more amazing twenty minutes per pair, but he used a grinding machine to flatten the skates.

Like many of the bone skates found in archaeological contexts, mine are not extensively worked. In the

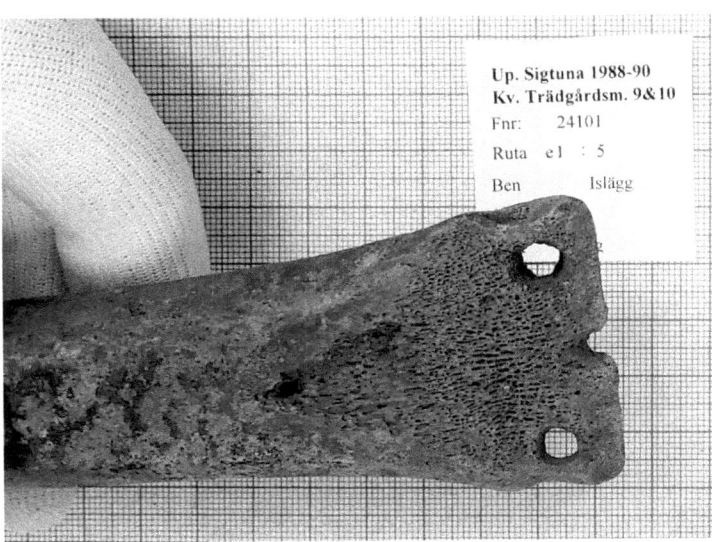

Figure 10: A medieval horse metacarpus skate from Sigtuna, Sweden, with two holes at the distal end. This skate is also upswept at both ends (Rune Edberg and Johnny Karlsson, *Isläggor från Birka och Sigtuna. En undersökning av ett vikingatida och medeltida fyndmaterial*, 27, fig. 4.7, courtesy Rune Edberg).

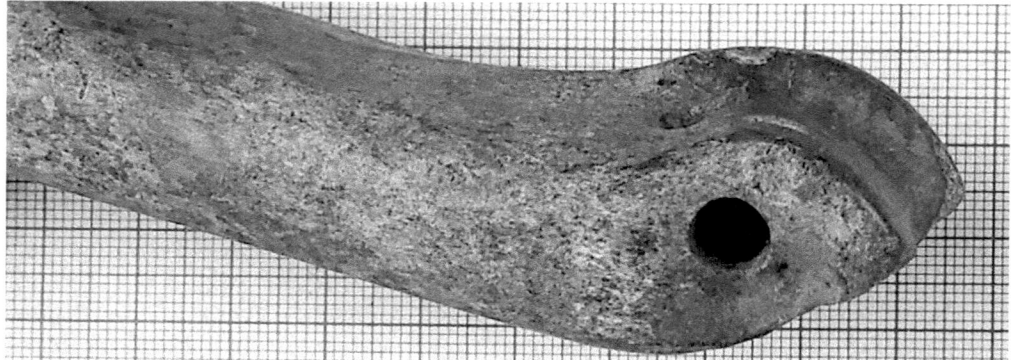

Figure 11: A fragment of a medieval horse metapodium skate from Birka, Sweden, that features an upturned distal end with a transverse hole (Rune Edberg and Johnny Karlsson, *Isläggor från Birka och Sigtuna. En undersökning av ett vikingatida och medeltida fyndmaterial*, 21, fig. 3.14, courtesy Rune Edberg).

skates that exhibit more extensive working, the next most common class of modification described by Küchelmann and Zidarov[70] includes several different ways of shaping the ends of the bones. The ends of some skates are pointed and/or upswept (see figure 11), presumably to help them travel over rough ice that may be covered with a thin layer of snow. Overall, bone skates are generally not elaborately worked, though there are some exceptions. They were made from the most suitable bones that were available, and became better with use—to a certain point. Eventually, they wore out or broke and were discarded or were simply lost, perhaps through a hole in the ice.

2.4. Attaching the Bones

Today, skaters must remove their shoes and put on their skates before getting on the ice, but skaters using bone skates were able to skip this step. They did not need to attach their skates to their feet, much less take off their shoes. Many bone skates lack holes through which to run cords for attachment, which confused some of the first people to study bone skates—if skaters did not attach their skates to their feet, how could they skate? The technique for skating, discussed in section 2.6, makes attachment unnecessary.

Although it was not necessary, most modern experimenters and some past skaters did attach their skates to their feet. The task turns out to be more difficult than expected. It is tricky to get the bones to remain firmly attached, especially for experimenters wearing period footwear rather than modern boots. Herman[71] includes drawings of two binding methods that were used in Hungary in recent times in his article about bone skates; they are reproduced in figure 12. Allen[72] depicts skates attached "exactly as were the sandals formerly worn by lades in England up to thirty or thirty-five years ago" in the drawing shown in figure 13. This method involves wrapping the laces around the skater's ankle and may resemble the one used by William fitz Stephen's twelfth-century skaters, who "singuli pedibus suis aptantes et sub talaribus suis alligantes ossa, tibias scilicet animalium"[73] (put on their feet the shin-bones of animals, binding them firmly round their ankles[74]).

Modern experimenters and skaters have generally used these or similar methods to

2.4. Attaching the Bones

Figure 12: Two methods of tying bone skates to the feet (Otto Herman, "Knochenschlittschuh, Knochenkufe, Knochenkeitel: Ein Beitrag zur näheren Kenntnis der prähistorischen Langknochenfunde," 220).

attach their skates to their feet. Küchelmann and Zidarov[75] found Herman's single-lace method "quick and easy." Schmidt[76] and the skaters at the Lödöse Museum settled on a similar system (shown in figure 14) after some experimentation. The children had to put pressure on the skates to keep them on, but they managed just fine. A system that seems functionally equivalent to this but that uses an extra heel piece is shown in figure 15. Short[77] developed a method that seems to follow the same general concept, but is somewhat more complex. Fowler[78] uses a similar method with two laces to attach bone skates. One lace goes through the toe of the skate, and the other goes through the heel, but they attach in a way that resembles Herman's first method. Fowler's two laces make a figure-eight pattern, which is the type Frans van Liere[79] recommends.

Some skaters preferred to use two separate straps instead of a single-lace system. This type of attachment scheme is described by Petenyi's Romanian child:

> ...lássa az úrfi, amik ispárga elöl van azt a csidma vaj már a bocskor orrára húzom fel s amik hátul van, avval a bokámnál kötöm keresztül.[80]

(...you see, young mister, the string up front is pulled over my boots or moccasin while the ones in the back are tied across my ankles.[81])

Figure 13: Another method of tying the skates to the feet, "exactly as were the sandals formerly worn by ladies in England up to thirty or thirty-five years ago" (J. Romilly Allen, "The Primitive Bone Skate," 32).

2. How to Skate on Bones

Figure 14: The attachment system used at the Lödöse Museum's skating parties (courtesy Lödöse Museum, Sweden).

A method similar to the Romanian child's seems to have been used by Formenti and Minetti,[82] whose bone skate is pictured with two straps.

I tried a few different methods of attaching my own skates to my hiking boots. First I used two laces, one through the front hole and one through the rear. I put one lace through the front hole of each skate and wrapped it around my toes, then tied a half-knot. Then I slid one end under my bootlaces and tied a bow. This helped anchor the skates, keeping them from sliding forward or backward. Doing this and connecting the laces in Herman's pattern worked very well, though I actually had three pieces of string: one going over my toes, one going through the rear hole, up in front of my ankle (where I tied a half-knot), and down through the front lace, and a third going through the rear

Figure 15: A bone skate and a nineteenth-century bladed skate attached to a skater's boot. The attachment systems seem equivalent even though the top uses cords and the bottom uses leather straps (Henry Balfour, "Notes on the Modern Use of Bone Skates," 30).

lace and around my heel. The overall pattern was similar to the one pictured by Herman, but the system is more like one described by Klindt-Jensen[83]: "The binding on the two-holed bone skate was carried partly through the front hole and over the toes and partly through the back hole behind the heel, while a third cord held the two bindings together and was tied tight over the instep." This description may reflect what archaeologist F. Nordin was recalling when he told Klindt-Jensen[84] about his experience using bone skates as child after excavating one at a site on Gotland in Sweden.

Next, I tried Herman's single-lace method. I put the lace through the front hole, tied a half knot, then put the right end through the rear hole. I pulled it up to my ankle and looped the left end around it, then ran the first (originally right) end around my heel, looped it under itself, and tied it to the other end in front of my ankle. It was easier to tighten the laces with this method because pulling on the two ends tightened the whole contraption. This was my favorite attachment method. Like Küchelmann and Zidarov, I found it quite easy and effective once I got the hang of it.

These techniques work well because the holes are at the front and rear ends of the skates. This placement suggests that either the skates were not substantially longer than the skater's feet or that the holes were not used for attachment after all (they could have been used for carrying or storing the skates instead). A hole all the way at the front end of a skate suggests that if it was used for tying the skates on, either skaters used a method like the single-lace one Herman describes or their feet were longer than the bones. If they used two separate straps, one across the toes and the other at the heel, the skates must have been shorter than the skater's feet. With two straps, the skater's toes must extend past the front strap for it to hold a skate on firmly. With Herman's method, the lace is pulled back toward the skater's ankle, enabling the toes to stick out even if the hole is aligned with the ends of the skater's toes. Without this pullback, the attachment would not be effective—with two straps and skates as long as or longer than the skater's feet, the front strap would not go around the skater's toes firmly.

The placement of the rear hole is a bit trickier. Formenti and Minetti[85] "noticed that the hole at the back was found approximately under the ankle" in the Dutch skates they considered. However, this depends on the length of the skate relative to the foot and its placement under the foot. In figures 4, 12, and 13, the rear holes are quite close to the proximal (heel) end of the bone. Axial holes in the heels of skates for use with pegs, nails, or iron loops placed the fastening as far back under the skater's heel as the skate extended. The placement of the rear fastening device may have been selected based on the shape of the bone (it is easier to drill through a flatter surface) and its type (it is easier to insert a peg at the heel of a horse metapodium than a cattle metapodium because the former is hollow, while the latter is effectively two conjoined cylinders). MacGregor[86] notes that some of the cattle metapodium skates found in England have axial holes "set at an oblique angle … so that they penetrate the upper (posterior) surface of the bone rather than the medullary cavity." Skates with perforations of this type would not have required heel pegs since a thong or cord could have been strung through the hole.

In my experience, it was more comfortable to have the skates reach nearly all the way to my heel and let my toes hang over the front end than the other way around. Küchelmann and Zidarov[87] preferred longer horse bones to shorter ones because they were "more comfortable to stand on," but I did not notice much of a difference between my cattle metacarpus skates and my radius skate. The metacarpi were slightly more comfortable because the radius was so curved. This may be because I was wearing modern hiking boots with stiff soles; medieval turnshoes may produce different results.

There are no published measurements of how far from the ends of the skates binding holes typically are. Making such measurements on well-stratified skates may show that there is a geographical and/or temporal component to the location of the holes, which could reflect local skate styles. Or perhaps the holes were simply located near the ends of the bones in the places that were easiest to drill through. It is notable that the holes are always near the ends of the bones and never near the middle, except in one of the 160 Dutch skates Jacobi[88] analyzed. This skate had the usual transverse holes at the two ends and an additional transverse hole in the middle. Because central holes are so rare, it is unlikely that a strap was placed across the skater's instep (except perhaps in this unusual case).

The holes may also have been used for carrying or storing the skates, but lateral holes would work better than the more common transverse holes. I ran a single string through the front holes in both my metacarpus skates and found carrying them rather awkward. This method might work better with narrower skates (i.e., skates made from horse bones or laterally flattened). Of course, there are many other ways to carry skates.

2.5. The Pole

Although the focus of this book is on bone skates, the one (or two) poles that were normally used for propulsion cannot be left out. Skating poles were generally wooden sticks that ended in metal spikes. William fitz Stephen describes them as "palos ferro acuto supposito" ("poles shod with iron").[89] Poles have not survived in the archaeological record, perhaps because they were probably made of wood, which does not survive well under typical preservation conditions. The points—of metal or, perhaps, bone—are more likely to have survived, and indeed, some mysterious points have been described in the literature.[90]

Iron points that may have been used with skating poles in the ninth and tenth centuries have been found at Birka, Sweden.[91] Occasionally, bone points found at archaeological sites have been identified, more or less tentatively, as the tips of poles used for skating.[92] However, some archaeologists are skeptical of this idea. MacGregor[93] notes that

> No supporting evidence has been found for this theory, however, iron tips normally being used for this function. Although the stabbing action suggested by the microscopic wear patterns on the tips is not at variance with such a suggestion, some of the points seem too blunt to have gained any purchase on the ice, while others are so slender that they would almost certainly have broken under any stress of this kind.

Küchelmann and Zidarov[94] "doubt if the bone and antler points mentioned above ... assumed to be used as prickstick tips will be suitable for a high-duty task like this."

To determine whether it was possible to push with a bone point, I made one that is similar to the ones described by Lauwerier and Van Heeringen[95] from a piece of a cattle radius (see figure 16). On the ice, I found that I was able to propel myself using it. Although my point is much stouter than the archaeological specimens, it shows that it is possible to use bone points for propulsion.

However, that bone points work for skating does not necessarily mean they were used for skating. Becker[96] has identified many of the bone points found at Berlin-Spandau (a site with numerous bone skates) as basketry pins, perforators, and pegs, but has left approximately half of the points unidentified. MacGregor[97] suggests that similar bone points from a York leather-working ship were used in crafts.

The poles used in other experiments have varied, but all had metal tips like the two poles on the right in figure 16. I made these two poles from dowel rods purchased at Home Depot that were 48 inches long and one inch in diameter. I drilled a hole in one end of each rod and glued in a nail. The pointed end of the nail protruded approximately 5.3 cm beyond the end of the pole. I found that placing the nail at a slight angle rather than perfectly parallel to the pole made pushing easier, provided the pole was oriented so that the nail pointed backward.

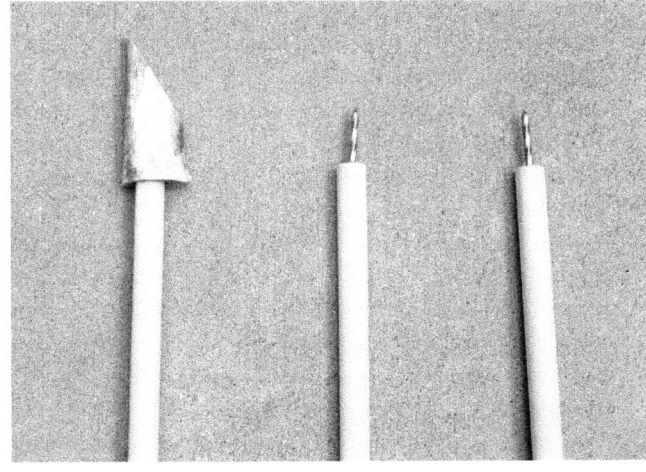

Figure 16: The tips of the skating poles made by the author. The one on the left features a bone point made from a cattle radius. The other two points are simply nails inserted into the ends of dowel rods.

Historically, poles could have been much more elaborate than these were. Berg[98] describes eighteenth-century skating poles in museum collections in Stockholm and Falun as "not seldom beautifully decorated." An example of a decorated pole is shown in figure 17. Von Luschan[99] calls this

> ...[ein] höchst interessantes Beispiel deutscher Hausindustrie den oberen Theil eines 130 Cm. langen Stockes abbilden lassen, den ich gleichfalls der Güte meines Freundes Kraus verdanke und der vorläufig noch meiner Sammlung angehört. Schnitzmesser und ein hohler Meisel sind die einzigen Werkzeuge gewesen, die zu der zierlichen und geschmackvollen Arbeit gedient haben. [...] Das untere Ende des Stockes, der die Zeichen Jahrzehnte langen Gebrauches aufweist, ist mit einem eisernen Stachel und einem schmalen eisernen Bande versehen, das in fünf Spiraltouren schützend um das selbe herumgeführt ist.
>
> (...a very interesting example of German home-industry, the upper part of a 130 cm long staff, for which I thank the kindness of my friend Kraus; it temporarily belongs to my collection. A carving knife and a concave chisel are the only tools that served in the dainty and tasteful work. [...] The lower end of the staff, which shows signs of decades-long use, exhibits an iron spike and a narrow iron band, which is wrapped around in five protective spiral turns.)

2.6. Skating!

Once the skates are made and on, it is time to get started, which may be the hardest part of skating. Standing on bone skates on ice for the first time is harder than it may appear. They are quite slippery, and it can be difficult to keep one's feet

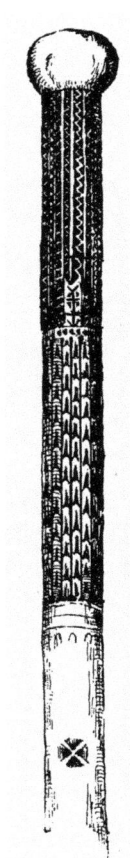

Figure 17: The top of a decorated skating pole (F. von Luschan, "Mitteilungen aus dem Museum der Gesellschaft," 142).

in place before beginning to skate. I recommend that beginners start with two poles instead of one because it is easier to balance with two things to lean on. Holding onto the barrier in a modern ice rink also helps, especially with single-pole skating. In this situation, von Luschan[100] advises beginning to skate "in halber Kniebeuge" (in a half squat) with the tip of the pole placed between the feet. This makes it unnecessary to move much and easier to stay balanced. Skaters simply stand on the skates, keeping their feet still relative to each other, and push themselves along with the pole or poles. The skates are never lifted from the ice.

One-pole skating seems to have been more common than two-pole skating, especially in and near Scandinavia. William fitz Stephen's skaters used single poles, and Olaus Magnus includes some images of skaters using two bone skates and a single pole in his 1539 *Carta Marina (Marine Map)* and his 1555 *Historia de gentibus septentrionalibus (Description of the Northern Peoples)*. They are all very similar; see figure 18 for an example. This technique is also shown in drawings in the works of Herman[101] (figure 19) and Söderbäck.[102]

The two-pole method, with one pole in each hand, is shown in figure 20. It is mentioned in a tenth-century Arabic source published by Markwart[103] and the account of Sharaf Al-Zamān Ṭāhir Marvazī, which dates to approximately 1120 CE.[104] Like Küchelmann and Zidarov,[105] I found it easier to balance with two poles. They became exhausted more quickly when pushing with two poles than with one, but in my experience, pushing with a single pole was much more tiring.

Regardless of which pole-pushing technique is used, the hardest part of skating on bones is keeping one's feet parallel and going forward. In my first attempts at skating on bones at an indoor ice rink, I found that I had to concentrate on keeping my feet aligned. It is easier with skates that are very flat on the bottom, as I found out during my first skate.

Figure 18: Skaters on bone skates using one pole below a reindeer-drawn sled in a woodcut from 1555. Note the fallen skater on the lower left, who has lost one skate but not the other (Olaus Magnus, *Historia de gentibus septentrionalibus*, 392).

2.6. Skating!

My right foot tended to go off on its own because the skate was not entirely flat. Küchelmann and Zidarov[106] encountered the same difficulty, and Herman[107] notes that

> [D]as Paar der Knochenschlittschuhe musste gegen einander parallel verblieben; denn sobald es in eine nach vorne divergirende Stellung kam, liefen die Füße auseinander und das unfreiwillige Niedersitzen war unvermeidlich, und zwar oft mit einer Gewalt, welche in das Eis einen sogenannten »Stern« schlug.
>
> Außerdem mussten beide Füße stets in gleicher Entfernung verbleiben, damit—bei Gebrauch nur einer Schiebstelze—in das Stoßen zwischen den Füßen leicht und sicher geschehen könne. Auch die Haltung des Körpers war wesentlich: man musste die Knie und den Leib etwas beugen und eben in dieser Stellung lag das ermüdende Moment, weil die betreffenden Muskeln dauernd in Spannung erhalten werden mussten.
>
> (The pair of bone skates had to remain parallel to each other; as soon as they went forward in different directions, the feet separated and an unwanted sit-down became unavoidable, often with a force that resulted in a so-called "star" in the ice.
>
> In addition, both feet had to remain at the same distance, so that when only one pole was used, it could be placed easily and securely between the feet. And the way the body was held was important: one had to bend the knees and the body, and the tiring part lay just in this stance because the relevant muscles had to be held in position for a long while.)

According to Herman, the next hard part is balancing in the right position for as long as possible. It is difficult to hold the sort of half-squat that works best for single-pole pushing. Küchelmann and Zidarov[108] "observed light muscular aches in the muscles of the thighs and seat" the day after experimenting with bone skates. With two poles, the need for a particular and tiring position is, in my experience, much less.

The ice plays some role in how fatiguing holding the right position is. I found it easier to balance on natural outdoor ice than on smooth indoor ice, perhaps because the outdoor ice was rougher and provided

Figure 19: A nineteenth-century Hungarian child on bone skates; this is the poster child for bone skates (Otto Herman, "Knochenschlittschuh, Knochenkufe, Knochenkeitel: Ein Beitrag zur näheren Kenntnis der prähistorischen Langknochenfunde," 220).

Figure 20: A person using two poles to skate (Otto Herman, "Knochenschlittschuh, Knochenkufe, Knochenkeitel: Ein Beitrag zur näheren Kenntnis der prähistorischen Langknochenfunde," 222).

the skates with a slightly better grip. Küchelmann and Zidarov[109] also found outdoor ice easier to skate on due to its roughness. This contrasts with Olaus Magnus's assertion that bone skates only work on smooth ice,[110] but he did not know about indoor ice, which is typically at least as smooth as the smoothest outdoor ice. While smooth ice certainly makes the skates slide faster, it also makes it more difficult to keep them from sliding apart, as I discovered in my experiments. I skated fairly slowly, which may have been a factor in my results.

A significant weakness of bone skates is that turning and stopping are very difficult. There is no easy curving as there is with modern skates, and stopping relies on the pole. Säve[111] describe this problem vividly:

> När isen är rätt glansk, går det i en brinnande fart; det farligaste för en ovan är att alldeles spräcka opp sig, ty isläggarna ränna gerna åt sidorna, och det att man i farten omöjligt kan vika åt sidan eller göra en lof, om man ock såge den djupaste vak för sig. Det enda medel i detta sednare fall är sätta iskäppen mellan benen, sätta sig vackert ned på den och låta den rispa uti isen och att sålunda något hejda farten; men då skall man ej vara för nära faran.
>
> (When the ice is very glassy, it goes at a burning speed; the most dangerous part of the above is splitting apart: the ice-legs easily run to the sides. And when sliding, one cannot turn aside or change course, if one sees a deep hole in front. The only thing to do in this case is to set the pole between the legs, lean back on it, and let it scratch the ice to slightly arrest the motion; but then, one should not go too close to danger.)

Küchelmann and Zidarov[112] were able to use their poles to adjust their motion slightly, which I was also able to do. The emphasis is on slightly, though. Jacobi[113] had more success with steering while skating without a pole. Due to concerns about damage to the ice, Jacobi was pushed by someone on modern skates instead of using a pole.

Stopping can be done by dragging the pole (as Berg suggests above) or by stabbing it into the ice in front. The latter may be inadvisable at high speeds but works tolerably well at low speeds. Küchelmann and Zidarov also had some success with turning their feet sideways. None of these methods are very effective, which underscores Antti Juvel's recollection that "if there was open water ahead, you had no choice but going into it, for it would be too dangerous to fall on the ice in such a great speed and turning was impossible."[114]

What makes it so difficult to stop and turn is that bone skates do not have edges that dig into the ice like metal-bladed skates do. Instead, they glide on the surface of the ice. This means that they slide equally well in any direction, making it possible to skate backwards and sideways as well as forwards, though I found it more difficult to get sufficient leverage with the pole when skating backwards and sideways. However, difficult is not impossible, and I was able to skate in any direction I wanted to. I was also able to spin by standing on both feet and pushing myself around with the pole. On my first attempt, I completed approximately one revolution per pole push. With some practice, I was able to complete two or three revolutions per push. I am not the only one to accomplish this; a nine-year-old girl at the Lödöse Museum's skating party was also able to spin on bone skates.[115] In subsequent years, other skaters also learned to spin.[116]

Some skates have binding holes and others do not, even within individual sites, which implies that some skaters attached their skates to their feet and others did not. Berg[117] reports that "only children and beginners had to bind on the skates, whereas the skilled could do as well without such arrangements." Without bindings, according to Ernst Friedel,[118] "stand der Eisläufer einfach auf dem Knochen aufrecht, wozu lediglich Vorübung und Gewandtheit gehöhrte, und trieb sich mit einem Stachelstock, besser mit zweien, in jeder Hand mit einem, vorwärts" (the skater simply stood upright on the bones, for which only preliminary practice and dexterity were necessary, and pushed forward with a prick-stick, or better with two, one in each hand). MacGregor[119] also had success using this method, which he explains as follows:

> Since the friction between the ice and the skate is lower than that between the skate and the foot (particularly when the upper (posterior) surface of the bone skate has been artificially roughened), the body weight of the skater is normally enough to keep the skates in position. On the indoor ice-rink where the recent experiments took place, even novice skaters had little trouble in staying on their skates in this way, but on uneven outdoor ice they would certainly have preferred those with a full set of bindings.

In contrast, Küchelmann and Zidarov[120] "did not reach the 'dexterity,' being the 'only requirement' described by Friedel in 1898 … as necessary for skating without the bones attached to the feet and we still cannot imagine how this works."

In my own experiments, I waited until I was fairly comfortable skating with bindings before I attempted to skate without them. At that point, skating without bindings was quite easy, and after some practice, I found it preferable. The balance was slightly trickier, but only slightly. The most difficult part was getting onto the skates the first time. I placed them on the ice and stepped onto first one, then the other, while holding a pole in one hand and the ice rink's barrier in the other. With some practice, I was able to dispense with the barrier. Once safely aboard, my hiking boots remained firmly on the skates, even when I hit a rough patch on the ice. On natural outdoor ice, I had no difficulty remaining on the skates even though the ice was rather rough, but that may have been due in part to my earlier practice on indoor ice.

Skating without bindings has certain advantages. It was hard to get up after I fell with the skates tightly bound to my feet because they slid away during the attempt, resulting in another fall. It was easier to remove them, stand up, and then put them back on. This process was much quicker without bindings to worry about. Removing the skates seems like a reasonable strategy for crossing rough patches of ice and ice-free areas between lakes. Skating on rough outdoor ice with many leaves frozen into the surface revealed another advantage of skating without bindings: when an attached skate hits a bump and stops, it is hard to remain upright, especially when skating at a reasonable speed. With unattached skates, falls can be avoided by simply stepping (or jumping) off a stuck skate. I suspect that this strategy would not work as well at high speeds but have not tested it.

My preference for no bindings is borne out by recent archaeological results. Edberg and Karlsson[121] found that some of the child-sized skates from Birka and Sigtuna lack binding holes, which implies that children were able to skate without bindings, and concluded that older and more skilled skaters (who presumably went faster) used longer bones and worked them more extensively, including adding a binding apparatus. However, Berg's statement is supported by the difficulty Küchelmann and Zidarov[122] had standing on the skates and my first attempts at stepping onto unattached skates on slippery ice.

All these experiments, except Jacobi's, were conducted using one or two poles for propulsion. These seem to have been the most common ways to skate on bones, but there are other methods. During my experiment on rough outdoor ice with leaves frozen into the surface, I developed another effective one: with a pole in each hand, I slid my feet forward alternately, each accompanied with the opposite pole (right pole/left foot, left pole/right foot). This helped me keep up a continuous slow motion despite all the obstacles. There is no historical evidence for this technique, but Short[123] also found it effective.

Pole-free skating is also possible, but it is not at all like the technique used by skaters on metal-bladed skates. There are two pole-free skating techniques that I am aware of: skating on one skate and pushing with the other foot (like a modern skateboarder) and skate-sailing. In both, the skates remain flat on the ice all the time. Thorsteinn Einarsson[124] observed the skateboarding technique in Iceland in about 1950. He describes "a girl us[ing] a single ice-leg, using her free foot to push with from time to time, in between which she rested it on her other foot or moved it out sideways to keep her balance." This method has also been used in Russia[125] and nineteenth-century Berlin, where children skated in gutters on single bone skates.[126] I tried skating with a bone skate on my right foot and pushing with my left foot. It was very difficult to keep the skate going straight and to stay balanced over it. This seems like a technique for advanced skaters.

Using a sail for propulsion is also fairly well-known in the literature. Ernst Friedel describes it clearly as follows:

> Auch wurde zwischen den beiden Stöcken wohl ein Tuch als Segel befestigt. Setzte sich der Wind dort hinein, so war man imstande, Außerordentlich schnell auf glatter Eisbahn vorwärts zu kommen.[127]
>
> (Also, a cloth was sometimes fixed between the two poles as a sail. When the wind blew into it, one could go forwards exceedingly fast on a smooth ice surface.)

A drawing of this technique is shown in figure 21. Brückner[128] describes a version of sail-based skating that requires less sophisticated equipment:

> Wenn das Eis vollkommen eben war und die "Knochen" durch langen Gebrauch eine spiegelartige Glätte erhalten hatten, konnten wir uns durch Ausspannen unserer Jacken vom Winde treiben lassen.
>
> (When the ice was completely even and the "bones" had gained a mirror-like smoothness through long use, we could allow ourselves to be driven by the wind by spreading out our jackets.)

I have not tried this on bone skates, but it works very well on modern figure skates. It is easy to imagine building up a great deal of speed with a large ice surface on a sufficiently windy day. A more complex version of ice-sailing is described by Heathcote[129]: skaters would take large sails onto the ice to catch the wind (see figure 22). Buck[130] goes even

Figure 21: A skater using a sail with bone skates. (Otto Herman, "Knochenschlittschuh, Knochenkufe, Knochenkeitel: Ein Beitrag zur näheren Kenntnis der prähistorischen Langknochenfunde," 222).

further and describes the equivalent practice with yachts on runners. More recently, Ann Bancroft and Liv Arneson used sails to ski across Antarctica.[131]

2.7. How Fast Did They Go?

The first question that comes to mind after mastering skating on bones is "how fast can I go?" This section attempts to answer this question quantitatively. Medieval authors answered it in the following ways:

- William fitz Stephen (late twelfth century) describes skaters as "tanta rapacitate feruntur, quanda avis volans" ("swift as a bird in flight or a bolt shot from an engine of war").[132]
- Saxo Grammaticus (early thirteenth century) calls Ullr's motion on his magic bone "nec eo segnius quam remigio" ("as swiftly as with oars").[133]
- William of Rubruck (thirteenth century) says skaters travel "cum tanta velocitate ut capiant aves et bestias" ("with so much speed that they may catch birds and beasts").[134]
- Olaus Magnus (1555) describes Ullr as traveling "nec segnius eo, quam celerrimo velorum, ventorumque vsu" ("as speedily as ... with wind-filled sails") on his magic bone[135] and other skaters as skating "velocissimum" (at "a very great speed").[136]

These comparisons are qualitative, but more recent authors provide measurements:

Figure 22: A skater using a sail with metal-bladed skates (J. M. Heathcote, "Skating as a Recreation," 214).

- The skaters in the study conducted by Formenti and Minetti skated on bones at 1.13 ± 0.40 meters per second (2.53 ± 0.89 miles per hour). This was a "low" speed that skaters were supposed to be able to sustain for eight hours. However, these former competitive speed skaters "declared that they would not feel safe at a higher speed" on bone skates after less than one day of practice.[137]
- Küchelmann and Zidarov[138] were able to skate 400 meters in 3 minutes 5 seconds or at approximately 2.16 meters per second (4.83 miles per hour) after practicing for three days. They suspect that this was slow, because there were "a lot of details to improve in locomotive technique as well as in the shaping of the skates."
- Brückner[139] claims that in his youth expert skaters were able to traverse

half a German mile in 15 minutes. In Prussia, before the advent of the metric system, a German mile was 7,532.50 meters,[140] which gives a speed of approximately 4.18 meters per second (9.36 miles per hour).
- Säve[141] describes men and women[142] skating across Lake Bästetrask on Gotland in Sweden to the church at Fleringe in 11 minutes. This lake has a water surface area of approximately 681 hectares[143] and appears roughly circular, which suggests that its diameter is approximately 1.5 km, giving a skating speed on the order of 2.3 meters per second (very roughly 5 miles per hour).

Brückner's claim seems quite high compared with the others, which could be due to his greater experience or an exaggeration of his friends' youthful accomplishments. To identify a reasonable speed for an experienced skater, I took a simple physics approach. First, I created a mathematical model that included the following three forces:

- *Friction:* Although bone skates slide very well, there is friction between the ice and the skate. The usual simple model for friction is $F_{fric} = \mu N$, where μ is the dynamic coefficient of friction and $N = mg$ is the normal force (here the mass of the skater times the acceleration of gravity). I assumed that the skater weighs 60 kg and that the acceleration of gravity is 9.81 m/s². For μ, I used 0.013 based on Formenti and Minetti's[144] measured value of 0.013±0.0022.
- *Air resistance:* Generally, air resistance is modeled as $F_{air} = kv^2$ where k is some constant and v is the velocity. Although k varies with a variety of factors (the skater's posture and clothing, for example), the data for modern speed skaters presented by Ingen Schenau[145] shows values of k between 1.2 and 2.1 kg/m; I used 2.1 because that value corresponds to a more upright body position, matching the pictures of skaters on bones from various sources. It may even be too low because ancient and medieval fashions were less aerodynamic than modern speed-skating clothing.
- *Pole-pushing:* I developed a model for the pole-pushing force based on the experimental data of Millet et al.,[146] who measured the force of a pair of poles used by someone on roller skis. They were interested in determining how the force varies with different skiing techniques. The technique that is most applicable here is the double poling method, where a skier's only source of propulsion is two poles. This is very similar to how skaters on bones skated with two poles; data were not available for the more common single-pole technique. I assumed the pole is 1 meter long, which is very roughly the distance from my hand to the ice when I skate.

Following Millet et al., I assumed that the pole touches the ice for half a second and then, the skater glides for nine tenths of a second for a rate of approximately 0.71 strides per second. This is similar to the poling pattern used by the skaters in Formenti and Minetti,[147] who skated at a rate of 0.50±0.01 strides per second. My rate seems reasonable because the roller skiers were probably more accustomed to their equipment, and the skaters in Formenti and Minetti's study were not attempting to skate fast.

I chose a maximum force of 400 Newtons directed along the pole. Only the horizontal component of this force goes into pushing the skater forward. I assumed that the skater plants the pole at a 90 degree angle, then picks it up when it reaches 45 degrees with respect to the ice, which is a very rough estimate based on my own pole-pushing experience. I also assumed that the angle varies linearly with time, which is probably not quite correct because the skater accelerates while pole-pushing.

$$F_{pole} = F_{str} \cos \theta$$

$$\theta(t) = \frac{\pi}{2}\left(1 - \frac{t}{t_{push}}\right)$$

F_{str} is fitted to the data, giving approximately

$$F_{pole} = \begin{cases} F_{max} \sin(\pi t / t_{push}) \cos \theta & \text{During each push} \\ 0 & \text{between pushes} \end{cases}$$

Combining these forces yields the equation of motion $m\ddot{x} = F_{pole} - F_{fric} - F_{air}$, which I solved using ode23 in GNU Octave. The resulting steady state velocity was approximately 3.55 m/s (7.94 miles per hour). This is closer to the speed reported by Brückner than to those reported by modern experimentalists, which suggests that extra practice can make a significant difference in how fast a skater can go. It also suggests that skating can be significantly faster than walking. This speed compares well with sea travel, as Saxo Grammaticus and Olaus Magnus note. In a 2003 trial voyage, the replica ship *Ottar* traveled at about 4.5 knots (5.18 miles per hour) under good weather conditions.[148] Davis[149] reports a top speed of 14 knots (16.11 miles per hour) for a Scandinavian longship. However, William fitz Stephen's comparison to birds and crossbow bolts appears to be exaggerated.

This model does not account for what Küchelmann and Zidarov[150] call "the limiting factor": "the power needed for the constant tension necessary to keep the legs in the parallel and bent position." That may explain why their speeds were so much lower than the model's results: the model implicitly assumes that the skater can keep this tension up at whatever speed is necessary. According to my experience learning to skate on bones, this is a muscular ability that can be built up over time; Brückner and his friends seem to have done so.

2.8. Wear and Discard

Eventually, every pair of bone skates must be replaced due to loss, excessive wear, breakage, or another problem or is discarded when conditions are no longer suitable for skating. It is at this point that bone skates have the opportunity to enter the archaeological record. The human actions that give bone skates and other artifacts this opportunity are what Schiffer[151] calls "cultural formation processes." These processes also include other human activities that affect the archaeological record, including excavation. They contrast with "noncultural formation processes," which are environmental factors events that affect an artifact's survival. Weather events and animal activities are two examples. This section is about the cultural formation processes that affect how bone skates enter the archaeological record. Interpreting this record is dealt with in the next chapter.

Of these factors, loss is perhaps the simplest. It is easy to imagine a pair of skates disappearing when a skater falls through the ice, especially if they are not tied on. Because bone skates are easy to make, it is unlikely that skaters would have gone to great lengths to retrieve lost skates. Instead, they would have simply made new ones, and indeed, skates have been found in places that were once under water.[152] An example is the pair of skates found in the company of a female skeleton in what was once a river in Ipswich, England.[153]

Sometimes, bone skates remained in use long enough to wear out. Nobody has yet figured out how long this process took. If a reasonable correspondence between the amount of wear on a bone skate and the amount of time it was used on the ice could be developed, it will become possible to determine how long a given skate was used. This could provide important information about the local climate, such as how many skating days there were in an average year. Such a scale will be difficult to develop because ice conditions and inclusions may change the rate at which skates wear down. A worn-out skate is shown in figure 23.

Although there is no such scale, a few archaeologists have commented on the rate of wear to bone skates. MacGregor[154] was able to observe striations under a microscope after three hours of skating on a juvenile bone; apparently, it takes longer for the harder bone of an older animal to show visible wear. Jacobi[155] also found that wear patterns emerged quickly—within 400 meters of skating. Similarly, my radius skate began to show signs of use after only fifteen minutes. Choyke and Bartosiewicz[156] note that the wear on Late Bronze Age skates is generally heaviest under the heel and can even result in exposure of the medullary cavity, as is the case in with the medieval skate in figure 23. Edberg and Karlsson[157] rate the wear of each of the 679 Swedish bone skates in their study on a four-point scale. Each skate is assigned one of the following four values: unassessable (because the skate is fragmentary), protruding parts of the gliding surface removed but no traces of use, light traces of processing or use, and heavy traces of use. They assign 13 skates to the first category, 62 to the second, 189 to the third, and 415 to the fourth.

Most of MacGregor's skates seem to have been in the third category, and the ones described by Choyke and Bartosiewicz sound like they are in categories three and four. These results suggest that wearing out was indeed a problem with bone skates. Brückner[158] and his friends seem to have gone looking for new skates every year, which attests to their ephemeral nature. Overall, the wear process seems to have been gradual but not unobservable.

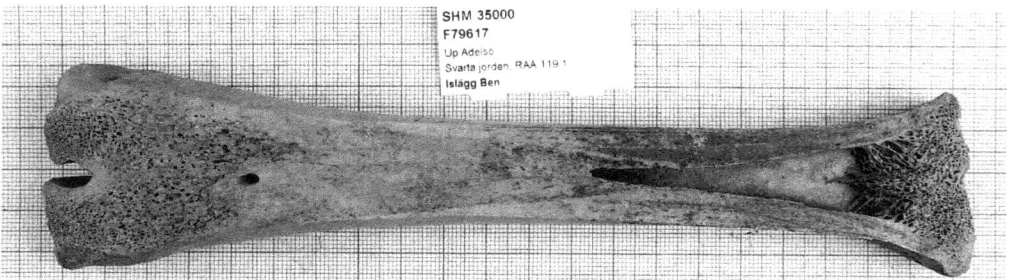

Figure 23: A worn-out medieval bone skate from Birka. This skate is made from a cattle metatarsus; the dorsal side, which was in contact with the ice, is shown. Note the exposed medullary cavity on the right. This skate has been worn all the way down by excessive use (Rune Edberg and Johnny Karlsson, *Isläggor från Birka och Sigtuna. En undersökning av ett vikingatida och medeltida fyndmaterial*, 19, fig. 3.5, courtesy Rune Edberg).

2.8. Wear and Discard

Contrasting evidence that supports the durability of bone skates comes from a modern account. In a story told by W. H. Barrett and collected by Enid Porter,[159] a man skated along a river on a pair of bone skates that were two hundred years old and still in use in about 1900. Unfortunately, it is not clear how frequently these skates were used. This story reflects my own experience the best. I found that filing down a bone to make a flat surface was hard work; it would take a lot of skating to wear down a bone, especially since metapodia are very dense and ice is generally smooth. This implies that skating sessions must have been long and frequent indeed to wear out a pair of bone skates.

An experimental study of the rate of wear of bone subjected to friction against ice would be helpful in assessing the wear on skates and determining how frequently they had to be replaced. In fact, a study of the rate of wear of bone tools in general would no doubt be of interest to those studying such tools, especially if the results could be used to provide an estimate of the number of hours a given tool was in use. With skates, one factor to consider is the width of the facet along the gliding side, which, as Choyke and Bartosiewicz[160] note, becomes wider with use.

Breakage may also have been a problem with bone skates, especially after they became badly worn. Many broken skates have been found. Some of them undoubtedly broke after being discarded, but others were probably discarded because they had broken. When one skate of a pair broke, the other could have been kept and a single replacement made; Choyke and Bartosiewicz[161] suggest that this occurred in a pair of Late Bronze Age skates from Törökbálint, Hungary (near Budapest). These skates show different amounts of wear, which they interpret as evidence that the newer skate replaced a skate that had broken.

Skates could also have been discarded for other reasons. A child's skate could have been outgrown, or the local environment could have changed in a way that made it less suitable for skating. People could have abandoned their skates when they moved to a new place. Sometimes skates were placed in graves, presumably with their owner. Regardless of why it was discarded, once a skate was no longer in use, it had the opportunity to enter the archaeological record. Once it was found again, archaeologists had to identify it as a skate, which is not as easy as it may seem. This is the subject of the next chapter.

3

The Study of Bone Skates

3.1. Skaters and Scholars

The modern era of scholarship on bone skates began in 1834 with a presentation by C. A. Rethaan Macaré to the Zeeuwsch Genootschap der Wetenschappen (Zeeland Society for Sciences). He identified a set of artifacts found in Serooskerke as bone skates, but was not taken seriously.[1] The next step was taken in 1842 with a description of a bone skate found in Moorfields, London in 1840.[2] Charles Roach Smith described the skate's provenance and listed some of the literary references that have become well-known, and Alfred Smee described the condition of the bone.[3] The skate was a well-preserved horse metatarsus with a pointed toe and holes at both ends. The hole at one end is drilled into the axis of the bone. Smith set the stage for the interpretation of bone skates by connecting them with references made by William fitz Stephen and Olaus Magnus and in Bishop Percy's translations of Old Norse poetry, discussed below. These references connect bone skates to the North, which was a very popular topic at the time.

In the following years, bone skates drew the attention of antiquaries across Europe. The *Verhandlungen der Berliner Gesellschaft für Anthropologie, Ethnologie und Urgeschichte* published in the *Zeitschrift für Ethnologie (Journal of Ethnology)* include at least one transcript of a presentation about bone skates in most of the years between 1870 and 1890. Some of the presenters described artifacts they had found or texts they had read. Others, including Brückner, Virchow, and Kuckuck, reminisced about skating on bones as they grew up.[4] They described bone skates as toys for children and adolescents growing up in small towns in the country. The Anthropologischen Gesellschaft in Wien (Anthropological Society in Vienna) also discussed bone skates frequently during these years, though not quite as often as its Berlin counterpart did.

One contributor was Dr. Heinrich Kraus, who told his friend, F. von Luschan, about schoolboys collecting bones (much like Brückner and his friends) and asking an adult for help shaping them before skating on them. He describes skating as an experience with "kaum in aesthetischer Beziehung zu vergleichen" (hardly anything to compare in the aesthetic realm).[5] The British Archaeological Association also showed interest in skating; J. W. Grover "exhibited some fashioned bones, used probably as skates" to the society at its meeting of 14 February 1872, and, two weeks later, E. Roberts showed the society three British bone skates from Broad Street.[6] No additional details of the skates are given in the proceedings.

All this work led up to two important publications about bone skates. In 1894, Robert Munro summarized the state of scholarship on bone skates.[7] His article focuses on the archaeological evidence and provides a list of the skates found at various sites and housed in museums. Less than a decade later, Otto Herman wrote about prehistoric bone tools, including both skates and runners, with emphasis on their continuing use.[8] These papers laid the foundation to the subsequent study of bone skates. They came near the peak of interest in bone skates in English-language scholarship. A graph made using Google's n-gram tool (figure 24) shows that bone skates were most frequently discussed in the English-language Google Books corpus shortly before 1900.

It is not surprising that people became interested in bone skates during the nineteenth century. This was a time when the North was being both explored and romanticized and, perhaps not coincidentally, figure skating was emerging. Explorers tried to reach the North Pole, while scholars began to translate medieval Scandinavian literature, including the Icelandic sagas, from Old Norse to English. Recent books such as Andrew Wawn, *The Vikings and the Victorians: Inventing the Old North in Nineteenth-Century Britain* (Woodbridge: Boydell and Brewer, 2000) and Annette Kolodny, *In Search of First Contact: The Vikings of Vinland, the Peoples of the Dawnland, and the Anglo-American Anxiety of Discovery* (Durham, NC: Duke University Press, 2012) describe the enthusiasm of the Victorians for everything northern. These two books focus on the reception of the Icelandic sagas. Wawn describes how the Victorians used medieval sources to effectively invent a Viking Age, and Kolodny focuses on why the Scandinavian discovery of North America was important to Americans. During this time, many of the stereotypes about medieval Scandinavians, often called vikings, arose. One of the best known is that they drank from the skulls of their enemies. This is based on a mistranslation—the text actually refers to the horns of animals.[9]

At the same time, figure skating was developing both in England and on the Continent. The earliest extant book on skating, Robert Jones's *Treatise on Skating*, first published in 1772,[10] ushered in a new era of figure skating. Following him were, most notably, H. E. Vandervell and T. Maxwell Witham, *A System of Figure-Skating: Being the Theory and Practice of the Art as Developed in England, with a Glance at Its Origin and History* (Lon-

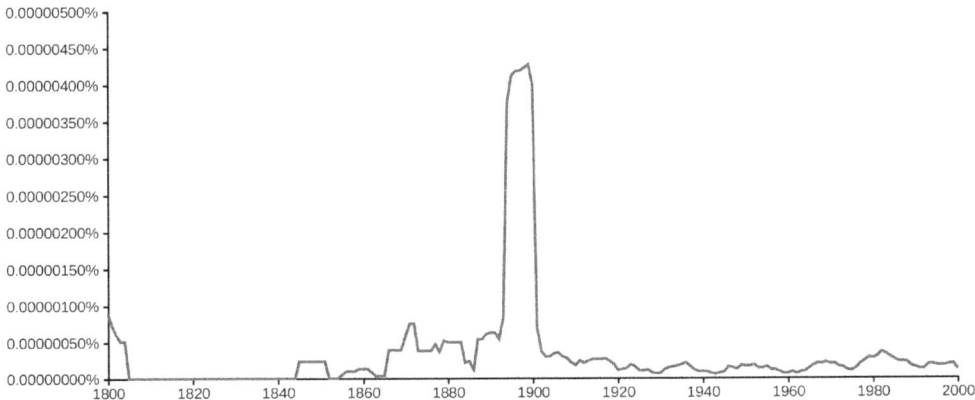

Figure 24: Results for "bone skates" and "bone skate" in Google's English-language n-gram corpus. The horizontal axis represents the publication year and the vertical axis shows the phrase's frequency (created using the Google Books Ngram Viewer at http://www.google.com/ngrams/).

don: Horace Cox, 1869) and Demeter Diamantidi, C. von Korper, and Max Wirth, *Spuren auf dem Eise: Die Entwicklung des Eislaufes auf der Bahn des wiener Eislauf-Vereines*, 2nd ed. (Vienna: Alfred Hölder, 1892), which cataloged all the known figure skating moves. All these books and many more recent ones begin with references to bone skates. Jones quotes William fitz Stephen, and the others provide more detail based on the scholarship available when they were written.

Figure skating and medieval Scandinavians were linked by Romantic literature, which was popular at the time. The clearest example of this link is the popularity of *Frithiofs saga* by Esaias Tegnér. First published as a whole in 1825, this was a modern Swedish reworking of the Old Norse *Friðþjófs saga ins frækna (The Saga of Frithiof the Bold)*. *Frithiofs saga* was translated into English, German, and other languages many times during the nineteenth century. It was so popular that Wawn devotes an entire chapter of *The Vikings and the Victorians* to its reception. The original saga was also popular; in 1894, John Sephton's reading of his translation at the meeting of the Liverpool Philosophical and Literary Society attracted more listeners than any subject other than Darwin's theory of evolution.[11] Since then, its popularity has declined, and today, Frithiof's story has mostly been forgotten.

The link to skating appears in canto 18. Frithiof accompanies the king and queen as they travel across a frozen body of water. They are in a sled pulled by horses with him on foot beside it. Suddenly, the ice breaks and the sled falls through. Frithiof heroically saves the day by pulling them, sled and all, back into the surface of the ice. He does this while wearing a *stålsko* (steel shoe), which has been interpreted as a pair of ice skates by translators and illustrators even though crampons are more historically and physically appropriate.[12] Bone skates would not have been useful in this situation. Had Frithiof been on them, he would have simply followed the sled through the hole in the ice.

The nineteenth-century enthusiasm for both medieval Scandinavia and figure skating spurred on the study of bone skates. However, the scholarship of the time was not error-free, and mistakes made early on were repeated by others who knew no better. This was known even to some contemporary authors, including Fowler,[13] who justifies his parts of his book on skating history with the remark,

> Some points of antiquarian and philological interest are dealt with superficially in the following pages. Little pretence at originality is made in respect of them, yet it seemed worth while to print them for two reasons:—The first, that every skating handbook but one either omits them altogether, blunders over them, or makes sweeping and unsupported statements thereanent; the second, that they may perhaps serve to warn the future historian of the Art of Skating from one or two less obvious rocks.

One example of a simple error is Guido Weiss's report to the Berlin Society for Anthropology, Ethnology, and Prehistory on John Stow's *Survey of London*.[14] Stow's book includes an English translation of William fitz Stephen's twelfth-century *Description of London*, but Weiss did not realize how old this text was. He thought it was contemporary with Stow's translation. Therefore, he concludes that people skated on bones in England in the sixteenth century.[15] Although this conclusion is probably correct, it does not follow from fitz Stephen's description. A more grievous example is the claim that metal-bladed skates were invented in 200 CE, which still occasionally appears in histories of skating. This seems to have originated with one Mr. Kreuger; a translation of his Swedish text is printed in Vandervell and Witham.[16] The actual text begins, "The origin of skates in their present form of a wooden shoe with iron runners cannot be reckoned further back than the so-called Iron Age, or about two hundred years after the birth of Christ, because iron first came into general use then throughout the North." This does not imply that metal skates were invented in 200 CE, merely that it would have been impossible for them to have been invented before 200 CE because metal was not available then. However, Kreuger finishes his brief account with the claim that "On these grounds the origin of the present form of skates and skating may be attributed to the Northern people about two hundred years after the birth of Christ," falling into the error of considering a lower bound the greatest lower bound without proof. Other authors simply copied this assertion without examining it carefully. In fact, there is no evidence for metal-blades skates until a thousand years after this date. That the technology to make something is available does not necessarily mean it gets made.

As Fowler notes, mistakes like this have propagated through histories of ice skating as the author of each new skating book repeats what was said in previous books. These errors are not always the result of carelessness; it can be hard to tell whether an artifact is a skate or when a text refers to skating, skiing, or snowshoeing. A serious problem in the prehistory of ice skating is trying to figure out what is and is not about skating. In this book, I rely on both literary and archaeological evidence. Each type of evidence presents its own problems. In literature, it can be hard to tell whether skating or another activity is being described. In archaeology, determining whether an artifact was actually used as a skate or as something else can be difficult, and some results are still controversial. The next section of this chapter is about translations of Old Norse literature, especially errors in older translations that have made their way into skating history. The final section deals with the archaeological problem of identifying skates. This problem has been considered repeatedly for over a century as archaeologists refine their understanding of bone skates and the societies that produced them.

3.2. Identifying Bone Skates in Written Records

The enthusiasm for both skating and the North led to the appearance of skating where it should not have been in translations of Old Norse literature. These early mistranslations highlight the difficulty of identifying references to skating in medieval literature. A good grasp of the language and a careful analysis of the text are necessary to extract information about skating from these texts. Because skating historians are typically not scholars of Old Norse, translation errors were repeated in histories of skating long after they lost the support of scholars. This was clear even in the late nineteenth century; Fowler[17] describes the problem aptly:

> One is frequently advised to look in the Scandinavian sagas for the earliest references to skating; but we must attribute to the neglect, with which the study of the old Scandinavian tongue has met until quite recent times, the extraordinary mistranslation of these passages from Icelandic sagas and the like, which were supposed to refer to skating, which have been copied and recopied from one skating manual to the next, but which now prove to have nothing whatsoever to do with the art.

The following two examples from Thomas Percy's popular translations of Old Norse literature were noted by Smith and Smee[18] and have spread through the literature on skating history. The first is from Percy's *Translations of Runic Poetry*, first published in 1763 and one of the earliest translations of Old Norse poetry into English. In a translation of a poem he calls "The Complaint of Harold," Percy includes the following verse:

> I know how to perform eight exercises. I fight with courage; I keep a firm seat on horseback; I am skilled in swimming; I glide along the ice on skates; I excel in darting the lance; I am dexterous at the oar; and yet a Russian maid disdains me.[19]

"I glide along the ice on skates" appears to the modern reader to be a clear reference to skating. The Old Norse from which this line is translated is "skriða kannk á skíðum," which actually means "I can slide on skis"; it occurs twice in the Old Norse corpus, in *Morkinskinna*[20] and *Orkneyinga saga*.[21] Percy calls both poems "The Complaint of Harold"; the former is quoted above, and he translates this line as "I traverse the snow on scates of wood" in the latter.[22] The dissimilarity of these two translations of the same line of poetry shows that Percy considered skating and skiing interchangeable.

The difficulty is compounded by changes in the English language between Percy's time and today. As English continues to change, the meanings of words shift. The current distinction between skates (for use on ice), skis (for use on snow), and snowshoes (for use on snow in a different way) is quite recent. Although Smith[23] explains the difference between "snow skates of wood … in the North called *skier*," which are "about six feet long, and, of course, very different in construction from those of bone," the conflation of skates and skis survived into the twentieth century. In his discussion of these passages, Brown[24] suggests that the "skates of wood" Harold used on the snow were skis or "snow skates in the manner of Axel Paulsen," a Norwegian speed skater in who competed in the late nineteenth and early twentieth century.[25] Adding snowshoes to the mix compounds the problem. Two popular Old Norse dictionaries, Geir T. Zoëga's *Concise Dictionary of Old Icelandic* (1910) and Richard Cleasby and Gudbrand Vigfusson's *Icelandic-English Dictionary* (1874), include "snowshoe" as a translation of the Old Norse *skið*, which means "ski." Both were written over a century ago, when skis were sometimes called snowshoes, but now a snowshoe is something else.

Percy made a more egregious translation error in an episode from *Gylfaginning* in Snorri Sturluson's *Prose Edda*. Thor and his friends go to visit Utgarda-Loki and are challenged to compete in a variety of events. They seem to fail, but when the true nature of those events is revealed, it turns out that they have done remarkably well. The Old Norse is as follows:

> Þá spyrr Útgarða-Loki, hvat sá inn ungi maðr kunni leika, en Þjálfi segir, at hann mun freista at renna skeið nökkur við einhvern þann, er Útgarða-Loki fær til. Þá segir Útgarða-Loki, at þetta er góð íþrótt, ok kallar þess meiri ván, at hann sé vel at sér búinn of skjótleikinn, ef hann skal þessa íþrótt inna, en þó lætr hann skjótt þessa skulu freista. Stendr þá upp Útgarða-Loki ok gengr út, ok var þar gott skeið at renna eftir sléttum velli. Þá kallar Útgarða-Loki til sín sveinstaula nökkurn, er nefndr er Hugi, ok bað han renna í köpp við Þjálfa. Þá taka þeir it fyrsta skeið, ok er Hugi því framar, at hann snýst aftr í móti honum at skeiðsenda.
>
> Þá mælti Útgarða-Loki: "Þurfa muntu, Þjálfi, at leggja þik meir fram, ef þú skalt vinna leikinn, en þó er þat satt, at ekki hafa hér komit þeir menn, er mér þykkja fóthvatari en svá."
>
> Þá taka þeir aftr annat skeið, ok þá er Hugi kemr til skeiðsenda ok hann snýst aftr, þá var langt kólfskot til Þjálfa.
>
> Þá mælti Útgarða-Loki: "Vel þykkir mér Þjálfi renna skeiðit, en eigi trúi ek honum nú, at hann vinni leikinn, en nú mun reyna, er þeir renna it þriðja skeiðit."
>
> Þá taka þeir enn skeið, en er Hugi er kominn til skeiðsenda ok snýst aftr, ok er Þjálfi eigi þá kominn á mitt skeið. Þá segja allir, at reynt er um þenna leik.[26]

3.2. Identifying Bone Skates in Written Records

Percy's full translation is:

> Then the king asked, what that young man could do, who accompanied Thor. Thialfe answered, That in running upon scates, he would dispute the prize with any of the courtiers. The king owned, that the talent he spoke of was a very fine one; but that he must exert himself, if he would come off conqueror. He then arose, and conducted Thialfe to a "snowy" plain, giving him a young man named *Hugo* (Spirit or Thought) to dispute the prize of swiftness with him. But this Hugo so much outstript Thialfe, that in returning to the barrier whence they set out, they met face to face. Then says the king; Another trial, and you may exert yourself better. They therefore ran a second course, and Thialfe was a full bow-shot from the boundary, when Hugo arrived at it. They ran a third time; but Hugo had already reached the goal, before Thialfe had got half way. Hereupon all who were present cried out, that there had been a sufficient trial of skill in this kind of exercise.[27]

Percy relied heavily on the Paul Henry Mallet's 1755 French translation in *Introduction à l'Histoire de Danemarc (Introduction to the History of Denmark)* for his translations of Old Norse,[28] and, in general, the results are "far from perfect."[29] The word for a footrace, *skeið*, is quite similar to *skreið*, the singular past tense of the verb *skríða*, which was used for skiing and, as I argue in chapter 6, would have been appropriate to use with bone skates. As Fowler[30] notes, *skeið* is also extremely similar to *skið*, which means "ski." The similarity of these words confused Mallet and, through his work, Percy. Another problem with his translation is that Percy describes the race as occurring on "a 'snowy' plain"; this seems to be a misunderstanding of *sléttum velli* (level ground). There is no snow in the Old Norse.

By the end of the nineteenth century, the passage was understood correctly.[31] A more accurate translation is

> Then Útgarða-Loki asks, what the young man could do, and Þialfi says, that he will try to run a footrace of some type with anyone who Útgarða-Loki chose. Then Útgarða-Loki says that that is a good skill, and he expects that he will be very nimble, if he should win this game, and they will try him at this. Then Útgarða-Loki stands up and goes out, and there is a good racecourse over level ground. Then Útgaða-Loki calls to a certain one of his boys, who is called Hugi, and asks him to run a race against Þjálfi. Then they run the first race, and Hugi is so far ahead, that he turns back to meet him at the end of the race.
>
> Then Útgarða-Loki says: "Þjálfi, you will need to apply yourself more if you are to win the game, but it is true that no men have come here who seem to me more swift-footed than you."
>
> Then they run another race, and when Hugi comes to the end of the route and turns back, it is a long bolt-shot to Þjálfi.
>
> Then Útgarða-Loki says: "It seems to me that Þjálfi runs the footrace well, but I do not believe now, that he wins the game, but now let us see how they run the third race."
>
> Then they run a race, but when Hugi comes to the end of the race and turns back, Þjálfi has not reached the middle of the race. Then, everyone says that the game is decided.

Hugi, the Old Norse word for "thought," turns out to be just that, and clearly unbeatable. Yet Þjálfi did well, which attests to his great athleticism. However, he was not a skater despite the popularity of Percy's mistranslation, which formed the basis for Friedrich Gottfried Klopstock's 1767 poem "Die Kunst Thialfs" (The Art of Thialfi),[32] one of the three odes with which he popularized ice skating in Germany.[33] Klopstock's poems were reprinted in one of the first German books on skating, Christian Siegmund Zindel's *Der Eislauf* (1825).

Old Norse is not the only language with passages that are difficult to interpret, and it was not only Victorian scholars who had trouble. Middle Dutch has also presented some difficulties. An entry from 1398 in an account book from the court of Duke Albert of Bavaria in the Hauge is occasionally cited as evidence that the earliest metal-bladed skates were used with poles. It records the purchase of "XII piecstaffen voir mijn here mede te vliegen mit ijsers aen de voeten."[34] This has been translated as "for my lord twelve pikestaffs for flitting with irons under his feet," with the suggestion that it may refer to "a medieval cross between skates and skis."[35] Janse[36] disagrees with this interpretation and proposes a different one: these were staffs meant to be stuck into the ground, not used for skating. Mulder[37] provides a more detailed explanation: pegs (*piecstaffen*) were used to secure nets used in bird-hunting (*vliegen*). Iron ends seem likely to be helpful in securing the pegs in the ground, and this interpretation also helps why so many were ordered. A better translation is "twelve pointed sticks with irons on the feet (i.e., shod with iron) for my lord for bird-hunting."

Illustrations add another layer of difficulty. The well-known pictures of skaters in Olaus Magnus' works are not entirely accurate, but the difference between skates and skis is clear (skis are long and skates are short; see figure 25, which shows both). Not all artists made this distinction. In figure 26, the footwear of the people depicted is similar to the bone skates and skis shown in figure 18, but longer than skates and shorter than skis. Are these skates or skis? Foster[38] was not sure. But the text, which includes the phrase "ligneæ soleæ longitudinem trium cubituum"[39] (wooden shoes three cubits long), suggests that they are actually skis.

Figuring out whether an image depicts skates or skis is an example of what Straten[40] calls "pre-iconographical description," the first step in iconography, the field of art history concerned with identifying objects in works of art and determining what they mean. It consists of cataloging the items. Later steps involve finding relationships among these objects to identify the work's theme and looking for deeper meanings. Where bone skates are concerned, the seemingly trivial task of identification can be daunting at times.

Despite these problems, it is possible to find skating in medieval literature. The archaeological evidence (Chapter 7) shows that medieval Scandinavians were enthusiastic

Figure 25: Skaters and skiers using single poles. Note how similar the skates (right, short) and skis (left, long) are (Olaus Magnus, *Historia de gentibus septentrionalibus*, 713).

3.3. Identifying Bone Skates in the Archaeological Record

skaters, and they did indeed mention it in their literature, but the references are often oblique. The evidence for skating in Old Norse literature is discussed in detail in Chapter 6.

3.3. Identifying Bone Skates in the Archaeological Record

While some scholars were reminiscing about their childhoods on bone skates, others began to question whether these artifacts were actually skates. This signaled a large disconnect between the archaeological and ethnographic evidence for bone skates. Part of the problem was that people who had not skated on bones, including enthusiastic figure skaters, did not understand how bone skates worked. The concept of skating without firmly attached blades was foreign to people raised with metal-bladed skates. The lack of binding holes on many bone skates made Virchow[41] look for another interpretation of these artifacts. Because a friend from Lithuania told him that similar bone tools were used for weaving there, he suggested that bones that looked like skates but did not have any

Figure 26: Are these people using long skates or short skis? It is difficult to tell in this woodcut (Alessandro Guagnini, "Arma Magni Ducis Moschouiæ," folio 16v).

holes for attachment were used in the textile industry instead. Similarly, Tergast[42] identified an object from a mound in Grimersum that looks very much like a skate made from a cattle metapodium as a "Knochen, an einer Seite poliert, zum Glätten des Gewebes" (bone, polished on one side, for smoothing textiles).

The confusion continued for over a century. Fowler[43] cites "a recent manual of skating" as denying that bone skates were used for skating because it would have been impossible for skaters to push with their feet as they do with metal-bladed skates. In response, he made a pair of bone skates and showed experimentally that it was possible to push with the toe, though not the side, of a bone skate. Fowler's goal was to settle the matter by showing that bone skates could have been used in the same way as modern skates, but this only addresses part of the confusion. For foot-pushing to work at all, the skates have to be attached to the skater's feet. The question of why many bone skates lack binding holes remained.

Nearly half a century later, Kjellberg[44] tried, unsuccessfully, to skate on bones. Because he could not get them to work, he followed Tergast in supposing bone skates were actually tools from the textile industry. Shortly after that, Semenov[45] argued that common artifacts from eastern European sites such as Olbia, Sarkel, and Staraya Ladoga dating to the first millennium BCE that had been identified as bone skates could not have been used for skating. His argument includes five main points:

1. There was no way to attach these artifacts to the skater's feet.
2. The edges of the bones ought to have been rounded off by skating.
3. Because skaters skate by pushing with their feet, the wear patterns are not parallel to the blades; they are at an angle.
4. If they were skates, they should exhibit wear without a definite pattern due to imperfections in the ice surface.
5. Some skates do not have upswept toes, and it would be impossible to skate over bumps without them.

All these criticisms are based on a single underlying assumption: that bone skates were used in the same manner as modern skates are, i.e., that skaters pushed with their feet. The problem is summarized well but unintentionally by Brown,[46] who describes Fowler's experiment as demonstrating "the possibility of these primitive bone runners having been used centuries ago as skates proper." The final phrase of this sentence brings out the significant difference between bone skates and modern metal-bladed skates. With "skates proper," skaters push with their feet. With bone skates, this was not done. If the definition of skating requires foot-pushing, bone skates are not skates, which is why Allen,[47] Balfour,[48] and others proposed different names for them. They are also not textile smoothers.

If Semenov had visited one of the rural areas in which bone skates were still in use, he would have been seen how they were used and realized that the wear patterns on bone skates ought not to be the same as those on metal-bladed skates because they were used differently. This is an instance of people in one field (archaeology) not knowing what was going on in another (ethnography). Arthur MacGregor, who eventually settled the matter, considered himself "fortunate to discover a more extensive and persuasive ethnographical literature which showed the same objects to be ice skates, in regular use in certain communities up to the present century."[49]

3.3. Identifying Bone Skates in the Archaeological Record

In his review article, MacGregor[50] attributes the confusion in part to the fact that the use of bone skates was becoming increasingly uncommon and in part to the popularity of other interpretations in earlier works. A year earlier, in a direct response to Semenov, he settled the matter by making and using two pairs of bone skates.[51] He showed that the wear patterns that result from skating match the wear patterns on archaeological artifacts. The wear patterns on both MacGregor's bone skates and the archaeological specimens he compared them to consisted primarily of longitudinal striations along the gliding surface with random secondary scratches.[52] A typical wear pattern is shown in figure 27. Since MacGregor's experiments, others have continued evaluating the wear patterns on var-

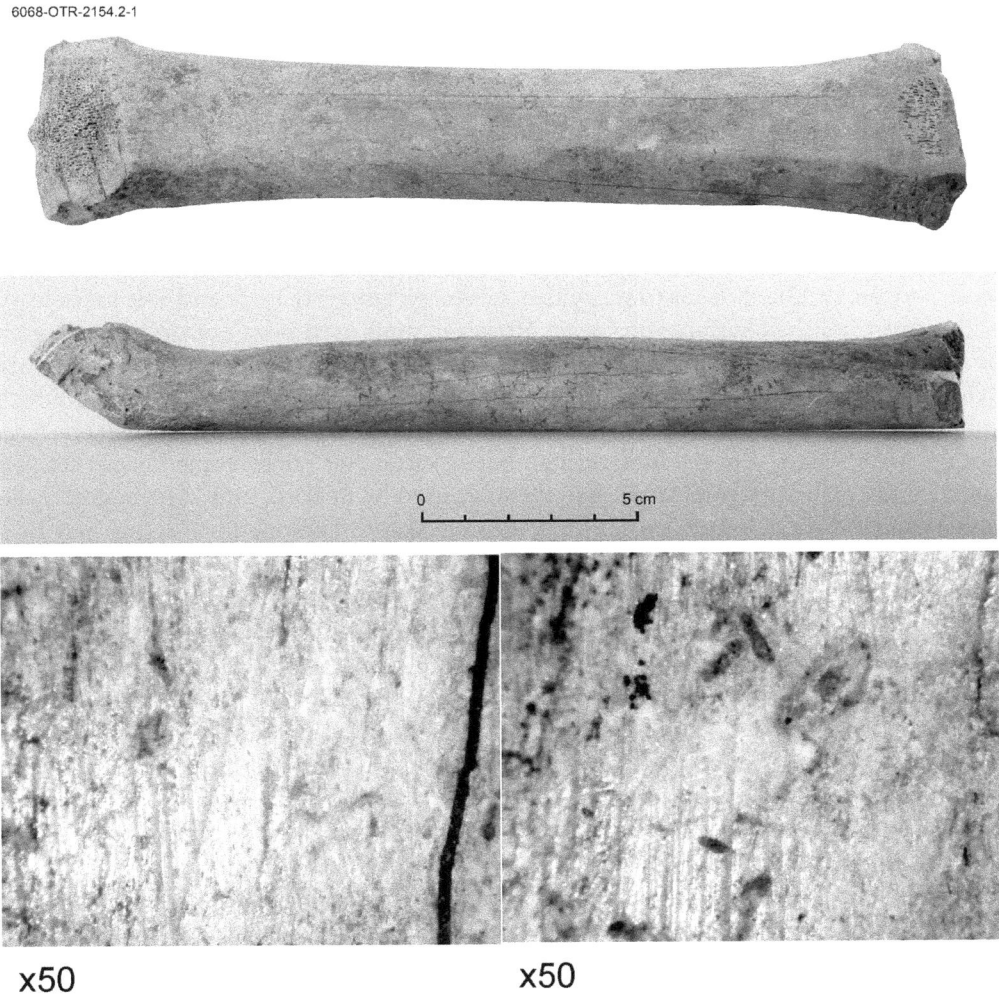

Figure 27: An equid metacarpus skate from Altorf, France, that dates to the fourth century CE. Shown are the dorsal (sliding) side (top), a side view (middle), and two images of the wear pattern (bottom row) (Olivier Putelat, "Les relations homme-animal dans le monde des vivants et des morts: Étude archéozoologique des établissements et des regroupements funéraires ruraux de l'Arc jurassien et de la Plaine d'Alsace de la fin de l'Antiquité Tardive au premier Moyen Âge" 615, fig. 633, courtesy Olivier Putelat).

ious bone artifacts that have been identified as skates. All agree that the primary wear pattern on bone skates ought to be parallel to the axis of the bone. Barthel[53] stresses the importance of striations running parallel to the long axis of a bone skate; bones used in the tanning process show striations perpendicular to this axis instead. This has remained the primary factor in the identification of bone skates. Küchelmann and Zidarov[54] require bones to exhibit noticeable wear to be considered skates.

The remaining questions are about the details of the wear pattern, such as exactly where on the bone it ought to occur. Becker[55] suggests that the location of the wear pattern can be used to differentiate between textile smoothers and skates. On smoothers, the wear pattern is more toward one end, as if the bone did not lie evenly on the surface (perhaps due to a hand holding onto one end while it was in use), whereas a well-worn bone skate is expected to show wear along its entire surface. "Well-worn" is the key here; my own experiments have shown that the shape of the bone affects the location of the wear pattern. Because my metacarpus skates are arched in the middle, only the ends, where the bone touches the ice, have developed the characteristic longitudinal striations. My radius skate also exhibited wear along only part of its length after a moderate amount of skating due to its curvature. However, skates become flatter with use, which means that the wear pattern should eventually cover the length of the bone.

When a skate has not been used long enough to develop the characteristic full-length wear pattern, can be difficult to recognize. LeMoine[56] notes that ice and soft hides leave with similar polished surfaces on bones. However, tools used in wet environments, such as antler ice picks and bone snow knives, are characterized by visible osteons. Osteons are cylindrical structures made up of compact bone tissue around haversian canals, which supply blood to the bone. LeMoine did not examine bone skates, and visible osteons have not yet been reported in the literature on bone skates. Her microscopic images are at a significantly higher magnification than the ones used by Becker[57] and MacGregor,[58] who only examined scratches on the surfaces of bone skates. Looking for osteons may be a way to differentiate between bone skates and tools used in the textile industry, provided the tools were used to work dry textiles. Archaeologists could also look recent advances in engineering science for more ideas about how to evaluate wear patterns. Stemp, Watson, and Evans[59] describe some new methods, including engineering techniques based on fractals, interferometry, and laser scanning confocal microscopy. These methods bring discoveries in metrology and tribology into the context of archaeology and allow wear patterns to be quantified.

Despite these advances, little has changed in the years since MacGregor noted that "their identification ... is still challenged from time to time,"[60] and it remains difficult to identify bone skates in some archaeological contexts, especially prehistoric ones. Kustár and Tugya[61] suggest that archaeologists can be overly hasty in identifying skates and that the definition of a skate needs refinement. They propose that bone skates be identified in a holistic manner using all the information available, not just the wear pattern. Wear patterns remain the most common method of identifying bone skates but are not always sufficient. There are some cases where using wear patterns as the only criterion can be limiting: skates do not always show wear, either because they had been shaped but were never used or because the gliding surface has not survived. Even when it has survived, wear to the gliding surface may not always be visible because of poor preservation due to soil type or other natural factors.

3.3. Identifying Bone Skates in the Archaeological Record

To improve on the use of wear patterns to identify bone skates and because many of the skates they examined appeared new, Edberg and Karlsson[62] required that each artifact classified as a skate have a flattened (if only rudimentarily) gliding surface. Visible wear was not required. They rated the amount of wear to each skate on a four-point scale. Skates that scored two on this had a prepared gliding surface but no signs of use. Skates that scored four were extensively worn, like the skate in figure 23.

Another difficulty is differentiating between bone skates and bone sled runners. They were used in similar ways and therefore, have similar wear patterns. This is where the presence of holes becomes an appropriate criterion for identifying bone skates. The presence of one or more vertical (dorsal-palmar) holes suggests that an artifact is a sled-runner instead of a skate.[63] Some sled-runners may have been bone skates of a different type: Schuldt[64] proposes that some such bones may have been used as skates in the form of foot-sized sleds, i.e., with a small wooden platform attached to the bone via the hole. MacGregor[65] calls this idea "unsupported by any evidence," but Thunig[66] recalls that "Söhne von Zimmerleuten oder Tischlern befestigten auf diese Knochen Klötze resp. Bretter, so dass ein schlittschuhähnlicher Gegenstand zum Vorschein kam. Ich kann mich noch sehr gut daran erinnern, dass mir meine Mutter zu Weihnachten 1820 oder 21 ein Paar dergleichen Schlittknochen mit Brettern schenkte" (sons of carpenters and joiners attached to these bones [i.e., bone skates] pads or boards, so that a skate-like object appeared. I can remember very well, that for Christmas 1820 or 21 my mother gave me a pair of such bone skates with boards). This type of skate may have been a local innovation. It complicates the distinction between bone skates and sled runners.

Questions about whether artifacts are actually skates continue to the present day. The rest of this chapter is devoted to a discussion of two particular classes of artifact whose identification as skates has recently been questioned: objects made from metapodia found at Sabatinovka culture sites near the Sea of Azov and objects made from radii found at various locations in the steppes and central Europe. In this book, I leave the possibility that all these artifacts are skates open because I do not consider the evidence against their identification as skates strong enough and because there is no need to identify them as other tools. However, I also leave the possibility that they are not skates open because I have not had the opportunity to examine them for myself.

The artifacts that were discussed in the literature first are from Sabatinovka culture sites near the Sea of Azov. Gerškovič[67] describes skate-like objects from three sites (Sabatinovka, Novokievka, and Zlatopol'). Pankovskiy[68] reports finding additional horse metatarsus and cattle radius skates at Sabatinovka. Skates or skate-like objects have also been found at other sites associated with the Sabatinovka culture. Gerškovič[69] reports that they were first identified as smoothers that may also have been used to drag small loads across ice by B. H. Peters in 1986. A. N. Usacuk examined the wear patterns and thought they showed signs of use on "weichen, elastischen Stoffen (Wolle, Leder, Filz usw.)" (soft, elastic materials (wool, leather, felt, etc.)). These comments are similar to previous thoughts on the use of bone skates in the textile industry. Peters and Usacuk may have known about Semenov's work but not MacGregor's refutation of it.

Gerškovič[70] suggests that Peters' idea that the bones could have been used as sled runners is untenable because the bones were too weak. This contrasts with Cornwall's description of a horse metapodium as quite strong and sturdy.[71] An experimental study of the strength of metapodia may provide insight into the maximum load they could bear. In any case, the artifacts in Gerškovič's figures look like skates, except that the skate from Zlatopol' in his figure 10.1 seems slightly curved. However, this need not rule out its identification as a skate. MacGregor[72] describes similar curvature in skates from Trondheim, Norway:

> Mr Clifford Long informs me that since 1971 at least sixty-five bone skates have been found during excavations in Trondheim. Some of these display a certain concavity of the bone which could hardly be the result of abrasion on a flat surface, a feature which led the excavators at Århus to suggest that the wear must have been produced by contact with some elastic material such as skins or textiles....[73] After consultation with Mr James Rackham (but without having handled the skates in question), I would suggest that this longitudinal curvature is probably the result of distortion, perhaps during drying out. Since the anterior surface will often have been seriously weakened by abrasion while the posterior surface remains relatively intact, the columnar bone will tend to distort due to internal stresses.

Based on this, it seems best not to rule out the notion that these artifacts were skates yet.

The other class of questionable artifact consists of tools made exclusively from radii that date primarily to the Middle Bronze Age and the beginning of the Late Bronze Age and have been found between the Black Sea and Hungary. The only modification made to the bones used to create these artifacts was to flatten the proximal end of the dorsal side by removing some of the surface and any protrusions.[74] Examples include Middle Bronze Age artifacts made from cattle radii found in Hungary[75] and near the Black Sea[76] and Early Iron Age artifacts found at Biskupin in Poland.[77] The Middle Bronze Age artifacts from Veselé and Nitriansky Hrádok in Slovakia[78] and objects found at Pécel near Budapest[79] may also belong to this category. Similar artifacts have been found at Smuszewie,[80] Słupcy,[81] Jankowie,[82] and Łagiewnikach.[83] Additionally, Choyke and Bartosiewicz[84] note that the Middle Bronze Age artifacts identified as skates by Furmánek, Veliácik, and Vladár[85] are based on radii of various animals and exhibit wear facets. Details are not provided, but these sound similar to the mysterious artifacts in this class.

Lukasiewicz and Rajewski,[86] who were the first to interpret the artifacts from Biskupin, identified them as skates. Then, Rajewski[87] suggested they were used in the textile industry, and Mogielnicka-Urban[88] interpreted them as tools for honing ceramics. Finally, Drzewicz[89] identified them as spatulae used for working hides based on an analysis of 75 artifacts (73 radii, 1 tibia, and 1 humerus) from different types of animal (42 red deer, 1 roe deer, 2 horse, 6 sheep/goat, 13 cattle, 1 dog, and 7 pig/wild boar). All these artifacts are between 25 and 34 cm long and between 3.2 and 4.5 cm wide.

The evidence Drzewicz uses to identify these artifacts as tools used in the leather-working industry is that they are not worn along their whole length like skates are. Instead, "[t]he working surface covers from 1/2 to 3/4 of the length of the bone.... Its appearance suggests that the objects were used in working soft organic materials, eg [sic], cloth and pelts."[90] This description is consistent with the wear on both skates and textile smoothers described by Becker,[91] who shows a drawing of a skate with a wear facet extending just over half its length. The presence of mainly longitudinal striations, as opposed to striations in various directions, seems to be the key to identifying a skate, but the angle at which a tool is held affects the wear pattern, resulting in shorter facets on some textile smoothers. The natural curvature of the bone also affects the location of the wear facet. The more common and flatter metapodium skates naturally have wear facets that extend along their full length (or focused at the ends of bones with concave dorsal sides). Curved radii would not necessarily have lain flat on the ice; therefore, the facet need not extend along the full length of the bone.

3.3. Identifying Bone Skates in the Archaeological Record

Drzewicz's interpretation of the artifacts from Poland has led to the re-interpretation of similar artifacts from other parts of Europe. Sofaer, Jørgensen, and Choyke[92] refer to this class of artifact as "[a]nother specialized tool, often incorrectly published as skates" that "spread rapidly over large areas of central and eastern Europe at the end of the Middle Bronze Age." In a recent paper, Choyke rules out the skate interpretation because these artifacts are made from the bones of animals that differ widely in size. She now classifies them as tools for "some kind of widespread and important craft activity, perhaps leather or felt-working" even though they exhibit wear patterns that are consistent with skating.[93] Currently, the general archaeological consensus is that these are not skates.

Several artifacts based on cattle radii found at Százhalombatta-Földvár are very similar to the artifacts based on horse radii found at the Early Bronze Age site Albertfalva, Hungary, that Choyke and Bartosiewicz[94] call skates. This similarity has been noted by Sofaer,[95] and the Albertfalva artifacts seem to fall into the mysterious radius-based artifact category. These four fragments are made from horse radii and exhibit flat facets with longitudinal striations, which is the characteristic wear pattern of bone skates. However, unlike the Middle Bronze Age radius-based artifacts from Százhalombatta-Földvár and many of the other artifacts in that class, at least one of the Albertfalva artifacts features a hole drilled through its distal end.[96] Choyke[97] notes that such holes occasionally appear on the mysterious artifacts.

Choyke and Bartosiewicz[98] also describe a Middle Bronze Age artifact based on a horse radius with vertical holes near both ends of the palmar side that was found at Százhalombatta-Földvár. They originally identified it as a sled runner because of the holes' orientation and placement, but Choyke is no longer certain of its function.[99] It and the other skate candidates from this site have wear patterns that match the patterns on the Albertfalva artifacts and other artifacts from other sites that are nearby in both space and time: Bell-Beaker sites in the Csepel-Háros group. Other candidates for this class include an unknown number of Early Bronze Age skates made from unidentified bones found in southwestern Slovakia described by Točik,[100] two horse radii from Höngeda, Germany,[101] two radius-based artifacts from Buch, Germany (near Berlin)[102] and Bronze Age skates or runners made from unidentified bones found in Switzerland.[103]

A careful study of all the Bronze Age radius-based artifacts, including the ones from Albertfalva, is needed to ascertain exactly what their function was. Because this question affects the earliest evidence for bone skates and is highly relevant to the origin of ice skating described in the next chapter, I experimented with a radius skate of my own devising. I began by purchasing the radius of a young steer from a local butcher. I left it in my backyard for several months for cleaning, then removed the ulnar protrusion and filed down the proximal end of the dorsal side. This resulted in a skate that felt very stable standing on concrete, except that a slight shift in weight could cause it to tilt toward the ulna—I had to balance on one or the other of the two fused bones. Balancing on the radius itself kept my ankle straighter. Removing additional material may have helped alleviate this at the expense of historical accuracy.

This skate, shown in figure 28, was approximately 28.9 cm long, which made it slightly longer than my boot. Because of its curvature, it was slightly uncomfortable to stand on; a longer skate would have helped alleviate such discomfort by keeping my foot toward the flatter middle of the bone. I found the skate more comfortable to use with the distal end in front. After skating on bumpy outdoor ice for approximately fifteen minutes,

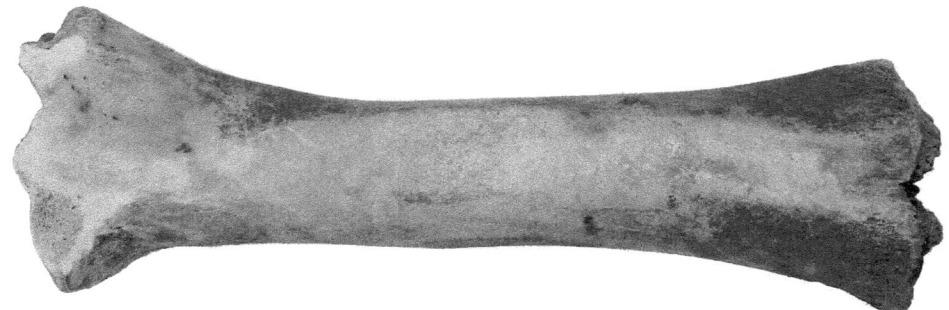

Figure 28: The author's radius skate. The wear facet described in the text is the light area on the right.

I saw signs of use running from the distal end to roughly the midpoint of the bone. The facet gradually lengthened and widened with additional skating time.

The outcome of this experiment and discussion is that none of the reasons given for why these radius-based tools cannot be skates entirely disqualify them. Longer facets may correspond to more use, but a short facet does not mean that the bone was not used as a skate. Bones from different species could have been chosen based on local availability or to meet the needs of different-sized skaters, and the facets need not have extended the full length of the bone. Perhaps some of the new methods discussed by Stemp, Watson, and Evans will shed more light on the use of these artifacts in the not-too-distant future. In particular, they note that scale-sensitive fractal analysis has promise because it makes it possible to quantify the characteristic wear patterns at different scales and see how the patterns resulting from plants and hides differ.[104] They do not discuss patterns formed by contact with ice, but the same principles should apply. Perhaps this will provide a way to differentiate between hide smoothers and ice skates. Determining how isotropic the wear patterns are may also help: ice skates mostly slide in one direction (forward), while hide smoothers may have been slid back and forth by their users.

Despite many questions and refutations over the last century, it remains indisputable that animal bones were used as ice skates in the manner described in Chapter 2. In what follows, I am cautious about using the Albertfalva artifacts as evidence and leave most of the radius-based artifacts out completely, except to note where they could fit in if they become accepted as skates. The identification of artifacts in that group remains open, as far as this book is concerned: maybe they are skates, and maybe they are not.

4

How Ice Skating Came to Be

4.1. An Origin Story

This chapter presents a story about how ice skating may have come to be. This story fits with the evidence currently available, but more evidence needs to be found before it can be considered more than a hypothesis. It provides a direction for archaeologists and other researchers to go as they continue looking for the first ice skaters. According to this hypothesis, ice skating was probably first done in the Eurasian steppes sometime between approximately 3300 and 2600 BCE and no later than 1500 BCE. The steppes provide the elements necessary for the invention of ice skates: a cold, dry climate and animals with sufficiently large bones. This location makes it possible that skating was originally connected with skiing, for which the earliest evidence so far comes from the Altai Mountains. Either skating was inspired by skiing or the idea of sliding across a frozen surface was developed independently in two relatively close areas.

Most of the scholars who have considered the origin of ice skating in the last half-century place it in or near the steppes. MacGregor[1] suggests that bone skates were invented in central Europe during the Bronze Age. Küchelmann and Zidarov[2] qualify this statement on the basis of additional archaeological evidence and suggest that bone skates appeared "sometime during the second millennium BC in the steppe zone stretching from the Northern Pontic Area in the east to the Great Hungarian Plain to the west." Choyke and Bartosiewicz[3] speculate that "skating may already even have been a rather important form of getting around for a number of Migration Period peoples, in their original homelands, the plains of the Eurasian steppe belt," which echoes the sentiment of Bartosiewicz.[4]

The outliers are two biomechanics researchers at Manchester Metropolitan University, Federico Formenti and Alberto E. Minetti. In 2008, they published a paper describing a mathematical calculation that they claimed showed that ice skating is most likely to have been invented in Finland to decrease the amount of energy required by winter travel.[5] This paper was met with enthusiasm by publications such as *National Geographic News*[6] and *Wikipedia*.[7] The study used data on the metabolic costs of skating and walking in Finland, the Netherlands, Germany, Sweden, and Norway, countries selected on the basis of Munro's statement, made over a century ago, that bone skates were "invented by the early Teutonic races who inhabited the shores of the Baltic."[8] Traveling by skating across lakes, as opposed to walking around them, had the lowest metabolic cost in Finland, and on that basis, Formenti and Minetti suggested that bone skates were invented there. Unfortunately, this result conflicts with the archaeological evidence. No Finnish bone skates that have been securely dated to before the fourteenth century CE are included in the literature on bone skates or Küchelmann's database, but prehistoric bone skates have been found in central Europe—a region Formenti and Minetti did not consider. Their analysis would be much more relevant to the invention of ice skating if it included the regions archaeologists have proposed as most likely. As it stands, it says more about the topography of the countries they evaluated than about where ice skating could have been invented.

The place to start is with archaeological evidence, but the earliest evidence for ice skating is difficult to pin down. The earliest skates included in Küchelmann's database are the four radius-based artifacts from Albertfalva, Hungary, discussed in the previous chapter. They date to approximately 2500 BCE[9] and were originally identified as skates, but later discussion of the class of mysterious radius-based artifacts has made that identification questionable. Similar artifacts from the Bronze Age have been found in southwestern Slovakia[10] and at Höngeda, Germany.[11] Kiekebusch[12] reports that two skates or runners made from radii (one horse and one cattle) were found at Buch, a Bronze Age site near Berlin, and Tschumi[13] describes Bronze Age skates or runners from Switzerland. Any or all of these may not actually be skates; they may belong to the class of mysterious radius-based artifacts. These sites, and others mentioned in this chapter, are shown in figure 29.

The next-earliest skates are the ones involved in the other recent controversy described in section 3.3. These artifacts are more generally accepted as skates than the

Figure 29: Sites where artifacts that have been identified as bone skates have been found (circles), sites with skis or rock art depicting skis (triangles) and horse-related sites (stars). The Sabatinovka culture area is cross-hatched, and bodies of water are shaded. The inset (lower right) shows an enlarged view of the cluster of sites in the Carpathian Basin (created with QGIS).

radius-based artifacts, in part because they are made from metapodia, which has been the norm for skates since the Late Bronze Age. They are from Ukraine and have been attributed to the Sabatinovka culture, which is generally thought to have flourished in the Dneiper basin just north of the Black Sea, in the region shaded in the map. This culture may date to 1500–1100 BCE,[14] but Gershkovich[15] notes that dates ranging from 2000 to 1700 BCE may be more accurate.

The controversy ends with the appearance of clear metapodium-based skates in central Europe roughly 1000–1500 years after the Albertfalva artifacts. Although the identification of some of these skates has been questioned from time to time, they are now generally uncontroversial. These include skates found at sites in the Czech Republic,[16] Slovakia,[17] Hungary,[18] and Italy[19] as well as two skates made from horse metapodia found in Vráblé, Slovakia.[20] From the Late Bronze Age onward, skates made from metapodia—usually of horses and cattle—are recognized and accepted as skates. They become fairly common in the archaeological record in the subsequent millennia. The later finds that are of most interest for a steppe origin are from two Sarmatian sites in Hungary: Gyoma 133, which contributes 48 skates (though one is questionable) from the Roman Iron Age,[21] and Endrőd 170, which contributes four skates from the Migration Period.[22]

Gyoma 133 is the site of prehistoric, Sarmatian (Roman period) and Avar settlements, but most of the evidence is from the Sarmatian settlement; in fact, this site is both large and early for a Sarmatian settlement in this region.[23] It dates to the second or third century CE.[24] The Sarmatians living there definitely had horses: Bartosiewicz[25] describes horse bones that are fragmentary or show marks of butchery and concludes that the Sarmatians ate horse meat at this site. Given the easy availability of horses, it is not surprising that skates were made from horse bones. The only other thing people living here needed was ice to skate on. The site was located near a stream, but the presence of a large well suggests to Vicze[26] that there was a prehistoric drought. A smaller Sarmatian well was found within it, which may or may not be taken as evidence of another, less serious, drought during Sarmatian times. The stream may have provided ice for skating, but drought may have made ice less readily available at times.

The Sarmatians were known for their horseback riding skills and for their steppe background; Herodotus describes them as the product of Amazon women and the Greek men who captured them and took them to sea. The women turned on their captors and returned the ships to land, where they captured a herd of horses and continued their invasion. When peace had been made, Sarmatian women continued dressing like men, riding horses, hunting, and fighting.[27] The archaeological evidence shows that the Sarmatians can also be remembered as skaters. Steppe peoples such as them are most likely to have invented ice skating.

The presence of bone skates at Sarmatian sites in Europe ties skating to the steppes, and perhaps more skates will eventually be found there. It is also possible that bone skates have already been found in the steppes, but have not been identified correctly, or that skates from the steppes have simply not made their way into Western archaeological literature. As Anthony[28] notes, much of the literature on the prehistory of the steppes is still difficult for Western archaeologists to access. As this literature becomes more accessible and steppe archaeology advances, it should become possible to trace the development of bone skates and skis in more detail. Küchelmann and Zidarov[29] looked forward to "interesting new results" from the East, which may be on the way.

This chapter explores the possibility that the first ice skates were an adaptation of skis to the conditions of the steppes. Why the steppes were a suitable place for bone skates to develop, including how they could have been linked to skis, is explained in the next section. Section 3 details the link between skates, skis, and horseback riding. The chapter ends with a description of how bone skates spread from the steppes to Europe and their distribution at the dawn of the Middle Ages.

4.2. The Steppes As a Homeland

Two factors combine to make the Eurasian steppes a suitable homeland for bone skates: the climate and the availability of suitable animal bones. The steppes are cold during the winter, and steppe-dwellers are known for domesticating the horse. Because the vast majority of the earliest ice skates were made from horse bones, the latter consideration makes the pastoral nomads of the steppes good candidates for their inventors. Furthermore, the archaeological and linguistic evidence for migrations within and out of the steppes connects steppe peoples with the first appearance of bone skates. Finally, the proximity of the Altai Mountains, where skiing appears to have been invented, allows a link between skating and skiing to be made.

A climate cold enough for bodies of water to freeze regularly in winter without too much snow is a prerequisite for ice skating. Dedicated skating enthusiasts could, of course, shovel snow off the ice before skating, but the archaeological evidence suggests that people living in climates with heavy snow preferred skiing to skating. Clark[30] maps the locations of early ski finds in Scandinavia with a line designating the boundary between regions covered by snow at least 150 days per year. The ski finds are all to the north of or very close to this line. In contrast, bone skates have generally been found south of this line. My own experiments on bone skates have shown that they are unusable on soft, powdery snow. When I tried to skate on it, I simply sunk into the snow, just as I do when wearing ordinary shoes or boots. The presence of an icy surface underneath does not help because the snow makes the skates stick. It may have been possible to use bone skates in areas with lots of snow as long as a thick crust formed; my attempts to skate under such conditions were more successful. Having the right type of climate is so important that Addyman et al.[31] suggest using bone skates as indicators of local climate. Bartosiewicz[32] suggests that bone skates show how the steppe environment influenced the cultures that arose there.

The steppe biome is generally characterized by cold winters, with nighttime temperatures below freezing for three to six months, and hot summers. Steppes average 10–20 inches of rain per year.[33] Today, the Eurasian steppes exhibit extreme temperature variations. Winter temperatures average 0°C in Tashkent (Uzbeckistan) and −12°C in Semipalatinsk (Kazakhstan), and the steppes are normally covered with snow, with forested regions receiving more precipitation than grassland regions.[34] There are a few large lakes and several major rivers that would make good ice rinks. Winters in Kazakhstan sound good for skating, as long as the snow is not too deep, but the climate of the steppes has not always been like this.

The climate of the steppes was warm between 6300 and 4800 BCE, then became milder, with warmer winters and cooler summers. Sometime between 3500 and 3000 BCE, the climate of the steppes cooled and dried.[35] This cooler climate improved skating conditions across the steppes; Bartosiewicz[36] suggests that cold winters froze smaller bodies of water and made skating possible. Climate considerations put the earliest date for the invention of bone skates between 3500 and 3000 BCE. Before then, the weather would not have been as well-suited to skating.

The use of horse bones for the earliest skates also points toward the steppes because horses were readily available there. They accounted for a significant part of steppe-dwellers' diet in about 5000 BCE. Domestication may have begun as early as 4800 BCE and was complete by 2500 BCE; horseback riding may date to between 3700 and 3500 BCE.[37] Horses need not have been domesticated before people began using their bones to make skates; people could have used the bones of horses dead of natural causes or hunted for meat. What was necessary was for horses to have been available in some way, and the Eurasian steppes were the place to find them. While horses were widespread in Europe during the Ice Age, their numbers dwindled and, between 10,000 and 3000 BCE, they could only be found in a few isolated places outside of the steppes. Nobody outside the steppes depended on horses as a food source at this time,[38] and horses were not used for non-meat purposes before being domesticated, which Anthony[39] puts shortly before the beginning of horseback riding.

After its invention, horseback riding spread across the steppes, and near 3300 BCE, perhaps as a result of the changing climate, the Yamnaya culture emerged. Yamnaya people were nomads who carried their possessions in wagons, rode horses, and probably spoke Proto-Indo-European.[40] They liked to spend the winter in marshes because the tall reeds offered food and protection.[41] Skating conditions could have been particularly good in these marshes, away from the reeds. These people, or people they were in close contact with, may have been the first skaters. If they were involved in the invention of skating, the first skating must have been done closer to 3300 BCE than 3500 BCE.

After 3200 BCE, the skating conditions got even better: the climate of the steppes dried up and the difference between summer and winter increased. Soil from the Kazakhstan steppe suggests that the climate was more arid between 3200 and 1900 BCE, and Lake Balkhash dried up somewhat in around 2990 BCE.[42] Kremenetski, Chichagova, and Shishlina[43] put the most arid phase of the climate of southern Russia and Ukraine at approximately 2700–2000 BCE. This is also about when the difference between winter and summer temperatures was greatest.[44] During this time, the amount of grassland increased while the area covered by forests and marshes decreased. Alekseeva et al.[45] report that the climate of the Russian steppe began to dry in approximately 3000 BCE and reached a minimum between 2000 and 1500 BCE.

This is when bone skates made from metapodia began to appear at Sabatinovka culture sites. This culture was the culmination of a period of settlement growth in the steppes.[46] There are some questions about its date, but it seems unlikely to have been earlier than the 2000 BCE or later than 1200 BCE. Unlike its predecessors and neighbors, the Sabatinovka culture was agricultural while retaining the pastoral characteristics of other steppe cultures, including the practice of transhumant herding.[47]

As settlements were growing, light horse-drawn chariots with two wheels emerged. Sintashta, a site in Chelyabinsk Oblast, Russia, slightly east of the Ural Mountains and rather farther east than the evidence for skates, is the site that is best known for them because of its elaborate chariot burials. Radiocarbon dates of several of these chariots indicate that they were probably invented by 2000 BCE.[48] This was not the only technological advance of the era. Dietrich[49] suggests that Late Bronze Age projectile points found in Eastern Europe underlie the new styles of warfare that may have emerged along with new, lighter bows. Later, the compound bow, invented around 1000 BCE, allowed people on horseback to shoot more effectively due to its short length and power.[50] Bone skates may have complemented these new inventions in the winter months, when ice made it difficult for horses to travel quickly without slipping.

Based on the discussed evidence so far, bone skates could have made their first appearance in the Eurasian steppes sometime after 3500 BCE. The very beginning of this period seems less likely because domestication would have made it easier for people to gain access to horse bones, which makes 3300 BCE a more reasonable date. The latest possible date is set by the archaeological evidence: 2500 BCE if the Albertfalva artifacts are skates, 2000–1200 BCE if the Sabatinovka skates are earliest (depending on which set of dates is correct), or 1500–1000 BCE if the first skates are the metapodium-based skates that began popping up around Europe.

4.3. Skates, Skis and Horses

Horse bones were the predominant material for all the early skate candidates. The first skaters may have considered skating on the bones of horses another way of riding those horses. Choyke and Bartosiewicz[51] suggest that there was "some kind of close cognitive relationship between skates and horse exploitation" based on the artifacts found at Albertfalva and Százhalombatta-Földvár. Even though the identification of these artifacts as skates is no longer supported, such a relationship need not be ruled out because of the strong preference for horse bones in the Late Bronze Age.

A parallel relationship between skis and horses has been noted by later authors. Markwart[52] notes that members of Finnish tribes "wore either skates from horse-bones or right 'wooden horses,'" and *Huan-jù kî (Description of Earth)*, a Chinese text that dates to between 976 and 984 CE, defines "wooden horses" as skis that were covered with horsehide.[53] This parallel goes back to approximately 5000–6300 BCE, the date of a ski found at Vis I, a Neolithic site near Lake Sindor in Russia. The front end of this ski has two small holes, which could have been used for attaching a belt, the other end of which was held in a skier's hand in a parallel to the reins used with horses.[54] This matches Abu-Hamid's twelfth-century description of "a long belt looking like a horse's bridle which is held in the left hand" by skiers beyond the Ural Mountains.[55] Another ski found at Vis I has an elk's head carved on one end, which may have kept the ski from sliding backwards and may have represented speed in a parallel to the carved heads attached to the fronts of ships.[56] It also suggests an association between skiing and swift animals, though not, in this particular case, horses.

These skis were found much farther north and somewhat farther west of the Altai Mountains than Omsk, and a bit north of the direct path from Omsk to Scandinavia. Although these skis are very early, there is even earlier evidence for skiing. Newly found rock art that depicts skiers in a cave in the Dundebulake Valley near Altai City, Mongolia, has placed the geographic origin of skiing in the Altai Mountains, which form the boundary between the eastern and western steppes. Although this rock art is difficult to date, it is thought to be Paleolithic,[57] which makes it old enough to win the Altai Mountains

4.3. Skates, Skis and Horses

the distinction of being home to the first skiers[58] even though no ancient skis have been found in the Altai Mountains yet. The Altai skiing tradition has continued to the present; traditional Altai skiing can be seen in various online videos, such as "First Skiers: Ski Altay Style" on the *National Geographic* website and Nils Larsen's 2011 documentary *Skiing in the Shadow of Genghis Khan*. These skiers use a pair of horsehide-covered skis with a single pole as a rudder.[59] The use of horsehide is reminiscent of the connection to horses proposed by Choyke and Bartosiewicz.[60]

Skis found in bogs and marshes in Scandinavia also provide strong evidence for early skiing. The oldest bog find, the Kalvträsk ski from Västerbotten in northern Sweden, dates to between 3623 and 3110 BCE. This ski's relationship to the skis from the Altai Mountains is clear enough that Berg[61] considers its connection with them "impossible to disregard." The numerous other preserved skis found in peat bogs in Scandinavia, including one dating to 3343–2939 BCE found in Drevja in Norway, are summarized by Weinstock.[62]

These finds and their similarity to the Altai skis favor diffusion over independent invention. It seems more likely that from the Altai Mountains, skis dispersed northwestward toward Scandinavia along the path proposed by Davidson[63] than that skiing was invented twice, once in the Altai Mountains and once in Scandinavia. Allen[64] summarizes the evidence for skis found along this path, including the skis from Vis I, petroglyphs from Russia and Scandinavia, skis from bogs in Scandinavia, and a dagger depicting skiers found near Omsk, which is located just north of Kazakhstan. The locations referred to in the following paragraphs are shown on the map.

Skis and bone skates share some features and have both been connected to horses. This has led to connections between bone skates and skis in the literature. Markwart[65] conflates bone skates and skis by bringing up parallels to both to provide context for the tenth-century Arabic text he translates. Berg[66] describes both bone skates and skis and notes that each works best in a specific climate. Dresbeck[67] includes a paragraph on "the bone 'ski' or skate," but does not explain the relationship between bone skates and skis. These connections are made through parallels in their form and use. The techniques used for skiing in central Asia during the Middle Ages are similar to the ones used for skating. In particular, using the pole for propulsion on skis in the steppes is attested by the Persian historian Rashīd al-Dīn Hamadani (1247–1318) in the early fourteenth century:

> When he runs over a snow-covered plain, he brings a long pole with him, which he sticks into the snow from time to time, like pushing a barge forwards: in this way he is also able to succeed in catching up with the animals he follows. The same pole serves him as a foothold when he climbs up slopes.[68]

Similarly, Francesco Negri reports that in 1664–1665, skiers in Scandinavia never lifted their skis from the snow.[69]

These techniques are similar to one that is known to have been used with later bone skates. The connection between bone skates and skis is strengthened by later literary texts that describe the use of bone skates on snow. The earliest of these is a description of skiing on bones in Yugra, an area east of the Urals. It concerns the Bulghār, a group that was originally from near the Sea of Azov[70] and therefore could have been connected with the skaters from that region. The first version of this tale is from a tenth-century Arabic source:

> Twenty days' journey from their land is a land called Īsū, and beyond Īsū (toward the north pole) is a people who are called Yūra. They are a wild group. They do not mix with people and are afraid of their malice. The Bulghār people travel into their land and bring clothes, salt, and other

things that are their wares. To carry the wares they have produced contrivances such as a kind of wagon (sled) that dogs pull, because there is a lot of snow there, and no other living being of that land can pull through it. The people bind reindeer bones to the soles of their feet, and they take two poles with spear-points in their hands and push them against the snow behind themselves, and they glide on the surface of the snow from there and run as if with a rudder, so that they reach the goal of the way in [number missing] days.[71]

This account was repeated approximately 200 years later by Sharaf al-Zamān Ṭāhir Marvazī in his "The Nature of Animals" (c. 1120):

At a distance of twenty days from them, towards the pole, is a land called Īsū [Wīsū], and beyond this a people called Yūrā; these are a savage people, living in forests and not mixing with other men, for they fear that they may be harmed by them. The people of Bulghār journey to them, taking wares, such as clothes, salt and other things, in contrivances drawn by dogs over the heaped snows, which [never] clear away. It is impossible for a man to go over these snows, unless he binds on to his feet the thigh-bones of oxen, and takes in his hands a pair of javelins which he thrusts backwards into the snow, so that his feet slide forwards over the surface of the ice; with a favourable wind [?] he will travel a great distance by the day.[72]

The "thigh-bones of oxen" ought to be tibiae, but the tibia was rarely used for skating. The radius is another, and more reasonable, candidate; although radii are in the front legs of animals, they are in the thigh position. The "favourable wind" that the translator seems unclear about may be evidence that sails were used.

Bīrunī's description (c. 1030) of people traveling from Bulghār to visit the Īsū is similar. After describing dogsleds, he writes, "They also use skates made of bone, with which they can travel long distances quickly."[73]

Those three reports are clearly closely related, but there are others that seem to have come from different sources. In the following centuries, two Western travelers provided independent observations of bone skates in the East. First is William of Rubruck's thirteenth-century description of bone skates in the modern Irkutsk region of Russia between Lake Baikal and the Altai Mountains:

Sunt etiam ibi Orengai qui ligant ossa limata sub pedibus suis et impellunt se super nivem congelatam et super glaciem cum tanta velocitate ut capiant aves et bestias.[74]

(There are also Orengai who tie polished bones under their feet and impel themselves over frozen snow and over ice with so much speed that they may catch birds and beasts.)

In all these reports, it sounds as if people used bone skates to travel on snow, like skis, as well as on ice. If this is true, the snow must have been quite hard and smooth. My own experiments skating on bones on snow were not very successful. With my metacarpus skates, I was able to shuffle along in the tracks left by cars, where the snow was very compressed. I had more success using the skateboarding technique with my radius skate, which slid over the compressed snow enough to convince me that skating on snow was not impossible under the right conditions. This experience highlighted one advantage of radii over metapodia: the concavity of the radius's surface helped keep my foot from sliding off the skate when it was stuck in the snow in the absence of bindings. Losing skates because they got stuck in the snow was a problem with the flatter metacarpus skates. Wide temperature swings, i.e., repeated freezing and thawing, would help produce hard, icy snow, which may have been possible to skate on. Such conditions may have been present at certain times of year; Alexander von Middendorff notes that they occurred regularly in the spring in eastern Siberia during the nineteenth century:

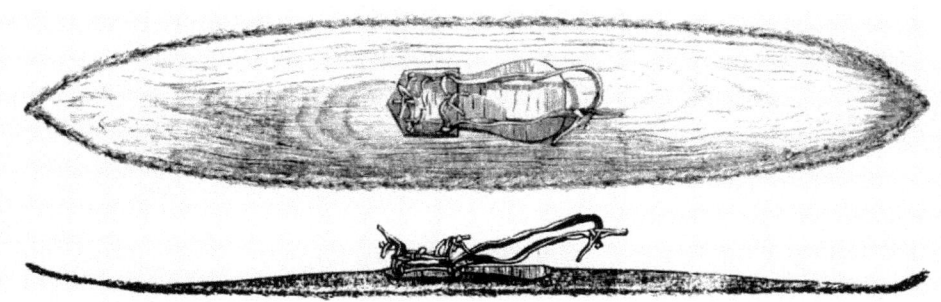

Figure 30: Alexander von Middendorff's "Schneeschuh" (snowshoe), which looks like a type of ski (Alexander Theodor von Middendorff, *Reise in den äussersten Norden und Osten Sibiriens während der Jahre 1843 und 1844*, 1349).

> Krustet der Schnee im Frühjahre so bindet der Tunguse ein paar Knochen-Plättchen unter seinen Schneeschuh, damit dieser besser gleite und doch nicht seitlich ausweiche.[75]
>
> (When the snow crusts in the spring, a Tungus binds a pair of small bone plates under his ski, so that it glides better and does not slide sideways.)

Although von Middendorff uses the word "Schneeschuh" (snowshoe), it is clear from the pictures he includes (see figure 30) and the context that a type of ski is meant.

Do these similarities suggest a common origin? Or were skates and skis invented independently and the similarities noted by later authors? In the former case, people from the steppes learned about skiing and wanted to do it themselves. They had to adapt what they observed to the resources available to them. This fits well into what Schiffer[76] calls the "cascade model" of technological development. It focuses on the need for new adopters to use local materials and perhaps even create new tools to adapt an observed technology to their conditions. This process can be quite complex, depending on the technology in question. Bone skates are quite simple, and it is not unreasonable to suppose that they were invented completely independently of skis; in this case, the similarities result from people in different places finding similar solutions to a problem.

In support of a connection between bone skates and skis, there is evidence for contact between the Yamnaya culture and people living in the Altai Mountains. Between 3700 and 3400 BCE, a group of people migrated from the Pontic-Caspian steppes to the western Altai Mountains and developed the Afanasievo culture, which has many similarities to the Yamnaya culture. Various cultural traits are known to have been transferred between the region near the Ural River and the Altai Mountains as people continued traveling between these two locations.[77] These migrants could have met skiers when they reached the Altai Mountains and learned to ski from them.

Contact could have been the first step in the cascade of innovation that led to bone skates. Bone skates could have arisen as an adaptation of skis the environment and resources of the steppes. If the Albertfalva skates and the other artifacts based on horse radii are skates, a link to skis may help explain why they were made from radii when metapodia work much better. Radii may have originally been used for skates because they were the bones that most closely matched skis in length while being flat enough to skate on. The early Altai rock art depicts skis as being much shorter than modern skis. It is hard to say how long the oldest skis were because the finds are fragmentary. The longest ski fragment from Vis I measures 45 cm by no more than 15 cm.[78]

Scandinavian rock art provides more information on prehistoric ski length. At approximately 4,000 years old, the Rødøy petroglyph (figure 31) is the oldest depiction of skiers in Scandinavia.[79] The skis shown are very long—much longer than the skis in other early works of art. In other Scandinavian petroglyphs, such as Bølamannen and the Alta skier, and in an image of skiers from the Karelia district in Russia (figure 32), the skis are closer in length to the ones in the Altai depictions than to the ones in the Rødøy petroglyph. Radii match the length of these early skis, as depicted in the works of art, more closely than metapodia do. Other examples of Russian rock art depicting skiing, mostly from the Neolithic, have been collected by Weinstock.[80] Most are from the area near Lake Onega in Karelia, Russia.

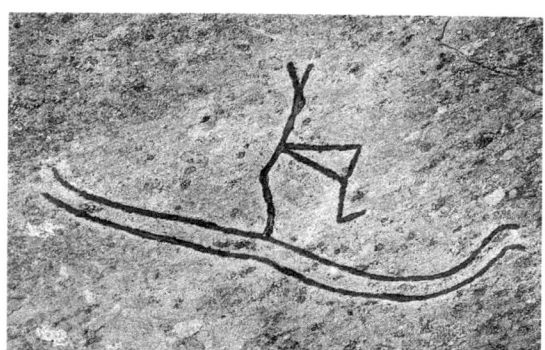

Figure 31: The Rødøy petroglyph. At 4000 years old, this is the oldest known depiction of skiing in Scandinavia (courtesy Ski Museum in Holmenkollen, Norway).

Figure 32: A petroglyph showing skiers from Zalavrouga on the Vyg near the White Sea in Russia that dates to the Stone Age, according to Gösta Berg, "The Origin and the Development of the Skis throughout the Ages," 14 (Semenov.m7, courtesy of Wikimedia Commons. Licensed under the Creative Commons Attribution-Share Alike 4.0 International License, which is available at https://creativecommons.org/licenses/bysa/4.0/deed.en).

Another place to look for a possible early connection is in the way early skis and bone skates were shaped. Some early bone skates and skis featured upturned toes. Skis found at Vis I, which remind Burov[81] of the skis used by the Yakut in Siberia, feature upturned toes. The accompanying pictures show that the upturned ends are also pointed. Another artifact from Vis I has an upturned end that is not pointed. The skis in the Rødøy petroglyph also have strongly upturned toes.

Upturned toes began to appear on metapodium skates during the Late Bronze Age. The pair of Urnfield skates from Ivanovice na Hané 7 in the Czech Republic features upturned toes along with several other features, as do one other skate from the same site and two skates from Törökbálint, Hungary. The skates from Ivanovice na Hané are shown in figure 33. Of the Sabatinovka skates, the one from Zlatopol' and one of the three from Novokievka have upturned toes. This could be evidence of a link between skates and skis or could simply show that people in two different places came up with the same solution to a problem. Presumably, upturned toes help skates and skis slide over imperfections in the gliding surface. This is not a high-tech solution to the problem of gliding, and there is no reason to think that people in two different areas could not have come up with it separately. Not all bone skates have upturned toes, which means it was not essential. It could have been a tradition or an innovation that was being tested.

Similarities and differences in form and technique, such as the use of bindings and poles, are examples of how the general idea of sliding can be adapted to specific surfaces. The modern conception of the difference between skating and skiing hinges more on the surface on which the activity happens—ice or snow. But these surfaces often go together, and William of Rubruck[82] and Middendorff[83] suggest that bone skates worked on hard snow as well as on ice. The modern distinction between skates and skis may not have been as strong during the Bronze Age.

There is not enough evidence to say for sure whether bone skates were an adaptation of skis or an independent invention. The former hypothesis would be supported by re-identification of the Albertfalva artifacts and the other mysterious radius-based artifacts as skates. If these artifacts are accepted as skates, a link to skis seems more likely because of their length. Further archaeological work could conceivably uncover early bone skates close to the Altai Mountains or early skis for comparison purposes. Sadly, this may remain an unsolved problem because, as Schiffer[84] notes, it is un-

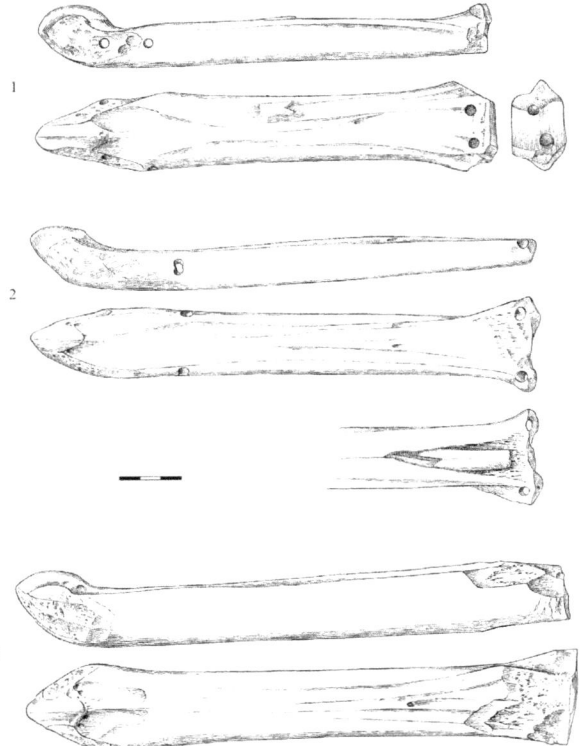

Figure 33: Three Late Bronze Age horse metatarsus skates from Ivanovice na Hané, Czech Republic. Note the pointed and upswept distal ends (on the left) and the placement of holes David Parma, et al., "Netradični materiál, neobvyklý předmet Opomíjený segment kostěné industrie mladší doby bronzové (Non-Traditional Material and a Non-Traditional Object: A Neglected Sort of the Late Bronze Age Bone Industry)," fig. 5.1-2, courtesy David Parma).

usual to find enough information in the archaeological record to figure out who came up with an idea first.

The argument for a separate invention is based on the simplicity of bone skates. They were not difficult to make—a child could do it! And sliding on ice is quite easy to do; in fact, it is often hard to avoid. Even the physical similarities of skates and skis could be due to people finding similar solutions to the same problems. The two factors that seem most suggestive of a link are the similarities between them and their close geographical ranges.

If skates and skis were separate inventions, their similarity could have been because they were developed to solve the same problem. What this problem was depends on the answer to the question of use: were the first skates for recreation or practical purposes? The next chapter addresses this question. The rest of this chapter follows the spread of bone skates across Europe from the Late Bronze Age until the onset of the Middle Ages.

4.4. Skating Across Europe

Regardless of whether bone skates were derived from skis, the steppes provide a suitable place for them to have emerged. Then, they are likely to have traveled westward toward Hungary along a route very similar to the one proposed by Davidson[85] for bearpaw snowshoes. This route was also followed by tools made from mandibles, possibly used for smoothing, that originated in Kazakhstan, and the radius-based artifacts discussed in section 3.3. The mandible-based tools spread into Hungary and surrounding areas during the Early Bronze Age.[86] The radius-based artifacts followed them slightly later. If these artifacts become accepted as skates, they add strong support for the argument of this chapter.

Between 3000 and 2800 BCE, a group of people left the Yamnaya cultural center in the steppes and migrated up the Danube Valley. This group has been associated with the Italo-Celtic branch of Indo-European[87] and may be also connected with the Germanic, Baltic and/or Slavic branches,[88] but these linguistic groupings have been contested. Isaac[89] describes objections to the Italo-Celtic branch of Indo-European and suggests that Celtic is more closely related to Baltic, Slavic, and Indo-Iranian than Italic or Germanic with references to more detailed discussions of the question.

Manco[90] favors the hypothesis that these migrants were speakers of Old European, a language ancestral to both Proto-Italo-Celtic (if it existed) and unattested languages, including Ligurian and Lusitanian. Anthony[91] and Haak et al.[92] put the end of this migration between 2800 and 2600 BCE, which is when the Indo-European languages began to split apart and spread across Europe. This migration was part of what Mallory[93] describes as "a phase of exceptional mobility throughout Europe when very different cultural traditions came into contact and learned much from each other." The "exceptional mobility" could have enabled bone skates to spread throughout the parts of Europe that had suitable climates. It is tempting to leave it at that and simply say that people shared and the concept of bone skates traveled around until, eventually, some were preserved well enough to be found by archaeologists. But it is possible to go a bit further into the problem and figure out a little bit more about how bone skates got to the places in which they first appeared.

In approximately 2500 BCE, the Bell Beaker culture arrived in the region surrounding Csepel Island in the Danube River near Budapest. This is the only Bell Beaker area in the region.[94] Choyke and Bartosiewicz[95] mention regular finds of skate-like radius-based artifacts at Bell Beaker sites in this area, including Albertfalva. Anthony[96] links this group of Bell Beaker sites to the Yamnaya culture by suggesting that they connected the Yamnaya in the steppes with the people living in what is now Austria and southern Germany. Heyd[97] notes that many bones from horses have been found at sites in this area, which indicates that the people living there kept horses. That they were kept instead of hunted is evidenced by their size: the horses at Csepel-Háros were smaller and more variable in size than horses at other sites in Europe.[98] Bökönyi[99] links these horses to horses from the steppes. This links people living in the Carpathian Basin during the Bronze Age and the pastoral nomads of the Eurasian steppes.

This association becomes stronger during the Late Bronze Age and the Iron Age. In the Late Bronze Age, skates made from the metapodia of horses and cattle began to appear. The Sabatinovka skates are probably the oldest metapodium skates found to date, but similar skates began to appear across Europe between roughly 1500 and 1000 BCE at

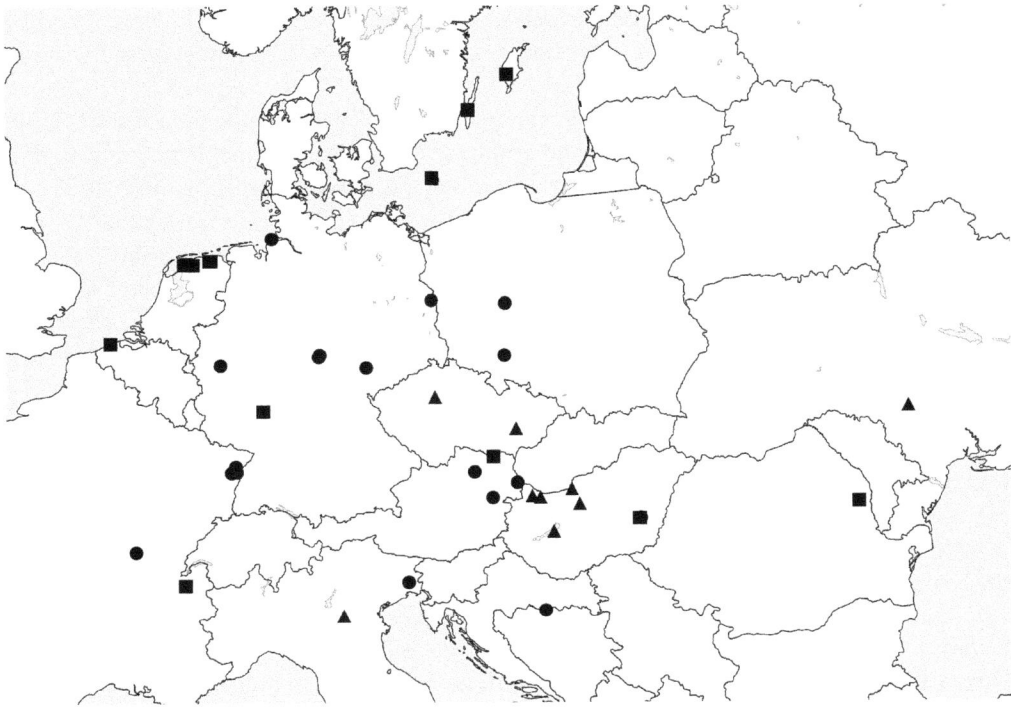

Figure 34: Sites with skates from the Late Bronze Age (triangles), Iron Age (circles), and Migration Period (squares). Bodies of water are shaded (created with QGIS).

the sites shown in the map in figure 34. At least 24 skates dating to the Late Bronze Age have been found in Italy, Hungary, the Czech Republic, Ukraine, and Slovakia.[100] During the Iron Age and the Migration Period, metapodium skates spread across Europe. Over 120 skates from these periods have been found at sites in Austria, Belgium, Bosnia, Denmark, France, Germany, Hungary, Italy, the Netherlands, Poland, Romania, and Sweden.[101] These skates are discussed in detail in the next chapter.

This is the time when, according to Jordanes, the Germanic peoples "burst forth like a swarm of bees from the midst of this island [Gothiscandza] and came into the land of Europe."[102] By 98 CE, according to Tacitus,[103] they had spread throughout much of northern Europe. They could have learned about bone skates from the Romans, the Celts, or the Sarmatians; they were in contact with all three during the Iron Age. The Romans seem like the least likely source because the climate of much of the Roman Empire was unlikely to have been sufficiently cold in winter. Contact with Celtic groups may have endured the longest; Green[104] describes late prehistoric contact between Germanic and Celtic groups that was primarily due to the expansion of the Celts between the middle of the first millennium BCE and the first century CE and notes that culture traveled primarily from south (Celtic) to north (Germanic), which is correlated with the motion of bone skates. They are a possible source of bone skates, but the Sarmatians seem more likely.

Germanic groups, including the Bastarnae, were living in the lower Danube region and the southern part of the Russian steppes by the early third century and were probably linked to their northern relatives, as shown by the spread of their loan words.[105] The Goths and the Gepids reached the Great Hungarian Plain and were in contact with the

Sarmatians by 300 CE, as shown by Sarmatian adoption of some special objects[106] and by Jordanes' book, which dates to 551 CE (though according to Jordanes, the contact was not always friendly).[107]

In addition to the places shown on the map, bone skates have been found in Gepid workshops alongside combs, spoons, and amulets.[108] It seems reasonable to suppose that the Gepids learned to skate from the Sarmatians because the two groups were allied and both lived in and around the Carpathian Basin in the fifth and sixth centuries CE.[109] A single skate included in the above list that dates to the fourth or fifth century CE and is associated with the Sîntana de Mureş/Černjahov culture has been found in Romania.[110] Green[111] notes that this culture was mostly Gothic and that Goths dominated this area, according to various literary sources. Combined with the Gepid skates, this is evidence that some of the East Germanic tribes encountered skates during their migrations.

Other skates have also been linked to Germanic tribes. Two fragmentary skates have been found in Kliestow, now a part of Frankfurt an der Oder in eastern Germany, which was occupied by Burgundians, who lived on the edge of the western part of the Roman Empire near the Franks and Saxons, from the end of the second century until sometime during the fourth century CE. One is made from the tibia of a cloven-hoofed animal, and the bone used for the other is not specified. Both skates are fragmentary.[112]

Lamb[113] suggests that the root cause of the migrations that brought the Goths and Gepids into contact with steppe peoples may have been drought in central Eurasia. Duncan-Jones[114] describes "worsening conditions" (improving conditions from the perspective of skating) that led to a decrease of approximately one third in the number of sites in the Rhineland during the last quarter of the third century CE. Büntgen et al.[115] describe greater variations in the climate between approximately 250 and 600 CE, which may have been related to the events described by Todd.[116] They depict the climate during the Migration Period as highly variable, with generally lower than normal temperatures but more precipitation than usual during the central part of the period. Part of this variability may have been due to the switch from a warming period to a cooling period. Lamb[117] describes general warming until the fourth century CE, a century or so of high variability, and then, a generally cold climate. There was strong volcanic activity during the first half of the sixth century CE, which may have contributed to the cooling effect; in particular, a large volcanic eruption in approximately 535 CE had a significant effect on the climate.[118] In addition, glaciers advanced in the Alps at some point between 600 and 850 CE and in Norway between 450 and 850 CE.[119] In about 700 CE, the ratio of warm to cold winters in central Europe began to decrease significantly.[120] Around 750 CE, winters were milder again, but then they became significantly colder in both central Europe and the British Isles until about 1000. Lamb[121] points to tree-ring data suggesting that tenth-century winters were cold and dry in some areas. Gräslund and Price[122] suggest that the dust veil resulting from the 535 eruption caused the Scandinavian climate to deteriorate and inspired the *fimbulvetr* ("'mighty' winter") that heralds the coming of Ragnarök (the end of the world) in Norse mythology. *Gylfaginning (The Tricking of Gylfi)*, part of Snorri Sturluson's *Prose Edda*, includes the following description in response to a question about how to tell when Ragnarök is coming:

> Mikil tíðendi eru þaðan at segja ok mörg, þau in fyrstu, at vetr sá kemr, er kallaðr er fimbulvetr. Þá drífr snær ór öllum áttum. Frost eru þá mikil ok vindar hvassir. Ekki nýtr sólar. Þeir vetrfara þrír saman ok ekki sumar milli.[123]

(There are many and important tidings to tell about it. First, a winter shall come, that is called *fimbulwinter*. Then snow will drift out of all directions. There will be great frosts and strong winds. The sun will be useless. There winter will come three times with no summer between.)

These conditions sound terrible for some, but avid skaters could have rejoiced at having a three-year skating season. It seems like this should increase the prevalence of bone skate finds, and indeed, the number of bone skate finds increased substantially in the Middle Ages. Thousands of medieval bone skates have been found at sites all over the parts of Europe with sufficiently cold winters. The next chapter addresses the question of use as bone skates spread across Europe: Were they primarily used by adults or by children? For practical purposes, such as travel, hunting, and fishing, or for recreation?

5

Tools or Toys?

5.1. The Question of Use

What did people use bone skates for? Were they tools for important adult activities, like travel, hunting, and fishing? Or were they toys for children and adolescents to play with? This chapter examines the archaeological and ethnographic evidence for how bone skates were used in Europe before the Middle Ages; skating in the Middle Ages, especially Scandinavia, is discussed in detail in the next two chapters. Because ethnographic evidence is only available from the Middle Ages onward, it is necessary to extrapolate backward. The main focus is on the archaeological evidence from the Late Bronze Age through the Migration Period (that is, skates dating very roughly to between 1300 BCE and 800 CE plus the Sabatinovka skates, which may be somewhat earlier). The earlier skate candidates are discussed in the final section.

Both practical and recreational uses have been proposed for bone skates. Berg[1] describes people using bone skates to travel both across bodies of water and along frozen roads, to hunt, and to fish. The practical use mentioned most frequently in modern ethnographic sources is travel. Bone skates seem to have been an effective mode of travel for the people described in medieval Arabic sources quoted in the previous chapter. Additionally, Olaus Mangus describes races that were 12–18 km long.[2] Although these races were recreational, their length suggests that using bone skates for travel was reasonable. In *Fljótsdæla saga*, an Icelandic saga that was probably written around 1500 CE, a character makes a journey of unspecified length on what are probably bone skates.[3]

There are also modern examples of people traveling on bone skates in England and Scandinavia. The example from England is a story told by W. H. Barrett, a collector of folklore from the fens and reportedly took place in approximately 1900:

> When the rivers were frozen a 90-year-old man from Soham Fen used to take the opportunity to visit his out-lying relations. One day he arrived at his niece's—she lived next door to us [in Brandon Creek in Norfolk]—having skated from Soham Lode into the Ouse, then on to Denver Sluice and back through Littleport. After a huge meal with his niece and a short rest he called to borrow from my father a file to smooth down a chip in one of his skates. These skates, which the old man boasted were two hundred years old, were simply a pair of sheep bones, beautifully shaped and polished. Holes had been bored through them for the leather thongs which he passed criss-cross style over his boots and up his legs.[4]

Modern maps suggest that this journey was roughly 19 km (12 miles), making it comparable to the races described by Olaus Magnus. In Sweden, Berg[5] reports that bone skates "were used when travelling over the lakes or along the coast, but also on the ice

5.1. The Question of Use

roads on land shore" in Småland, Bohuslän, Uppland, and Hälsingland. One objective for such travel was to attend church services; Säve[6] describes men and women skating to church and hiding the skates in the bushes during services. Similarly, Christina Marcusson describes "män som unga flickor" (men and young girls) using bone skates to get to church.[7]

Küchelmann and Zidarov[8] discuss the possibility of traveling on bone skates in the context of their trip to Finland. They found that frozen bodies of water effectively create new roads in winter, making travel much easier than it is otherwise because people can simply go across rather than dealing with a boat or going around. Such changes in topography due to ice could have made it easier to get from place to place but did not necessitate the use of skates. The impact of these changes is underscored by the results of Formenti and Minetti[9]: they compared the energetic cost of traveling across lakes (walking and skating) with going around them (walking only) and found that going across was significantly better in Finland and advantageous in Sweden and Norway. Because they assumed skating and walking were equally taxing, the result is due to topography rather than skating. People could have crossed the frozen lakes in rough-soled shoes or crampons just as easily. However, the speed of skating may have given it an advantage over walking, and the metabolic cost of skating could have been decreased significantly by using sails.

There are also some medieval and modern references to hunting and fishing on bone skates. William of Rubruck's thirteenth-century description of skaters moving "with so much velocity that they may catch birds and beasts"[10] suggests that bone skates may have been used in hunting. Herman[11] mentions the use of bone skates by modern Hungarian fishermen, and there is evidence for such use in medieval Scandinavia: Stenberger describes a boy buried with four bone skates, an ax, and a fish hook, among other items, in the tenth century on Gotland and interprets this as evidence that bone skates were used for ice fishing: "One is led to assume that the young man died in the winter time as a result of drowning while fishing in the neighbouring lake. He thus used the skates for travelling across the ice and the socketed axe, in that case, would have been used for hacking holes in the ice."[12]

Berg[13] and Säve[14] also note that bone skates were used for ice fishing in Sweden. Edberg and Karlsson[15] identify a pair of bone skates made from horse metatarsi that had been used in a fishery in Sweden until the early 1900s, when they were donated to the local history museum in Tjörn. Berg[16] provides more details of how ice fishing was done:

> When in the autumns [sic] pike and burbot were clubbed on the thin and transparent glassy ice, the bone skates certainly were preferable to the ice-prods, used in other connections, not the least due to their silent run. This fishery was realized in the following way. The shadow of the fish under the ice was watched, and then the fish was made unconscious with a strong hit by a club or an axe, an icy hole was rapidly cut out, and the quarry was hold [sic] up with a small fishspear.

Olaus Magnus provides a picture to go with this description (figure 35), with the important difference that the fisherman in the foreground is wearing crampons; there is no sign of bone skates. Edberg and Karlsson[17] express skepticism about the use of bone skates in ice fishing, and it is possible that using crampons was the norm because they would have provided more stability. Säve[18] and Berg[19] also mention the use of crampons but note that skates were better because their silence allowed people to sneak up on fish.

Recreational uses of bone skates include the races described by Olaus Magnus

Figure 35: Ice fishing in sixteenth-century Scandinavia. Here, the fisherman is wearing spiked shoes (crampons) instead of bone skates (Olaus Magnus, *Historia de gentibus septentrionalibus*, 709).

and the jousting described by William fitz Stephen. Tricks, such as spinning, were also possible but are not mentioned in the literature. Numerous recent accounts describe skating as a fun pastime. Kuckuck,[20] Petényi,[21] and János Makkay,[22] among the others listed in the Appendix, note that bone skates were popular among boys in Hungary and nearby areas in the nineteenth and twentieth centuries. Von Luschan[23] describes skating on bones as incomparable in terms of "Schnelle und Annehmlichkeit" (speed and pleasantness).

The most recent and detailed discussion of whether bone skates were primarily for fun or for practical purposes was written by Edberg and Karlsson.[24] They present a persuasive argument that bone skates were primarily used for the amusement of young people in medieval Scandinavia. It is based on medieval literary references and modern ethnographic analogues as well as an analysis of 679 medieval bone skates found at two sites in Sweden. They assume that to be usable with medieval shoes, bone skates had to be least as long as the skater's feet. According to their argument, soft-soled medieval shoes did not provide the support necessary to skate on shorter bones. Furthermore, all the images of bone skates in use depict them as about as long as the skater's feet (see figures 12, 13, 15, and 19). They measured all the whole skates (505 of the 679 skates; the others were fragmentary and therefore, their full length could not be determined) and found that 29 percent were shorter than 18.9 cm, 54 percent were between 19.0 and 23.9 cm, and 17 percent were longer than 24.0 cm. Comparison with a table of shoe sizes shows that 19 cm corresponds to a European size 30 shoe, which typically fits modern seven-year-olds. The second group, containing 54 percent of the skates, corresponds to shoe sizes 30–38, and the third group matches larger shoes.[25]

Corroborative evidence is provided by detailed studies of human foot development.

5.1. The Question of Use

In particular, Howard Meredith has collected information on the growth of modern children's feet. He reports average lengths of 19.7 and 19.4 cm for 450 seven-year-old boys and girls, respectively,[26] which supports Edberg and Karlsson's supposition that the smallest group of skates were sized for such children. Meredith quotes data showing that the feet of boys averaged 23.9 cm at age 14 at the Orphan Asylum of Brooklyn and at age 16 at Letchworth Village, both in New York. The feet of average girls at the former institution had only reached 23.0 cm by age 16.[27] This also supports Edberg and Karlsson's supposition: the second group of skates seems to have been sized for teenagers. Naturally, there is some variability, and modern feet are probably slightly larger than medieval feet were.

The third group of skates, the 17 percent that were longer than 24.0 cm, were approximately adult-sized. Meredith[28] found a median foot length of around 26 cm for adult men reasonably consistently across all the studies he collected and a slightly smaller size for women, whose median foot size was 23.3 cm in one study. Since people were slightly smaller in medieval Scandinavia,[29] their feet were probably smaller, which means that these skates are likely to have been usable by most adults. They could also have been used by children—the assumption was that the skates need to be at least as long as the skater's foot, so short skates rule out tall skaters, but long skates do not rule out short skaters.

Edberg and Karlsson[30] conclude that bone skates could have been used by children as young as two or three years old and all their family members, regardless of age, but that skating was probably most popular among children and teenagers. Because medieval Scandinavian literature (discussed in detail in the next chapter) and modern reminiscences mostly describe recreational uses of bone skates, those probably predominated during these periods. Bone skates were probably mostly (but not necessarily exclusively) used for recreation by children and teenagers in medieval Scandinavia and modern times.

Was it always like this? In the rest of this chapter, the archaeological evidence from the Late Bronze Age through the Migration Period is evaluated within the framework of Edberg and Karlsson's analysis to see how far their argument can be extended. Because few skates from this period have been found, the skates from all sites are combined for a general statistical analysis. The result is a broad overview. The uses of skating and the ages of skaters could have varied from site to site, as Edberg and Karlsson suggest was the case in Sweden—Sigtuna seems to have been home to more teenage skating enthusiasts than Birka. The main focus of the following sections is evaluating the archaeological evidence to determine whether skates were primarily tools or toys in antiquity.

Since the focus is on the artifacts themselves (no written evidence for this period is available), Choyke's[31] "manufacturing quality continuum" for bone tools is used as a basis for the analysis. This continuum is defined by four factors, but only the first two are applicable here. The four factors are:

1. The type of animal and bone;
2. How extensively the bone was worked;
3. How worn the artifact is; and
4. Whether the artifact was repaired or reworked.

Artifacts that score highly in these four areas (Class I artifacts) are, Choyke argues, more likely to have been planned in advance and the result of a long, complex manufacturing process. Such tools show extensive working and are generally cared for well. Artifacts on the other end of the continuum (Class II artifacts) tend to be made on the spur of the moment and may be difficult to distinguish from garbage. They are more likely to have been made opportunistically, for events like the "'urgent' competition" mentioned by Küchelmann and Zidarov.[32] Choyke[33] suggests that Class I artifacts are more likely to have been used in tasks that were important to the economy of the society that produced them, whereas Class II artifacts "were more likely to have been made to suit the individual needs of the moment such as repair or sport."

Only the first two points are discussed in detail in this chapter. The type of animal and bone selected is the point for which the most data are available and the subject of section 2. Section 3 discusses how extensively the bones were worked as part of the skate manufacturing process. The third point is not considered because no consistent data on use wear are available for prehistoric skates, but such data would add to this analysis. The fourth point is easy to set aside—there is no evidence that bone skates were repaired; they seem to have been discarded when they broke or wore out, which puts them on the Class II end of the continuum with respect to this criterion. In general, bone skates are closer to the opportunistic end of the continuum (Class II), which implies that their use was more likely to have been recreational than economically important. They may have been like bicycles are today: both practical and fun, with their exact place on the continuum varying from site to site.

5.2. Bone Type

The type of bone selected for skates may yield clues to their function by helping place them on the manufacturing continuum. Edberg and Karlsson base their analysis on size, which makes it difficult to follow exactly because measurements of many of the early skates have not been published. However, even without published measurements, it is possible to get an idea of how large a skate was from the type of bone it is made from. On a given cow or horse, the metacarpi are shorter than the metatarsi, which are shorter than the radii, and the bones of cattle are generally shorter than those of horses. This is very rough, and detailed measurements would be much better, but at this point, they are simply not available. However, size was not the only factor in bone selection; the shape and availability of particular bones also made a difference. Consistent selection of a single bone type points toward the Class I end of the continuum, whereas inconsistent choices and the use of waste bones point toward Class II.

In general, metapodia make better skates than radii. Küchelmann and Zidarov,[34] who made skates from both metacarpi and radii, found that the flatness of the metacarpi makes them easy to stand on. Balancing on radii takes more effort because they are curved. The ease of balancing on metacarpi suggests that they may have been more suitable for beginners, and their smaller size may have been perfect for children. The metapodia of horses are better for skating than those of cattle because they are flatter and longer; the length may have been unimportant to young skaters with small feet.

Additionally, Choyke and Bartosiewicz[35] note that metapodia were less likely to be used for food than radii. Compared with metapodia, radii have substantial amounts of marrow, which may have been an important source of food in cultures that practiced

5.2. Bone Type

hippophagy. Levine[36] argues that horse meat and milk were very important—perhaps as important as the mobility provided by domestic horses—for pastoral nomads in the Eurasian steppes during the late Eneolithic. It was necessary to break these bones to extract their marrow. Breaking them would have rendered them useless as skates, meaning that people had to make a choice: eat the marrow or skate on the bones? Choosing the latter implies either that enough food was available that eating the marrow was not necessary or that there was an important reason to make skates. This means that the selection of radii over metapodia may point more toward practical uses for skates. However, in general, radii would have become more readily available after the Roman period because people ate horses less frequently.[37] The prevalence of hippophagy depends on the culture and the site.

Their unpopularity as a food source is reflected by the fact that metapodia often remained with the animal's hide after slaughter. Head-and-hoof burials, in which the skull and lower leg bones of sacrificed sheep, goats, and cattle were buried (probably with the hide, but the hides have not survived) have been found at Khvalynsk, a site in the Pontic-Caspian steppes that dates to 4700–3800 BCE.[38] The practice of leaving the metapodia with the hide continued through the sixteenth century, as shown by a woodcut of a deerskin with the metapodia still attached in the upper right of the sixteenth-century woodcut *Maximilian's Triumphal Arch* by Albrecht Dürer and others. This practice is still the norm today. I was unable to obtain metapodia from the butcher shop that supplied the bone for my radius skate because they had been removed from the animal with the hide during processing.

Since they were not highly desirable, metapodia were good candidates for toys. However, radii may have been preferred by older and more serious skaters. They may have been more difficult to balance on because of their curvature, but once balance was learned, they may have worked well. Lambert van Es[39] found that radii made faster skates because their curvature leads to lower friction compared with metapodia. Küchelmann and Zidarov[40] found the longer radius skates more comfortable than the shorter metapodium skates because the former were longer than their feet, which is probably even more important for skaters wearing soft shoes than for those wearing modern stiff-soled boots. Children's feet could have easily fit onto the shorter metapodia, negating this advantage of radii.

The bone skates from the Late Bronze Age through the Migration Period are almost exclusively based on metapodia rather than radii. To build the list of skates the following analysis is based on, I used Küchelmann's online database with some additions and modifications.[41] This yielded a total of at least 151 skates from 46 sites.[42] The sites, shown in figure 34, are scattered across Europe in Austria, Belgium, Bosnia, the Czech Republic, Denmark, France, Germany, Hungary, Italy, the Netherlands, Poland, Romania, Slovakia, Sweden, and Ukraine. The data are dominated by the 48 skates found at Gyoma 133 in Hungary, including one whose identification is questionable.[43] A total of 18 skates on the list are neglected in the following analysis because the published information about them is incomplete or questionable.[44] The bone types of the remaining 133 skates are distributed as shown in table 1. Plus signs mean that the numbers are minima; there may be more skates. In particular, the numbers of Late Bronze Age skates found at Sabatinovka, Ukraine, are denoted by plus signs because Pankovskiy[45] does not include them. I assumed a single skate in such situations because there must be at least one. The "Other" skates are one made from the tibia of an even-toed ungulate (perhaps a cow) found at Kliestow bei Frankfurt (Oder),[46] two skates from donkey metapodia (one metacarpus and one metatarsus) found at Wiwersheim, France,[47] one skate made from a deer metatarsus found at the Pozzuolo del Friuli hill-fort, Italy,[48] and one very unusual skate made from a wolf radius found in Cologne, Germany.[49] The skates under "metapodium" were made from metacarpi or metatarsi; the sources are silent as to which.

Table 1

Bones used to make skates from the Late Bronze Age through the Migration Period.

Species	Bone	N
Cattle	Metacarpus	2
	Metatarsus	7
	Radius	6+
Horse	Metacarpus	69
	Metatarsus	21+
	Metapodium	17
	Radius	6
Other		5

Table 1 shows that horse metapodia predominate among the skates that have been described in the literature to date. They account for slightly over 80 percent of the skates in this analysis. This preference is even stronger than the preference for horse bones in Sigtuna, where Edberg and Karlsson[50] found that 235 (60 percent) of the 389 skates were made from horse metapodia. The preference for horse metacarpi seems particularly strong, with them accounting for just over 50 percent of the skates. Following Edberg and Karlsson, the strong preference for horse metacarpi, which generally fall into the middle size bin in their study, is probably indicative of teenagers being the most enthusiastic skaters during this period. Horse metatarsi were generally a bit longer, and radii even longer; such skates could have been used by adults. However, this size spectrum is based on broad generalizations that are not always true. For example, there is a medieval horse radius skate from Wallingford, England, that sounds like it should be long based on its bone type but that measures only 20 cm.[51] While the generalizations about bone sizes are enough for a general overview of skate sizes in the absence of numerous readily available measurements, they cannot replace such details.

Proceeding with caution, the small number of cattle bones is noteworthy. These skates, which are likely to have been shorter than skates made from horse bones, suggest that bone skates may not have been popular among young children in antiquity. The cattle metapodium skates are from five sites scattered across Europe: Gara Banca, Romania; Gyoma 133, Hungary; Ezinge, the Netherlands; and Feddersen Wierde, Germany. The last three of these four sites also feature skates made from horse bones. Ezinge contributes three skates from cattle metatarsi and one from a horse metacarpus; Feddersen Wierde contributes two cattle metacarpus skates and one horse metacarpus skate. The skates from Gyoma 133 are discussed in more detail below; that only three of the 48 skates found there are from cattle bones, which shows that there was a strong preference for horse bones. Only a single skate, made from a cattle metacarpus, was found at Gara Banca. That cattle bones outnumber horse bones among the skates at Ezinge and Feddersen Wierde could suggest that skaters got started at younger ages there, that cattle were more readily available, or simply that the archaeological record is incomplete.

Overall, the data from the bone types suggest the broad generalization that bone skates were probably mainly used by adolescents, since horse metacarpi, which generally fall into the middle bin, predominate. Naturally, skater demographics could have varied from site to site. If measurements for all the early skates were readily available, it would be easier to apply the method used by Edberg and Karlsson. But this is more difficult than it appears. The Sarmatian skates from Gyoma 133 show why: of the 48 skates described by Choyke,[52] only 15 are whole. The others were broken at some point—during manufacture, use, excavation, or that in-between time they spent in the ground. Despite plenty of breakage, measurements of skates from three different sites or localized groups of sites, Gyoma 133, a set of sites in France, and the Pozzuolo del Friuli hill-fort in Italy, make it possible to consider local variations.

The complete skates from Gyoma 133 range in length from 19.7 cm to 26.2 cm. Only two fall into the largest bin developed by Edberg and Karlsson (24.0 cm or greater). The shorter of those two skates, at 24.0 cm, is made from a horse metacarpus, and the longer, at 26.2 cm, from a horse metatarsus. In contrast, none of the measurable skates were shorter than 18.9 cm, the smallest bin in Edberg and Karlsson's analysis, but there are three fragmentary skates made from cattle metacarpi that might be candidates for that bin.

The picture seems somewhat different at the group of sites in France examined by Putelat,[53] who provides measurements for half of the 20 skates he describes. At 16.8 cm, one of these skates—one made from a donkey metacarpus—falls into the smallest of Edberg and Karlsson's bins. Only three of the skates (a donkey metatarsus and two horse metacarpi measuring 20.2, 21.2, and 21.9 cm, respectively) fall into the middle bin. The other six skates, at 25.0, 27.1, 28.4 (two skates), 29.9, and 35.0 cm, fall into the largest bin. These skates are all made from horse metatarsi except the longest, which is made from a horse radius and is the most elaborately worked of all. In this place and at this time (these skates date to between the second and the fourth century CE), skating may have been mainly for adults because the skates were so big.

Similarly, four of the nine skates found at the eighth- or ninth-century Pozzuolo del Friuli hill-fort in northern Italy were made from horse radii (all fragmentary). The other five skates were made from horse metapodia: two metacarpi (21.2 and 20.7 cm) and two metatarsi (23.05 and 24.71 cm). The ninth skate was made from a deer metatarsus (30.34 cm).[54] With the majority of skates in the largest bin, skating may have been primarily for adults in this region too. The popularity of radii here is unusual and may suggest a practical use for skating.

The different distributions of skates at these three sites may suggest that bone skates were primarily used by adolescents whose feet had not finished growing and possibly by children at Gyoma 133 but, in contrast, were more frequently, but not exclusively, used by adults at the sites in France and Italy. This variation parallels that noted by Edberg and

Karlsson for Birka and Sigtuna. They suggest that the prevalence of horse bones at the latter site in comparison with the former suggests that skating was mainly a pastime for children at Birka but more popular with teenagers at Sigtuna.[55]

Across Europe, the strong but inconsistent preference for horse metacarpi, which could have been thrown away, puts bone skates on the opportunistic (Class II) end of Choyke's continuum. This points toward recreational use and is not inconsistent with a primarily adolescent skating population. Class I artifacts with important practical uses are more likely to have been made consistently from a single bone type. Choyke[56] notes that the careful planning of economically important tools is often based on strong cultural traditions, including what type of bone is selected. This seems not to have been the case with bone skates. However, the reasonably consistent use of horse rather than cattle metapodia keeps bone skates from being solidly at the Class II end of the continuum as far as this criterion is concerned.

5.3. Complexity

The second factor in Choyke's manufacturing continuum is the amount of work put into an artifact, which is evident in the amount of wear due to manufacturing. Artifacts that show extensive working are on the Class I end, and those that are close to plain bones are on the Class II end. Baxter[57] suggests that artifacts that were made inexpertly or with little labor—those on the Class II end of the continuum—may have been made by children and are more likely to have been toys. Children may have been encouraged to make their own skates as a way of developing boneworking skills. Edberg and Karlsson[58] note that among the 679 bone skates from medieval Sweden they examined, those made from horse metapodia tend to be more extensively worked than those made from cattle metapodia. They suggest that this is because older children, who had larger feet, put more effort into making their skates.

Gyoma 133 provides a case study for this idea. Nine of the 48 skates from Gyoma 133 probably broke during manufacture, and seven fragments do not have clear signs of use, which means they could have broken during manufacture.[59] This high rate of breakage (9–16 of 47, or 19–34 percent) is remarkable given the strength of metapodia. It may be a sign that the people making skates were not skilled artisans, a group that includes children learning craft skills. One particular example is a skate found in square B13 at Gyoma 133 in Hungary. Unlike other skates from this site, which were shaped using metal tools, this skate was ground with sandstone.[60]

Of the 133 bone skates analyzed in this chapter, 48 (36 percent) do not have documented signs of working beyond the minimal preparation necessary to create a gliding surface. These may have been simple skates, but many are fragmentary (with parts that may have been modified missing) and others lack detailed published descriptions, which means the manufacturing process could have been far more complex than it appears. The other 85 skates show signs of having been worked, presumably to make them better skates. Of all the skates, 37 (28 percent) exhibit only one documented feature, 34 (26 percent) exhibit two or three documented features, and 14 (16 percent) exhibit four or more documented features in addition to the diagnostic surface.

All the horse radius and "other" skates and most of the skates made from cattle

5.3. Complexity

bones (4 of the 6+ radii and 7 of the 9 metapodia) exhibit one or more features. Horse metatarsi were more likely (17 of the 22+, or 77 percent) to exhibit features than horse metacarpi (44 of the 66, or 67 percent) and somewhat more likely (11 of the 22+, or 50 percent) to exhibit two or more features than the metacarpi (25 of the 66, or 38 percent). This may be due to inconsistent reporting of features rather than an actual preference to modify metatarsi more extensively than metacarpi, but it is consistent with Edberg and Karlsson's[61] conclusion that older skaters (with larger feet) tended to put more work into their skates.

The features of the skates are summarized in table 2.[62] The most common feature is an upswept front end; chopping and adding a binding apparatus are second and third most common, respectively. Although both upswept ends are grouped together in the table, 59 of the 63 skates with either or both ends upswept have the distal end upswept, and only 12 have the proximal end upswept. Eight of these skates have both ends upswept. Chopping refers to flattening of the sides or top surface of the skates, which was often done with an axe. Adding a binding apparatus was optional, which makes the relatively wide range of variations interesting. Pointed toes and preparation of the gliding and standing surfaces beyond what is absolutely necessary are relatively rare.

Table 2

The most common features of bone skates from the Late Bronze Age to the Migration Period. Percentages are expressed as the percent of skates with features.

Feature	Number of skates
Upswept end(s)	63 (78%)
Chopped (top or sides, either end)	31 (36%)
Binding apparatus (any type)	20 (24%)
Pointed toe	13 (13%)
Visibly flattened gliding surface	11 (13%)
Roughened standing surface	6 (7%)

Chopping to flatten the ends of the skate on the palmar side or to flatten the sides at both ends was performed on 31 of the skates. Of these skates, nine were chopped in two places, four were chopped in three places, and one was chopped in all four places. The single chops and two-chop combinations are shown in table 3. The skates with three chops are the wolf radius from Cologne and a horse metatarsus skate from Oberdorla, which were chopped on the sides at the toe and on the top on both ends, and an extensively worked Late Bronze Age pair from Törökbálint, Hungary. The only skate four with chops is a horse radius from Vallhagar, Sweden. These special skates are described in detail later.

Attachment systems are less common; only 20 of the skates exhibit some type of system. The most common of these is a transverse hole at the distal end of the skate, which 10 of the skates have. Each of the other types of fastener is only present on a few skates. The "other" holes can get quite creative. One skate from Lébény-Kaszás domb, Hungary, (shown in figures 36 and 37) features a cross-shaped hole in the palmar side. The combinations are shown in table 4. In addition to the skates in the table, three skates (the pair

76　　　　　　　　　　5. Tools or Toys?

Figure 36: Side view of a Late Bronze Age horse metapodium skate from Lébény-Kaszás domb, Hungary. Note the upswept distal end and the transverse hole (courtesy Gabriella Németh).

Figure 37: Another view of the Late Bronze Age horse metapodium skate from Lébény-Kaszás domb, Hungary. The palmar side is visible. Note the pointed distal end and the unusual hole at the proximal end (courtesy Gabriella Németh).

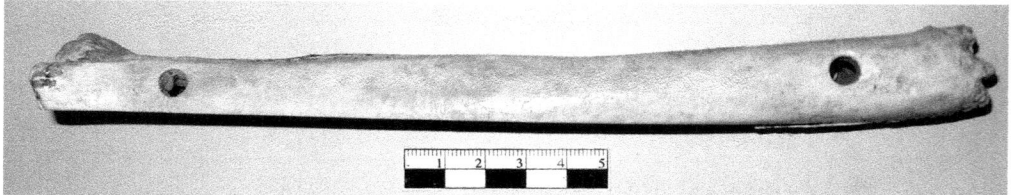

Figure 38: Side view of a Late Bronze Age horse metatarsus skate from Zámardi, Hungary. Note the placement of holes at both ends. Site: Zamárdi-Kútvölgyi-dűlő, Hungary; excavation leader: Zsolt Gallina (Ásatárs Kft.); archaeozoologist: Beáta Tugya (György Thury Museum, Nagykanizsa, Hungary) (courtesy Beáta Tugya).

Figure 39: Another view of the Late Bronze Age skate from Zámardi, Hungary, showing how worn it is. Site: Zamárdi-Kútvölgyi-dűlő, Hungary; excavation leader: Zsolt Gallina (Ásatárs Kft.); archaeozoologist: Beáta Tugya (György Thury Museum, Nagykanizsa, Hungary) (courtesy Beáta Tugya).

5.3. Complexity

from Törökbálint and one skate from Zámardi, Hungary) feature transverse holes at both ends plus other holes. The Zámardi skate is shown in figures 38 and 39.

Table 3

Combinations of chops in skates chopped once or twice. The diagonal entries indicate skates that were chopped only once. Off-diagonal entries indicate two chops, e.g., three skates were chopped at both ends of the palmar side.

Number of Skates		Lateral		Palmar Side	
		Distal	*Proximal*	*Distal*	*Proximal*
Lateral	Distal	1	1	0	2
	Proximal		2	0	3
Palmar Side	Distal			6	3
	Proximal				8

Berg[63] cites the "fact that almost only children and beginners had to bind on the skates, whereas the skilled could do as well without such arrangements." If true, this would mean that the presence of a binding apparatus points toward skates being toys. The data from Birka and Sigtuna contradict this supposition: Edberg and Karlsson[64] found that shorter skates, those sized for small children, tend to have been less extensively worked and were never attached. In contrast, longer skates are more likely to have been worked, including having holes drilled in them. Binding holes and other attachment systems push bone skates toward the Class I end of Choyke's manufacturing continuum. That these are only documented in 16 percent of the skates with features that are analyzed in this chapter is consistent with the idea that skates were primarily toys. This may also reflect cultural preferences. Edberg and Karlsson[65] suggest that "lösa" (loose, i.e., without bindings) and "fasta" (fast, i.e., attached) traditions may have existed in different places.

Table 4

Combinations of holes or other fastening systems in skates. The diagonal entries represent skates with only one hole. Off-diagonal entries represent skates with two holes. In addition, three skates exhibit transverse holes at both ends and other holes.

Number of Skates		Transverse		Two holes		Axial fastener	Other
		Distal	*Proximal*	*Distal*	*Proximal*	*Proximal*	
Transverse	Distal	3	2	0	3	0	1
	Proximal		0	0	0	0	0
Two holes	Distal			1	0	0	0
	Proximal				0	0	0
Axial fastener	Proximal					4	1
Other							2

Pointing the toe is one of the most basic modifications to bone skates, but only 12 of the skates analyzed in this chapter feature pointed toes. These skates were exclusively horse metapodia (6 metacarpi, 4 metatarsi, and 2 unidentified metapodia). All exhibit other features as well; all but one, a Migration Period skate from Ezinge, the Netherlands, also have the distal end upswept. Nine of the skates with pointed toes exhibit three or more features. These skates are extensively worked, as bone skates go, which shows that a pointed toe was an extra feature that may have made for premium skates.

Choyke and Bartosiewicz[66] suggest that horse metapodia were used more often than cattle metapodia because it is easier to carve them into points. This is because unlike horses, cattle have cloven hooves, and their metapodia are actually made of two bones fused together, whereas in horses, the second and fourth metapodials are vestigial and present only at the proximal end. As a result, cattle metapodia have narrow spaces between the two original bones at the distal ends. This space is exactly where a skater would want the toe to be pointed. This explains why no cattle metapodium skates are pointed and may be one reason why horse metapodia were more popular than cattle metapodia, but it is only partly borne out by the data.

Because the data on features are incomplete, it may be more fruitful to more closely examine some of the skates known to have been extensively worked individually.

The Pair of Equid Metacarpi from Törökbálint, Hungary

These two skates are from horses or possibly small donkeys.[67] Their distal ends are both pointed and upswept, and their sides (at the proximal end) and tops (at both ends) were flattened by chopping. Transverse holes were drilled at both ends, and extra transverse holes were drilled in the proximal part of the bone, with vertical holes drilled into these holes.[68] Choyke and Schibler do not say how long these skates are, but they look like they are probably at the long edge of the small bin or the short edge of the middle bin.

These are the most extensively worked skates of the set. Their small size means that they probably fit children, especially since their upturned toes would prevent a skater's foot from extending past the front end of the skate. Their size suggests that they were toys, but their extensive working pushes them toward the Class I end of the manufacturing continuum and suggests that they had some economic importance. They could have been made for an adult with small feet or for a very special child.

One of the Late Bronze Age Skates from Lébény-Kaszás-domb, Hungary

This skate (shown in figures 36 and 37), made from a horse metapodium, has a pointed and upswept distal end, a transverse hole near the distal end, and an unusual cross-shaped hole on the palmar side near the proximal end. Close examination of the photographs suggests that it is approximately 27 cm long, but the way the distal end is upturned makes it impossible for the skater's toes to reach the end of the bone. It is suitable for a skater whose feet are about 23 or 24 cm long. Like the skates from Törökbálint, this skate is closer to the Class I end of the manufacturing continuum. Its size suggests it may have been made for a small adult.

The Three Late Bronze Age Skates Made from Horse Metatarsi Found at Ivanovice na Hané 6 and 7, Czech Republic

These skates are shown in figure 33. The distal ends of are pointed and upswept. One has its proximal end chopped flat on both the sides and the top. It is 26.3 cm long, which puts it in the largest of Edberg and Karlsson's bins. The other two, which were found as a pair, have one transverse hole drilled in the distal end and two oblique holes drilled in the proximal end, like one of the two cattle radius skates from the same site. The skates in this pair are 21.6 and 24.2 cm long, which puts them in the middle bin. The longer skate is just long enough for the longest bin, but if they are a pair, their owner's feet must have been small enough to use the shorter one. The single skate is sized for an adult, and the pair of skates is rather smaller. The extensive working puts these three skates further toward the Class I end of the continuum, and their size supports use by adolescents and perhaps adults. These skates may have been used by advanced skaters for recreation or for a practical purpose.

The Roman Iron Age Skates from Oberdorla, Germany[69]

These skates, which have been described by Barthel,[70] exhibit extensive working. One, made from a horse metatarsus, exhibits an upswept distal end, three of the four possible chops (only the sides at the proximal end remain unchopped), and a roughened footbed. The other, made from a horse metacarpus, exhibits an upswept distal end with its sides chopped, has been chopped flat on the top of its proximal end, and has a roughened footbed. These skates are 27.5 and 22.3 cm long, respectively. The other two skates, both made from horse bones, are less remarkable. The metacarpus features an upswept and pointed distal end, and the metatarsus features an upswept distal end and a 17-mm hole drilled along the axis at the proximal end. These two skates are 22.6 and 28.2 cm long, respectively.

Two of these four skates are very long, which points to adult use, and the other two are sized for adolescents. The longest skate includes a binding apparatus (the hole drilled into the proximal end), which is consistent with Edberg and Karlsson's conclusion that attachment systems were used by older, more skilled skaters. These skates may have had practical uses or been used by adolescents or young adults for recreation. The pointed end of one skate shows that someone took the time to make a nice skate, but this skate falls into the middle bin, which may suggest that it belonged to an enthusiastic adolescent.

The Wolf Radius Skate from Cologne[71]

This skate dates to between 0 and 300 CE and is 18.0 cm long, which puts it in the smallest bin. Its sides and proximal end have been chopped flat, leaving a straight, narrow skate. This is one of two skates with the proximal end, but not the distal end, upswept. Berke[72] suggests that this is because the bone is naturally thicker at the proximal end. Due to its curvature, this skate only shows use wear in the middle. Because the striations are light, Berke thinks it did not see much use.

This skate is sized for a child but shows moderately extensive working. This may have been necessary to make this bone into an effective skate because its ends were probably excessively bulbous in its natural state. Berke compares it with a mastiff bone to show how much material had to be removed to make a reasonable skate from such an unusual bone. This bone selection seems to have been opportunistic: rather than seeking out one of the more common bones, whoever made the skate used whatever was at hand, in this case, a wolf bone. This places it toward the Class II end of the continuum and, along with its small size, points toward recreational use by a child.

The Four Migration Period Skates from Zwingendorf, Austria

Three of these skates are made from horse metacarpi, and the other is from a horse metatarsus. Two feature shaped toes: one is upswept and chopped on the sides, and the other (the metatarsus) is upswept and pointed. The latter, the most extensively worked of the set, was also deliberately flattened on the skating side and roughened on the top, and part of the bone was removed or broke off to expose the medullary cavity. Because the medullary cavity may have been exposed through breakage rather than deliberate removal, this exposure is not considered a feature in the analysis.

The other two skates are less extensively worked. The third has a shaped (upswept and chopped) proximal end and a flattened gliding side. The fourth skate shows only flattening of the bottom and roughening of the top; Küchelmann[73] notes that this skate was probably not used and may not be a skate after all. Sizes are not available because all these skates are fragmentary, which also means that the skates may have exhibited even more features when they were whole.[74]

The skate with the pointed toe is the most extensively worked of the set and, since it is made from a metatarsus, probably the longest. This puts it closer to the Class I end of the continuum than the other skates and is consistent with the idea that a pointed toe is an extra feature added by experienced skaters. The other skates show basic shaping and were probably smaller. Not one has a binding apparatus, which may mean this area followed the "loose" tradition. These skates may have been used by adolescents.

A Migration Period Horse Radius Skate
from Vallhagar, Gotland, Sweden

This skate is roughly 29 cm long, which puts it among the longest skates, and features transverse holes at both ends. It has also "been retouched and flattened at the ends,"[75] which seems equivalent to having been chopped in all four places considered. Its large size points to adult use. Although it was not extensively worked, the holes imply that it was attached to the skater's foot. This is consistent with the idea that older, more skilled skaters attached their skates. Because it is made from a radius instead of a metapodium, the skater may have carefully selected this bone rather than simply making use of waste bones. It could have been used for practical purposes, such as ice fishing.

A Very Long Horse Radius Skate from Geispolsheim, France

This last skate (figure 40) is the longest of the set at 35.0 cm and dates very broadly to antiquity. Both ends have been upswept, and the palmar side has been flattened and perhaps roughened by blows.[76] Like the Vallhagar skate, this skate is sized for an adult. Some

5.3. Complexity

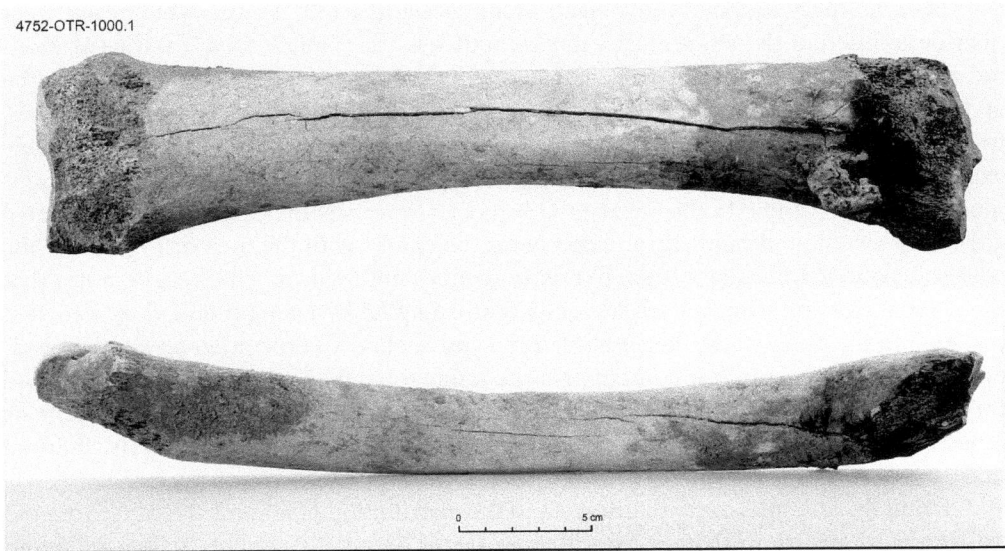

Figure 40: A large horse radius skate from Geispolsheim, France. Note the upswept ends (Olivier Putelat, "Les relations homme-animal dans le monde des vivants et des morts: Étude archéozoologique des établissements et des regroupements funéraires ruraux de l'Arc jurassien et de la Plaine d'Alsace de la fin de l'Antiquité Tardive au premier Moyen Âge," 615, fig. 633, courtesy Olivier Putelat).

care has been taken with its manufacture, but it was not extensively worked. Because this skate was large and functional, it could have had a practical use. The other two skates from this site, one made from a horse metacarpus (21.9 cm long) and one from a horse metatarsus (28.4 cm long) are not as extensively worked. They feature only upswept distal ends. The longer of the two is certainly adult-sized, and the shorter is large enough for an adolescent. These skates may be evidence that skating was primarily, but not exclusively, an activity for adults in Geispolsheim.

Most of the extensively worked skates described above generally tend toward the long end of the size range. Five of these skates are over 24.0 cm long, and some are significantly longer. Only the wolf radius skate and perhaps the pair of skates from Törökbálint are in the small bin. As Edberg and Karlsson observed for medieval Sweden, the more extensively worked skates also tend to be larger. The extensive working pushes them toward the Class I end of the continuum, which suggests they may have had practical uses. Their large size means they would have fit adults.

However, these skates were selected because they are unusual. Most of the skates discussed in this chapter are not extensively worked, or at least, extensive working has not been documented. The Sarmatian skates from Gyoma 133 are worth a closer look, not because they were extensively worked, but because of their sheer numbers. Of the 48 skates, 30 exhibit one or more features.[77] In general, skate-makers at Gyoma 133 chopped the bones to form flat gliding and standing surfaces and upswept one or both ends or (less frequently) made other modifications using metal tools.[78] The resulting skates feature upswept toes (the most common feature), upswept heels, side-chopping, top-chopping, and flattening in various combinations, but none have any type of binding apparatus. This means that Gyoma 133 is an example of a site where skaters followed the "loose" tradition.

Because so many skates have been found at Gyoma 133, it is possible to use their distribution across the site to learn more about how they might have been used. Evaluating the distribution of finds at a site is one technique for identifying children in the archaeological record. Schiffer[79] describes the role of children in moving trash from place to place, creating new features as they go, based on a summary of the literature. Notable experimental studies of this process have been conducted by Bonnichsen[80] and Hammond and Hammond.[81] In the former, the author uses archaeological methods to evaluate a recently-abandoned campsite and compares the results with the memories of someone who had occupied the site. Although this study does not focus on children, he notes that the presence of child-related artifacts in a tent occupied by young men led to incorrect results.[82] In the latter study, the authors, one a young child and the other an adult archaeologist, went to a vacant lot containing a trash pile to study how the child redistributed the trash around the lot while playing. They conclude that children tend to spread artifacts around a site when they might otherwise be concentrated in a single area, that is, children make messes.

Using these results to try to understand the distribution of bone skates at Gyoma 133 leads to more questions than answers. Of the skates found at Gyoma 133, 38 were found in features scattered across the site, seven were found in squares at the southern end of the site, and three were found on the surface with no specific provenance. The function of most of the features is unclear. Many were probably garbage pits, including some for workshops. Some were ovens, and one was a well. Only one was a house.[83]

This distribution seems consistent with the idea of children wandering around the site and scattering their belongings, which is not unexpected behavior. If these skates were made from bones that would normally have been thrown away, which is consistent with the exclusive use of metapodia and the account provided by Brückner,[84] this distribution is consistent with the archaeological studies of children. However, unlike other toys, skates could only be played with where ice was available, and therefore, might be expected to appear near the stream. It would be interesting to compare the distribution of bone skates at Gyoma 133 with the distribution of bone skates at other early sites that yield them in large numbers, if suitable sites are ever found.

Overall, the evidence points to bone skates being primarily used by young people for recreation. They tend to be on the Class II end of the continuum with respect to the first, second, and fourth of Choyke's criteria; data for the third are not readily available. In addition, the predominant use of horse metacarpi suggests that bone skates from this period were generally in the middle of Edberg and Karlsson's three bins. This means most skates were the probably about the right size for teenagers and young adults. Horse bones are even more prevalent from the Late Bronze Age through the Migration Period than they are in medieval Sigtuna, which suggests that rather than starting as young children and continuing into their teenage years, as suggested by Edberg and Karlsson,[85] people in this earlier period began to skate when they were a bit older. They were still not experts at bone-working and made mistakes that resulted in broken skates at Gyoma 133. Because bone skates are closer to the Class II end of Choyke's manufacturing continuum, they may have been primarily used for recreation. Some practical uses need not be ruled out, and early skaters may have used skates for travel and fishing, as some later skaters did. Naturally there was some variation from site to site, but the main point of skating seems to have been to have fun.

5.4. A Note on the Earliest Skate Candidates

The analysis of this chapter begins with the Late Bronze Age because that is where the evidence becomes uncontroversial. However, the earliest skate candidates found to date are the four radius-based artifacts found at Albertfalva in Hungary. These and other similar artifacts dating to the Bronze and Iron Ages are discussed in section 3.3. They have been identified as skates in the past, but current archaeological thinking places them in the textile industry instead based on their wear patterns and size variations. Their wear patterns have already been discussed. In this section, I show how the size variation that Choyke[86] takes as evidence that these artifacts were not skates can be explained without undermining the skate interpretation. If these artifacts are skates, they may have been used for practical rather than recreational purposes. The selection of radii rather than metapodia suggests that these skates were used by adults.

Although the Albertfalva artifacts are fragmentary, their size can be extrapolated from estimates of the size of horses in Early Bronze Age Hungary. Bökönyi[87] reports that Hungarian Bell Beaker horses were between 126.4 and 144.5 cm tall at the withers like other early horses from eastern Europe. May[88] notes that the lateral length of the radius times 4.317 gives the height of a horse at the withers, which means that skates made from these bones could have been between 29.28 and 33.47 cm long, which is more than long enough for an adult skater. In contrast, May also provides factors of 6.403 and 5.331 for the lateral lengths of metacarpi and metatarsi, respectively, which yield skate lengths of 19.74–22.57 cm and 23.71–27.11 cm, respectively. These are comparable in size to the longer of the medieval skates discussed by Edberg and Karlsson,[89] who suggest that such skates were used by teenagers, but horse metatarsi were also long enough to have been used by many Bronze Age adults.

Because metatarsi were probably long enough for many adults and generally made better skates, the selection of radii seems odd. Radii have much more marrow than metapodia, which makes metapodia a more logical choice if food value were a concern. Breaking the bones to get at the marrow would have relegated them to the garbage heap and ruled out their use as skates. Instead, they were kept whole; Choyke and Bartosiewicz[90] note that these bones "must have been deliberately and carefully selected for the manufacture of these objects." This pushes these artifacts toward the Class I end of Choyke's continuum, suggesting that they were used in an economically important activity. There must have been a strong reason to use radii for these artifacts; the entertainment of children seems insufficient to explain this choice.

The selection of radii points to the possibility that these artifacts were used by adults for practical purposes because people would have had to have been careful not to break them to get at the marrow. It would also have been preferable not to boil them, as discussed in Chapter 2, which means that the people who manufactured them had an application in mind that was more important than dinner. This is underscored by the fact that these artifacts were the only things made from horse bones in Albertfalva and Százhalombatta-Földvár during this period.[91] They are closer to the Class I end of the spectrum than most of the later skates are because of the careful selection of radii and only radii—first of horses, then of cattle, and finally, of other animals as well. If these are skates, this means that a utilitarian origin of skating, particularly one involving travel, cannot be ruled out even though it appears that in later times, bone skates were primarily used for recreation. The selection of radii may point to ski ancestry and support the hypothesis of Chapter 4. Skis are known to have had practical uses in antiquity—Dresbeck[92] and Allen[93] describe their use in travel, hunting, and warfare.

Choyke's second criterion, the amount of manufacturing wear, is also applicable. One of the Albertfalva fragments features a transverse hole at the proximal end,[94] and one of the cattle radius-based artifacts from Százalombatta-Földvár features has an upswept proximal end. The later artifacts in this class were normally merely flattened[95] with no binding apparatus and no features to improve their ability to slide across ice, if that is what they were used for. This level of manufacturing wear shows that some care was taken to make the earliest artifacts, which keeps them from being all the way at the Class II end of the manufacturing continuum, but they are not nearly as far toward the Class I end as some of the extensively worked skates discussed earlier in this chapter are.

With later artifacts in this class, the selection of the radius is consistent, but the animal is not, which results in significant variations in size. Lukasiewicz and Rajewski,[96] who identified these artifacts as skates, divided them into two categories based on size: small (12–16 cm) and large (20–30 cm). The small ones could have been made specifically for children. Instead of ruling out their use as skates, as Choyke[97] suggests, this variation can be seen as evidence for a process of miniaturization similar to the one that resulted in a miniature material culture for Inuit children.[98]

In a survey of artifacts found in houses at 31 Thule sites from the Arctic areas of Canada and from northwestern Greenland, Park[99] found numerous miniature tools associated with harpoonery, archery, fishing, and transportation, including three miniature sled runners, as well as miniature knives, slings, and darts. He argues that these miniatures were used by children playing at adult activities and describes Inuit children using miniature bows, arrows, and harpoons to hunt animals such as caribou and seals as their Thule ancestors could have.[100] Miniature artifacts in the transportation class seem to be over-represented (in comparison to full-sized transportation artifacts); Park[101] suggests that "'vehicles' like boats or sleds may simply have been seen as particularly enjoyable toys, much as miniature cars, trucks, trains, boats and spacecraft are today in Western societies." It is not a stretch to suppose that children would have wanted their own skates if they observed adults using skates for travel and hunting, especially in light of the popularity of other transit-based toys. The skates could have been sized in relation to the skater's foot, with smaller children using smaller skates, which would account for the wide variation in size. With the artifacts consistently made from radii, it seems as though using that particular bone may have been important. The need for large size variations explains the selection of radii from different animals. It seems plausible that such play began with a parallel to the Inuit experience: children learned to skate on miniature bone skates.

6

Skating and Skiing in Medieval Scandinavian Literature

6.1. Skates and Skis

The previous two chapters focus on bone skates from their invention through the Migration Period. This chapter and the next are about the skates from the Middle Ages, broadly and loosely defined as the period from the end of the Migration Period to about the sixteenth century. From an archaeological perspective, this is an amazing time for bone skates. Thousands of medieval bone skates have been found in Europe, especially along the northern coast. Many of these skates are from Scandinavia and areas visited, raided, or occupied by Scandinavians. Some Scandinavian sites have produced hundreds of skates.

Scandinavia is unique in having a large body of medieval vernacular literature that includes details of daily life. Bone skates, skating, and skiing are occasionally mentioned in medieval Scandinavian literature, which is the focus of this chapter. The Icelandic sagas, which generally take place during the Viking Age (793–1066 CE)[1] but were written down a few centuries later, the *Prose Edda* written by Snorri Sturluson in the early thirteenth century, and the *Poetic Edda*, a collection of poetry that is not easily datable, provide information about skating and skiing. The eddas and sagas were written in Old Norse, the language spoken in Scandinavia during the Middle Ages. It is quite similar to modern Icelandic, one of its children. Some of the sagas take place in Iceland, and others take place in Sweden or Norway, or elsewhere. They describe journeys to lands as far as North America (Vinland) and both historical and legendary events. Today, they are studied by scholars around the world and provide entertaining reading for anyone.[2]

Although skis are mentioned in the Icelandic sagas several times as important tools for winter travel and hunting, bone skates are almost entirely absent. Their apparent absence from the literature is puzzling because over a thousand bone skates have been excavated in Scandinavia. Birka, Sigtuna, and Oslo have each contributed several hundred bone skates, and other sites have contributed fewer skates. The relative absence of bone skates from the sagas can be explained using the hypothesis that they were considered close relatives of skis that were adapted to particular environments and uses. I argue that bone skates and skis were similar enough to be recognized as two variations on one concept: sliding across a slippery surface while standing on something that assists, rather than arrests, this motion. The difference between them is that skates were probably primarily used by children for recreation on smooth ice, whereas skis were for practical uses on snow, such as hunting.

The most prominent of mention of bone skates in the literature occurs during the boasting contest in *Magnússona saga (The Saga of the Sons of Magnus)*,[3] one of the sagas that make up Snorri Sturluson's *Heimskringla (History of the Kings of Norway)*. In it, King Eysteinn makes parallel boasts about his skating and skiing abilities. The structure of these boasts provides a connection between the two skills. In addition, the similarity of the techniques used for skating and skiing suggests that when Scandinavians began using bone skates, they could have identified them as a type of ski because of the similar motions of skaters and skiers. This is supported by the generic nature of the verb *skríða* (to slide), which can refer to a range of activities involving smooth gliding.

Bone skates have been found in large numbers at a few medieval sites in Scandinavia. This concentration contrasts with the situation across the rest of Europe, particularly central and eastern Europe. In those areas, fewer skates have been found at more sites. This may show that skating was extremely popular at the few Scandinavian sites with excellent skating conditions, or it may reflect the way archaeological research is conducted and how the results are disseminated. More in-depth studies of bone skates on both local and regional levels are needed. The model for such studies is the largest study of a set of finds published to date, the report on 679 bone skates from Birka and Sigtuna, two sites on Lake Mälaren in Sweden, by Edberg and Karlsson.[4] Medieval skates have also been found at Lund (47) and Lödöse (37) and on Gotland (10) and Öland (3). A handful of undated skates have been found at other locations in Sweden, including Dalarna, Uppsala, Småland, and Uppland. As for Norway, 402 medieval skates have been found in Oslo, 65 in Trondheim, and four in Bergen. In Denmark, fewer skates have been found; Küchelmann's database includes 16 medieval skates, eight from Ribe and eight from Trelleborg in Zealand, five from Århus, and one from somewhere in Denmark. To these can be added a skate from Viborg.[5] Iceland is only represented by three pairs of bone skates from the nineteenth and twentieth centuries.[6]

The connection between bone skates and skis relies on the fact that Scandinavians were already skiers when bone skates reached them. When they arrived in Scandinavia during the Migration Period, bone skates entered a region in which skis had been used for thousands of years. The earliest evidence for skis in Scandinavia comprises the Rødøy petroglyph in Norway and early bog finds; some of this evidence is discussed in Chapter 4. Medieval skis have been found in peat bogs in Finland, Norway, and Sweden. Many bog finds have been described by Berg,[7] and their locations have been mapped by Sørensen[8] and Clark.[9] Literary references to the *Skrið-Finnar* ("sliding Finns," variously spelled) begin with Procopius and Jordanes in the sixth century.[10] These references continue through Old Norse literature and have been summarized by Björn Bjarnason.[11]

The rest of this chapter details the literary evidence and relates it to archaeological finds when appropriate. The next section focuses on the boasting contest in *Magnússona saga* and other details about skaters and skiers. Then, the focus widens into a comparison of the motions involved in skating and skiing. These motions can both be described by the verb *skríða*, which can be used for any smooth gliding motion. Finally, the conclusions reached in this analysis are summarized.

6.2. Skaters and Skiers

There are few references to bone skates in the corpus of Old Norse literature. The clearest and most well-known is in the boasting contest in *Magnússona saga*. This contest was held by kings Eysteinn and Sigurðr, two brothers who ruled Norway jointly during the late twelfth century, at a dinner party at Eysteinn's house for the entertainment of their guests. Eysteinn uses skating and skiing as the finishing touches on two consecutive boasts with parallel structures. Their placements emphasize skating and skiing because they seem only loosely related to the other parts of Eysteinn's boasts. These activities tie the two boasts together, and their parallel placement suggests that skating and skiing were considered similar.

Eysteinn's first few boasts focus on speed and agility, whereas Sigurðr's boasts focus on strength and fighting ability. Right before Eysteinn boasts about his skating ability, Sigurðr boasts of his ability to best Eysteinn in a body of water:

> Mantu, hversu fór um sundit með okkar? Ek mátta kefja þik, ef ek vilda.[12]
> (Do you remember how it went when we went swimming? I could duck you, if I wanted.)

Eysteinn responds with another swimming-related boast and adds the boast about his skating ability:

> Ekki svam ek skemmra en þú, ok eigi var ek verr kafsyndr. Ek kunna ok á ísleggjum, svá at engan vissa ek þann, er þat keppði við mik, en þú kunnir þat eigi heldr en naut.[13]
> (I did not swim for a shorter time than you, and I was not worse at underwater swimming. I could also [slide] on bone skates, so that I did not know anyone who contended with me, but you could not do that better than a cow.)

In these two boasts, three different words for swimming are used: *kefja* (to dip, put under water[14]), *svima skemmra* (to swim shorter[15]) and *kafsund* (swimming under water[16]). Eysteinn is worse than Sigurðr at the first and at least as good at the other two. Eysteinn's boast about his skating ability continues the skill progression begun with swimming. He also progresses through the surface of the water: swimming is at the surface, underwater swimming is below it, and skating is on top of the surface of frozen water. Later in his boast, Eysteinn refers to bone skates clearly with the word *ísleggir*, which Cleasby and Vigfusson gloss as "ice-legs, shin bones of sheep used for skates."[17]

This is the only occurrence of *ísleggir* in Old Norse literature, but may not be the only reference to bone skates. There is another candidate in *Fljótsdæla saga*: "Síðan skýtr hann beinspýtum under sik, hleypr síðan upp eptir vatni, slíkt er fara má"[18] (Then he [Nollar] put *beinspýtum* under himself, and he ran up over the [frozen] water as quickly as he could). The meaning of *beinspýtum* is uncertain. Jón Jóhannesson[19] includes a footnote suggesting "grannur fótur, spóaleggur" (thin feet, bird-legs), which may refer to a type of foot disease in the Íslenzk fornrit edition of the saga. Porter[20] interprets this passage similarly and translates it as "[t]hen he stretched out his skinny legs and ran up along the lake shore as fast as he could go." Haworth and Young[21] concur with "then he took to his skinny legs and ran along to the end of the lake."

In contrast, *beinspýtum* has been interpreted a reference to as bone skates or similar footwear by Halldór Halldórsson and Thorsteinn Einarsson. Halldór Halldórsson[22] argues that the nominative form of *beinspýtum* ought to be *beinspita* on the basis of the Norwegian word *spita* (a type of bone, possibly the fibula). This could give it a meaning related to feet or legs, which he interprets as a reference to bone skates based on the narrative. This suggestion has been tentatively accepted in the new dictionary assembled at the University of Copenhagen, which glosses it as "?(ice)skate (or with the sense '(skinny) legs')" in this particular phrase.[23] Thorsteinn Einarsson[24] translates this word as "bone-pins" and writes, "this probably refers to split bones, as they are known from later sources. What was probably more common was to use the unbroken shinbones of cattle as skates." "Split bones" are bone skates of a particular type: Hyltén-Cavallius[25] mentions "is-läggor eller tvänne kluvna och inunder glättade läggben" (bone skates or two leg bones that have been split and smoothed on the bottom) and includes a drawing showing skates that have been flattened so far that the medullary cavity is exposed along its entire length, like some of the skates from Sigtuna described by Edberg and Karlsson (see figure 41). These differ from worn-out skates (figure 23) in that the palmar side of the bone has been deliberately removed to leave a nice place for the skater's foot to rest. The dorsal side, which is in contact with the ice, remains in place. On worn-out skates, the dorsal side is worn down until the medullary cavity is exposed.

Returning to the boasting contest, Eysteinn's skating boast echoes Snorri's description of Ullr as "skíðfærr svá, at engi má við hann keppast" (such a skier that none can contend with him) in *Gylfaginning*, part of Snorri's *Prose Edda*,[26] a description Eysteinn echoes in his skating boast. Snorri's application of this phrase to both skating and skiing in different texts provides a link between the two activities. One notable difference is that Ullr is an adult, whereas Eysteinn refers to his childhood. This draws a parallel

Figure 41: A pair of bone skates made from horse metatarsi found at Sigtuna that have been deliberately flattened enough to expose the medullary cavity ("split bones"). The palmar side (where the skater's foot rested) is shown (Rune Edberg and Johnny Karlsson, *Isläggor från Birka och Sigtuna. En undersökning av ett vikingatida och medeltida fyndmaterial*, 28, fig. 4.10, courtesy Rune Edberg).

6.2. Skaters and Skiers

between skating and skiing while putting skating in the context of childhood. This link between skating and skiing helps link this boast to Eysteinn's next one, which continues the pattern:

> Sigurðr konungr segir: "Hǫfðingligri íþrótt ok nytsamligri þykki mér sú at kunna vel á boga. Ætla ek, at þú nýtir eigi boga minn, þótt þú spyrnir fótum í." Eysteinn konungr svarar: "Ekki em ek bogsterkr sem þú, en minna mun skilja beinskeyti okkra, ok miklu kann ek betr á skíðum en þú, ok hafði þat verit enn fyrr kǫlluð góð íþrótt."[27]
>
> (King Sigurðr says: "A more noble and more useful skill, it seems to me, is to be able [to shoot] well with a bow. I expect that you can't use my bow, even if you press both feet against it."
> King Eysteinn answers: "I am not as bow-strong as you, but I remember less of a difference in our straight-shooting, and I can [slide] much better on skis than you, and that has previously been called a good skill.")

Sigurðr links his second boast with Eysteinn's skating boast by using the same structure: he begins with what he can do best and ends with an insult to Eysteinn. Eysteinn's response follows the same structure as his previous boast, beginning with *ekki* to admit his inferiority, then adding something they are equally skilled in. He ends with something he is better at, skiing. In this exchange, the kings move to activities requiring more skill, namely, shooting and skiing. Eysteinn's word for skis, *skíð*, is much more common than *ísleggir*. Degnbol et al.[28] list 46 references to *skíð* in the Old Norse corpus, and Zoëga[29] glosses it as a "billet of wood, firewood" or "long snow-shoes, 'ski.'" Cleasby and Vigfusson[30] list the first of these meanings and provide more information on the second: "[cp. Engl. *skid*, the drag applied to a coach-wheel], of *snow-shoes*, such as are used by the Finns, Norsemen, and Icelanders in the north-east of Iceland." Although the dictionaries use the word "snowshoes," both were published over a century ago, when skis were often called snowshoes in English.

As Eysteinn notes in his final clause, skiing has been included in two lists of accomplishments given in skaldic poems. The first of these is a poem attributed to Haraldr harðráði[31] in a manuscript called *Morkinskinna (Rotten or Moldy Parchment)* containing a compilation of sagas from c. 1220.[32] It is as follows:

> Íþróttir kannk átta
> Yggs fetk líð at smíða
> fœrr emk hvasst á hesti
> hefk sund numit stundum:
> skríða kannk á skíðum
> kýtk ok rœk svát nýtir
> hvártveggja kannk hyggja
> harpslátt ok bragþáttu.[33]
>
> (I have eight accomplishments: I know how to forge Yggr's [Óðinn] wine [skaldic poetry]; I am a swift horseman; on occasion I have practiced swimming. I can slide on skis; I shoot and row well enough; I have command of both harp playing and poetry.[34])

The three types of accomplishment in the boasting contest so far, swimming, shooting, and skiing, are included in this poem. Skating is not mentioned, but Eysteinn mentions several different types of swimming and shooting, and the place of skating suggests that it can be interpreted as a type of skiing. These accomplishments are echoed in *Orkneyinga saga,* when Kali Kolsson recites a poem listing his nine accomplishments:

Tafl emk ǫrr at efla,	(At nine skills I challenge—
íþróttir kannk níu,	a champion at chess
týnik trauðla rúnum,	runes I rarely spoil,
tíð er mér bók ok smíðir.	I read books and write:
Skríða kannk á skíðum,	I'm skilled at skiing
skýkt ok rœ'k, svát nýtir,	and shooting and sculling
hvárt tveggja kannk hyggja	and more!—I've mastered
harpslǫtt ok bragþǫttu.[35]	music and verse.[36])

Like Eysteinn, Kali and Haraldr are skilled at both shooting and skiing. Sigurðr and Eysteinn's use of *íþrott* to refer to both and its placement at both the beginning and end of the pair of boasts serve to link these two accomplishments, which are both used in hunting. *Konungs skuggsjá (The King's Mirror)* provides a detailed description of hunting on skis:

> En þat mun þykkja meira undr […] at sá maðr, er hann er eigi fimari á fœti en menn aðrir, meðan hann hefir ekki annat á fótum en skúa sína eða elligar bera fœtr sína, en jafnskjótt sem hann bindr fjalar undir fœtr sér, annattveggja sjau álna langar eða átta, þá sigrar hann fugla at flaug eða mjóhunda at rás, þá sem mest megu hlaupa, eða hrein er hleypr hálfu meira en hjörtr; þviat sá er mikill fjöldi manna, er svá kann vel á skíðum, at hann stingr í einni rend sinni 9 hreina með spjóti sínu ok þaðan af fleiri.[37]

> (But that may seem a greater wonder […] that the man, who is not more agile afoot than other men, as long as he has on [his] feet nothing other than his shoes or his bare feet, but immediately when he binds boards either seven or eight ells long under his feet, he surpasses birds at flight or greyhounds at running, or can run as fast as a reindeer which runs half more than a hart; for there is a great number of men, who can [slide] so well on skis, that he strikes in one run nine reindeer and even more with his spear.)

An example of a specific ski-hunter is provided in *Gylfaginning*: Skaði "fór […] upp á fjall ok byggði í Þrymheimi, ok ferr hon mjök á skíðum ok með boga ok skýtr dýr"[38] (went up on a mountain and lived in Thrymheim and she often goes on skis and with a bow and shoots deer). Following Skaði's example is Atti inn dœlski[39] a man who hunts on skis with a bow in winter and is called "mestan veiðimann" (the best hunter) in *Óláfs saga helga (The Saga of Saint Olaf)*.[40] Skis gave hunters a distinct advantage over their prey, which could have gotten stuck in the snow.

This close look at the structures of the boasts of Eysteinn and Sigurðr has shown that skating and skiing are positioned to link Eysteinn's two boasts. This positioning helps show that they were considered closely related accomplishments and possibly even facets of the same activity, like the different types of shooting and swimming used in the boasts. The difference could be that skates were primarily for recreation and mainly used by children and adolescents, whereas skis had important adult uses, such as hunting. The similar techniques used for skating and skiing, described in the next section, help strengthen the link between them.

6.3. *Skating and Skiing*

Skating on bone skates and skiing require similar motions. Pushing with a single pole while on two skates, the standard technique described in Chapter 2, seems to have been the most common method of skating used in medieval Scandinavia; Olaus Magnus is the primary source of guidance here because of his level of detail and placement in space and

6.3. Skating and Skiing

time. For comparison, some details of the techniques used for skiing can be gleaned from Old Norse literature. One source is Snorri Sturluson's description of skiing in *Óláfs saga helga* in *Heimskringla*:

> Síðan fekk Þórir skíð hvárum tveggja þeira. Arnljótr rézk til ferðar með þeim. Steig hann á skíð. Þau váru bæði breið ok lǫng. En þegar er Arnljótr laust við geislinum, þá var hann hvar fjarri þeim. Þá beið hann ok mælti, at þeir myndi hvergi komask at svá búnu, bað þá stíga á skíðin með sér. Þeir gerðu svá. Fór Þóroddr nærri honum ok helt sér undir belti Arnljóts, en fǫrunautr Þórodds helt honum. Skreið Arnljótr þá svá hart sem hann fœri lauss.[41]
>
> (Then Þórir got skis for both of them. Arnljótr prepared to go with them. He stepped on his skis. They were both broad and long. But as soon as Arnljótr struck with the pole, then he was very far from them. Then he waited and said, that they would come nowhere that way, he asked them to step on the skis with him. They did so. Þórodd went near him and held onto Arnljotr's belt, and Þórodd's companion held him. Then Arnljótr skied as fast as if he were unencumbered.)

Because it is a masculine noun, *geisl*, which refers to the ski pole, could be singular or plural in the form used here, *geislinum*. Singular seems more likely here because Norwegian skiers used a single pole until quite recently, and in 1908, Fritz Huitfeldt went as far as to write that the modern technique of skiing with two poles "is at complete variance with the very nature of skiing."[42] Poles were used singly by Þórðr and Eyvindr in *Þórðar saga hreðu* to help them slide down a snow-covered hill:

> Skafl var lagðr af hamrinum niðr á jöfnu ok ákafliga brattr; var þar in mesta manhætta ofan at fara. Síðan settu þeir spjótin í milli fóta sér og riðu svá ofan af hamrinum allt á jöfnu; komust þeir nú á Sviðgrímshóla.[43]
>
> (A snow-drift was lying from the crag down to the plain and it was extremely steep; it was very dangerous to go over. Then they set their spears in between their feet and rode down from the crag all the way to the plain, they come now to Sviðgrímshóla.)

The context implies that Eyvindr and Þórðr were wearing ordinary shoes rather than skis here, but their method of sliding down the hillside seems quite similar to skiing with a single pole. They could have used the pole for braking, as was done by the later skiers of Østerdal described by Bø.[44] A similar technique was used by people on bone skates in recent times, as vividly described by Säve and by Juvel.[45] The pole could also have helped skiers turn.[46] Additionally, skiers could have used their poles for propulsion. Arnljótr seems to have done this, since he outdistanced his companions when he applied it.

The use of a pole with a metal end for purposes other than skating or skiing is known from *Valla-Ljóts saga*, where Halli carries a *broddstǫng*. Zoëga[47] glosses this as "a (mountaineer's) staff, pole, with an iron spike." This is the only occurrence of the word reported by Cleasby and Vigfusson[48] and Degnbol et al.[49] Falk[50] suggests that this word refers to a type of spear, and Björn Bjarnason[51] uses *broddstaf* to refer to the pole used with bone skates. The use of *brodd* (spike) certainly suggests a sharp point rather than a blunt metal-covered end. Other types of spear, such as the ones used by Eyvindr and Þórðr, may also have worked as skating or skiing aids.

The single-skate technique observed by Thorsteinn Einarsson[52] also has a medieval Scandinavian parallel. Pairs of skis with different lengths were popular in medieval Scandinavia; each included a long *skíð* for gliding on and a short, fur-covered *andar* for pushing.[53] This is the Central Nordic type of ski, which Weinstock[54] thinks is a Scandinavian and/or Sámi innovation developed by combining skis of other types. The earliest secure archaeological evidence for such skis is from 1605 CE.[55] Sørensen[56] attempts to trace this type of ski to its roots through the poems of the *Kalevala* but is unable to document its route. It is possible that skis of this type are quite old; Berg[57] suggests that the laws of Magnús Lagabœtir (law-mender) about hunting, designed to protect elk from skiers, show that unequal-length skis were in use in the late thirteenth century:

> When Magnus Lagaböte makes his Norwegian land law, passed in 1278, to preserve elks[58] [sic] in private grounds from all those who run on skis, such an order can only be understood if it refers to that new type of hunting which the different-length ski type had then brought in.

This may refer to the following passage from Magnús Lagabœtir's new land-law from 1274[59]:

> Elgir allir skolu frið hafa firir þeim monnum er a skiðum renna allt innan takmarkar þers er iorð a.
> (All elk shall have peace in the presence of men who run on skis anywhere within the boundaries of an estate.)

There is no mention of what type of ski these men used; however, its sudden appearance suggests that there was something new. Berg's suggestion is that the new thing was an innovation in ski technology.

The literary evidence for uneven skis includes a proverb from *Morkinskinna* ("[s]nęliga snvGir sveinar qvoþo FiNar atto andra fala"[60] ("it looks like snow, the Lapps said: they had skis/snowshoes for sale"[61])) and Snorri Sturluson's statement, in *Gylfaginning*, that Skaði "heitir öndurgoð eða öndurdís"[62] (is called ski-god or ski-goddess). *Öndurás* (ski-god) is also given as a kenning for Ullr in *Skáldskaparmál (The Language of Poetry)*, part of the *Prose Edda*.[63] *Öndur*, the first part of both these names, is the singular form of *andrar*, which Cleasby and Vigfusson[64] gloss as "snow shoes" with a note that that this word is unique to Norway, where it is still used. In modern Norwegian, *ånder* refers to a "short right ski for climbing (used with long left one and usually having reindeer hide on the underside)."[65] Cleasby and Vigfusson[66] suggest a Finnish origin for the word.

Although there is no evidence for or against the use of the skateboarding technique with bone skates during the Middle Ages, its similarity to the technique used for skiing with skis of different lengths is evident. Each of the two prevalent techniques used for skating on bones has a corresponding skiing technique that resulted in a similar motion. This adds support to the idea that bone skates and skis were considered closely related. From a distance, it may have been difficult to tell which a person was using.

Another similarity between skates and skis is the difficulty of making quick turns. This necessitates bindings on skis because skiers going downhill quickly must make sharp turns that require a strong connection between the foot and the ski. Suitable bindings were not available until Sondre Nordheim invented them in the nineteenth century.[67] This problem was less significant for bone skates because they were primarily used on ice, which does not have a steep slope. Because of its relatively flat surface, skaters had somewhat better control over their velocity than skiers did. In addition, skating on lakes probably did not require as many tight turns, which is fortunate because skaters would not have been able to make them. The impossibility of turning quickly on bone skates is a significant weakness and described in section 2.6. Skiers solved this problem by using bindings and new skiing techniques. Skaters may have attempted to solve it in a similar manner; Edberg and Karlsson[68] note that the longer skates found in Birka and Sigtuna, which were used by "[o]lder, probably more skillful skaters," were more likely to have holes for bindings. The eventual solution was to use metal-bladed skates instead. These new skates were much more maneuverable and left skaters' hands free but were not known to the saga authors.

6.4. Skríða *As a Generic Verb of Motion*

Despite the differences between bone skates and skis, skating and skiing have similar gliding motions. This type of motion can be described using the verb *skríða* (to slide), which Snorri Sturluson uses to describe Arnljótr's motion at the end of the quote given above. *Skríða* is also used for other smooth gliding motions, including sliding across ice in ordinary shoes and ships sailing. Creeping, of both people and reptiles, is another common meaning of the word.[69] This range of meanings shows that *skríða* leaves the exact motion open to interpretation. The only requirement is that it be smooth, which suggests that the authors of the eddas and sagas considered smoothness a basic quality used for classifying types of motion.

Jan de Vries[70] defines *skríða* as "gleiten, kriechen; schreiten" (to glide, to creep; to stride) without referring to any particular type of footwear. Sveinbjörn Egilsson and Finnur Jónsson[71] also give the generic definition "bevæge sig glidende, krybe, især om skibe" (to move oneself gliding, to creep, especially of ships). In the modern Scandinavian languages, *skríða* has retained a broad meaning. In Swedish, *skrida* means to "slide, glide ... proceed" and a *skridsko* is an ice skate.[72] In his etymological dictionary, de Vries[73] connects the Icelandic *skriðna* and the Swedish dialectal *skrinna* (to ice skate) to *skríða*. In Norwegian, *skri* means "1 slide, slip. 2 glide, run (on skates or skis). 3 advance, proceed, walk with a slow, solemn gait; stride. 4 advance, proceed; develop, make progress."[74] The use of Norwegian *skri* with both skates and skis shows that the two motions are similar enough to be described using the same verb.

In Old Norse literature, *skríða* can be seen in action as people slide across ice and snow in ordinary shoes and in skis. Sliding in ordinary shoes on ice is described using the phrase "renna fótskriðu" (to run foot-sliding) in *Reykdæla saga*,[75] *Valla-Ljóts saga*,[76] and *Njáls saga*.[77] This activity is also mentioned by William fitz Stephen just before his description of ice skating in London:

> Hi ex cursu motu captato citatiore, distantia pedum composita, magnum spatium latere altero praetenso perlabuntur.[78]
>
> (Some, gaining speed in their run, with feet set well apart, slide sideways over a vast expanse of ice.[79])

He follows this with his well-known description of the "others" who skate on bones.

Of the occurrences in the sagas, the fact that the foot-sliding occurs on ice is most obvious in *Reykdæla saga*, where the sliding surface is explicitly mentioned: "Hann hafði rennt fótskriðu yfir ísinn [...] Hann snýr nú þegar aptr til árinnar ok rennir enn fótskriðu yfir ísinn"[80] (He had run foot-sliding over the ice [...] Now he turns back to the beginning and runs foot-sliding over the ice). *Valla-Ljóts saga* contains a similar story: "Ok íþví þrífr Eyjólfr til Bersa ok rennði fótskriðu með hann"[81] (and at that Eyjólfr grasps Bersi and ran foot-sliding with him). In *Njáls saga*, "Skarpheðinn [...] rennir þegar af fram fótskriðu.

Svellit var hált mjǫk, ok fór hann svá hart sem fogl flygi"[82] (Skarp-Heðinn [...] ran forward from there foot-sliding. The ice was very smooth, and he went as fast as a bird flies). The comparison of Skarp-Heðinn's smooth gliding and the motion of a bird in flight brings to mind the phrase "saman níðingar skríða" ("birds of a feather flock together"[83] or, literally, "shameful men glide together"). The smooth flight of a bird seems to be in keeping with *skríða*'s generic meaning of any smooth motion.

Skríða has also been applied to the smooth motion of ships sailing, as shown in a passage from *Grettis saga* in which Hafr includes all places where "skip skríðr, [...] Finnr skríðr"[84] (ships slide, [...] Finns slide) in his list of places from which a truce-breaker is outlawed. The fact that the same verb is used for both ships and skiers here shows that these two motions were considered similar. The link between the smooth motions of ships and skiers is strengthened by kennings in skaldic poems. Kennings are short phrases or compound words used in poetry to replace of the name of something; Barraclough[85] describes a kenning as "the Norse equivalent of a cryptic crossword clue." Weinstock[86] collects three examples where skis are used as kennings for ships.

In the first, a verse attributed to Skarp-Heðinn in *Njáls saga*, "Baldr skíða víðs hrings Hernar" (Baldr of the skis of the wide ring of Hern), the word *skíða* (skis) refers to a ship.[87] The second of the collection is attributed to Steinunn Refsdóttir and published by Kock.[88] Steinunn refers to "skíð Atals grundar"[89] (ski of Atall's green fields, i.e., the sea). The third of the kennings in Weinstock's collection uses *ǫndur* instead of *skíð*: "ǫndri andness" (ski of the headland) is a kenning for a ship.[90] All three of these refer to ships as "skis of the sea" in some manner, drawing a clear parallel between the two.

To this collection, a verse from *Haralds saga Gráfeldar (The Saga of Harald Greycloak)* in *Heimskringla* can be added[91]:

> Skyldak, skerja folder
> skíðrennandi, síðan
> þursa tœs frá þvísa
> þinn góðan byr finna,
> es, valjarðar, verðum,
> veljandi, þér selja
> lyngva mens, þats lengi,
> látr, minn faðir átti.

The Old Norse term *skíðrennandi* (ski-running) provides the link with skis. Samuel Liang's loose translation emphasizes the link between a ship sailing smoothly on the sea and the gliding motion of skaters or skiers[92]:

> I go across the ocean-foam,
> Swift skating to my Iceland home
> Upon the ocean-skates, fast driven
> By gales by Thurse's witch fire given.
> For from the falcon-bearing hand
> Harald has plucked the gold snake band
> My father wore—by lawless might
> Has taken what is mine by right.

In his revision of Liang's work, Anderson[93] calls Liang's supposed translations of the poems "original songs or ballads in modern measures" that "do not even paraphrase the thought of the original Icelandic texts," but in this particular instance, he captures the

Figure 42: Ullr on his magic bone, which is generally thought to be a bone skate but looks more like a surfboard here. The runes supposedly carved on it are not visible in the picture (Olaus Magnus, *Historia de gentibus septentrionalibus*, 122).

sense of the term in question adequately with "swift skating," assuming that "skating" is used in the older sense of "skiing." Hollander's[94] translation is more accurate and modern:

> Should I, ship's keen steerer,
> share thy favor henceforth:
> would that well befit thee,
> warrior, ruling Norway,
> seeing I give thee this goodly
> golden arm ring, dragon's-
> lair's rich treasure, liege,
> which long had owned my father.

Hollandar translates *skíðrennandi* by the phrase "ship's keen steerer," which emphasizes its use as a kenning for a ship in motion. A similar link between skating on bones and the motion of ships is provided by Saxo Grammaticus:

> Fama est, illum adeo præstigiarum usu calluisse, ut ad trajicienda maria osse, quod diris carminibus obsignavisset, navigii loco uteretur, nec eo segnius quam remigio præjecta aquarum obstacula superaret.[95]
>
> (According to one tale he [Oller] was such a cunning magician that instead of sailing in a ship he was able to cross the seas on a bone which he had engraved with fearful charms, and skimmed the waves that rose before him as swiftly as with oars.[96])

Oller is what Saxo calls Ullr, the famous skier of the *Prose Edda*. His magic bone is generally accepted as being a bone skate, though the one shown in Olaus Magnus's illustration (figure 42) looks more like a surfboard. A decorated bone skate that illuminates this idea,

Figure 43: A decorated medieval bone skate from Birka. This skate, made from a cattle metacarpus, is reminiscent of Ullr's magic bone (Rune Edberg and Johnny Karlsson, *Isläggor från Birka och Sigtuna. En undersökning av ett vikingatida och medeltida fyndmaterial*, 21, fig. 3.15, courtesy Rune Edberg).

even though it is not inscribed with runes, was found in Birka; it dates to c. 750–980 is shown in figures 43 and 44. The entry for another skate in the Swedish History Museum's catalog includes a note that it may have a boat-like carving. A few curved scratches can be seen in the picture.[97] Another parallel can be drawn to skate fragment from Viborg, Denmark, that may date to the Viking Age. It features a face carved at one upturned and pointed end using only a few strokes,[98] which is reminiscent of the carvings at the prows of ships, except that it faces the skater.

On his magic bone, Ullr was able to glide across ice smoothly as well as quickly. This links bone skates and ships explicitly by the speed and smoothness of their motion.

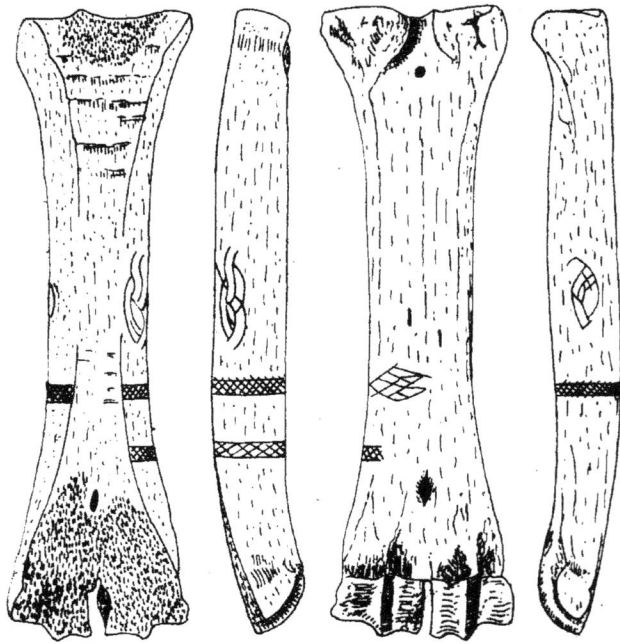

Figure 44: More details of the decorated bone skate from Birka (Erik Sörling, courtesy Swedish History Museum, Sweden. Accession number 270021).

Swimmers may also have appeared to glide smoothly through the water, making the link between Eysteinn's boasts about swimming and skating more than just the water involved. This contrasts with Steinþór's uncontrolled slipping in *Eyrbyggja saga*:

> Freysteinn bófi hljóp eptir Þorleifi; hann var á skóbroddum. [...] ok varð Steinþóri fallhætt, er jakarnir váru bæði hálir ok hallir, en Freysteinn stóð fast á skóbroddunum ok hjó bæði hart ok tíðum.[99]
>
> (Freysteinn bófi (rogue) ran after Thorleifr; he was on spiked shoes. [...] and Steinthor was in danger of falling, because the ice was both slippery and sloped, but Freysteinn stood fast on spiked shoes and struck both hard and long.)

In this passage, Steinþór slips uncontrollably while wearing ordinary shoes on ice while Freysteinn is able to control his motion because of his crampons. This motion is quite

6.4. Skríða As a Generic Verb of Motion

different from the smooth, straight gliding of a ship or a skater, and a different phrase, "varð fallhætt" (became fall-risky), is used to describe it.

Olaus Magnus[100] connects bone skates and skis when he refers to "[d]uo genera hominum ad stadia, & brauia celerrimè sub præsignatæ artis compendio currentium" ("[t]wo kinds of men rushing swiftly over a racecourse for a prize, by means of the art described earlier [skiing]"[101]). This statement emphasizes both the similarity (the same skill) and difference ("two kinds of men") between skates and skis. In one of Olaus Magnus's woodcuts (figure 25), the bone skates look like short skis. This suggests that the illustrator either did not understand the distinction between bone skates and skis or considered it minor. The latter supposition is supported by the example of Ullr as a skater (after Saxo Grammaticus) and a skier (after Snorri Sturluson), which emphasizes the similarity of the two accomplishments. Because there are two examples of people possessing both skills, Eysteinn in a single text and Ullr in two texts by different authors, skating and skiing are likely to have been considered closely related. This fits in well with the other evidence presented here: bone skates and skis were used in similar ways and produced similar types of motion. The sharp modern distinction between skates and skis was not made by Old Norse authors. Because this distinction was not sharp, Old Norse authors did not need to be explicit about which of the two they referred to. In *Magnússona saga*, it was important that the distinction be made so that Eysteinn would not repeat one of his boasts. In other texts, the distinction is less important, and the authors were vague. There are two places in the *Poetic Edda* in which *skríða* is used without further specification in contexts that could refer to skating or skiing: in *Hávamál* and in *Völundarkviða*.

Verse 83 of *Hávamál* begins with "[v]ið eld skal ǫl drekka // en á ísi skríða"[102] ([d]rink ale by a fire, slide when on ice). The latter advice could be to use bone skates, but it is not explicit, as Halldór Halldórsson[103] notes. Foot-sliding and skiing are also possibilities. Because the sliding is done on ice instead of snow, skiing can be probably eliminated, but foot-sliding cannot. Björn Bjarnason[104] suggests that this passage is about skating, and Hollander,[105] Dronke,[106] and Crawford[107] support this interpretation with their translations of *skríða* as "skate." Both Larrington[108] and Orchard[109] provide the more generic translation of *skríða* as "slide," which allows for both interpretations, but also leaves uncontrolled slipping as a possibility.

In *Völundarkviða*, three brothers are said to "skriðo ok veiddo dýr"[110] (slide and catch animals). In this passage, *skríða* is often translated as "to snowshoe."[111] This translation is in keeping with the glosses provided by Zoëga[112] and Cleasby and Vigfusson,[113] who use "snowshoes" in its older sense, which includes skis. The latter specify that the type of snowshoe referred to is *skíð* in Old Norse, and snowshoes (in the modern sense) were not common in medieval Scandinavia.[114] It seems likely that the three brothers hunted on skis because there are other examples of hunters on skis in Old Norse literature, such as Skaði. The translation given by Dronke,[115] "they traveled on skis and hunted wild beasts," is in keeping with this suggestion. Uncontrolled sliding does not seem like a reasonable interpretation of *skríða* in this context.

These examples show that *skríða* can refer to a range of activities; Old Norse authors were not always explicit about what they meant. The possibilities include people creeping, ships gliding across water, people sliding on ice in shoes, and skiers skiing. The common factor in all these different activities is a smooth gliding motion. Although *skríða* has not been explicitly used for skating on bones, the motion of a skater was probably similar enough to that of a skier for the word to apply.

Another verb used for skating and skiing in medieval Scandinavia is *hlaupa* (to leap, jump, or run[116]). This verb is used to describe skiers in the quotation from *Konungs skuggsjá* above and to describe Nollar's motion on *beinspýtum* in *Fljótsdæla saga*. It is also used to describe skaters in an unfortunate event that occurred at Lake Vättern on 18 January 1596, reported in *Calendaria Caroli*, the journal of King Charles IX of Sweden:

> [E]fter middagh drugnedde 4 drenger, såm wille löppe på islegger, gudh upueckke kråpperne till saligh upståndelse.[117]
>
> (After midday, four boys drowned, who wanted to run on ice-legs; may God reanimate their bodies at the holy resurrection.)

Here the Old Norse verb *hlaupa* appears in its Swedish form, *löpa*. The use of this verb with skis in one text, with *beinspýta* (which may refer to bone skates) in another, and with *islegger* (which clearly refers to bone skates) in a third helps tie these three things together. *Hlaupa* is also used to describe Freysteinn's motion on crampons in the passage from *Eyrbyggja saga* quoted above. Crampons are a different technology that is associated with a different type of motion. Cleasby and Vigfusson[118] explain that although its "proper meaning" is "to leap, jump," *hlaupa* is applicable to "any sudden motion," including, for example, water or ice rising and blood coagulating. It seems that the emphasis of this verb is not on the smooth gliding inherent in *skríða* but rather on the quick onset of motion.

Cleasby and Vigfusson add that *hlaupa* rarely occurs with the meaning "to run" "in old writers," which helps explain why it appears with bone skates in later texts, such as *Fljótsdæla saga*, which was written down in about 1500. As a diary, *Calendaria Caroli* was probably written not long after the event occurred in 1596. In contrast, *Eyrbyggja saga* was written down no later than the thirteenth century and *Konungs skuggsjá* dates to around 1250. It may be that *skríða* was more suitable for skating and skiing earlier on, and when *hlaupa* developed its "to run" meaning, it took over because skates and skis are used with the feet and, through them, linked to running.

6.5. *The Similarity of Bone Skates and Skis*

It seems as though skiing and skating on bones were considered closely-related activities in medieval Scandinavia. The key to this result is the structure of the boasting contest in *Magnússona saga*. Eysteinn's parallel boasts about skating and skiing allow the two accomplishments to be linked in the same way his boasts about different types of swimming and shooting are. Skating and skiing could not have been considered exactly the same because it would not make sense for Eysteinn to make the same boast twice. Although Eysteinn keeps skating and skiing separate, the techniques used for performing these two activities were similar. Both skaters and skiers had the option of propelling themselves with a pole or with one foot. This means that the motions of skaters and skiers would have looked similar enough that both could have been described using the same verb. The most suitable verb is *skríða*, which is used for skiing and a variety of other types of smooth motion, including ships sailing. The most important aspect of *skríða* is the smoothness of the motion it is used to describe, and both skating and skiing qualify. Although *skríða* is

6.5. The Similarity of Bone Skates and Skis

never explicitly used to refer to skating on bones in the surviving Old Norse texts, such use was possible.

The paucity of references to bone skates in Old Norse literary texts contrasts sharply with the numerous bone skates found at certain medieval sites in Scandinavia. They may not appear frequently in the sagas because they were primarily used for recreation. The four main references to bone skates do not conflict with this supposition. Eysteinn's boast about his skating ability harkens back to his childhood, and the line from *Hávamál* suggests a context of leisure. Although Nollar in *Fljótsdæla saga* and Ullr use bone skates for travel, such use could have been less common. For comparison, the literature includes descriptions of hunting and travel on skis that are more detailed and frequent than the mentions of bone skates.

That bone skates were considered similar to skis and were primarily used for recreation helps explain why there are so few references to bone skates in the sagas despite the number found in medieval Scandinavian archaeological contexts. This analysis of the evidence suggests that they were used, but perhaps not in a way that was worth remarking on. Skis had been used in Scandinavia for so long that their use was commonplace, and when Scandinavians learned about bone skates, they may have associated them with skis. This similarity may help explain their popularity at certain sites in medieval Scandinavia.

7

Skating on Bones in the Middle Ages

7.1. *The Scandinavian Expansion*

The archaeological record reflects a veritable explosion of bone skates during the Middle Ages.[1] Thousands of medieval bone skates have been unearthed to date. Many have been described in the literature, but many others languish in museum collections. This chapter is about what can be learned from the archaeological evidence that has been published to date. The main conclusion is that Scandinavians may have acted as a vector for the spread of ice skating across Europe. They seem to have brought bone skates to Great Britain and are likely to have contributed to the popularity of ice skating on the Continent. However, the evidence on the Continent is more difficult to sort out because of the long history of bone skates among the people living there—most notably, the Slavs. It may eventually become possible to differentiate between Scandinavian skates and those attributable to other cultures based on their features. The first steps toward a catalog are taken here.

The Old Norse word *víkingr* (pirate) is often applied to the Scandinavians who traveled the world between the eighth and eleventh centuries. This label was not unearned; they did inflict violent raids on many seaside towns. They also established trade routes throughout the known world and built permanent settlements. The best-known of these are their colonies in Iceland and Greenland, but they also moved into England and other areas of northern Europe. Most studies of the Scandinavian expansion, depicted in figure 45, have focused on westward movement from Denmark, Norway, and Sweden to Great Britain, Iceland, Greenland, and North America (Vinland). A friendly introduction to this subject that focuses on material from the sagas is Eleanor Barraclough's *Beyond the Northlands: Viking Voyages and the Old Norse Sagas*. Other recent studies include Grahame Davis's *Vikings in America*, which argues that Scandinavian activity in these areas was more extensive and longer-lasting than previously supposed based on evidence from a variety of sources, and Annette Kolodny's *In Search of First Contact: The Vikings of Vinland, the Peoples of the Dawnland, and the Anglo-American Anxiety of Discovery*, which describes the nineteenth-century interpretation and reception of stories of this spread. Studies specific to Great Britain include Dawn Hadley's *The Vikings in England: Settlement, Society and Culture* and Jane F. Kershaw's *Viking Identities: Scandinavian Jewellery in England*,[2] which focus on archaeological evidence, and *English: The Language of the Vikings* by Joseph Embley Emonds and Jan Terje Faarlund and Matthew Townend's

7.1. The Scandinavian Expansion

Figure 45: A rough map of the Scandinavian expansion (Max Naylor, courtesy of Wikimedia Commons, with slight modifications by B. A. Thurber).

Language and History in Viking Age England: Linguistic Relationships between Speakers of Old English and Old Norse, which evaluate the linguistic evidence. Anders Winroth devotes a bit more space to journeys to the east than the other sources do in his *The Age of the Vikings*.

According to Himstedt,[3] their preference for sea travel meant that Scandinavians settled near harbors and other resources for seafaring. Such areas were ideal for skating because they were near bodies of water that froze readily in winter. Perhaps it is no coincidence that period in which Scandinavians were exploring the world, approximately 800–1200, is also the period that has contributed the most bone skates. Finds are especially common in Great Britain and Scandinavia and along the northern coast of the Continent. The distribution of medieval bone skates is shown in Figure 46. The Czech Republic, Germany, Great Britain, the Netherlands, Norway, and Sweden have each contributed over 100 medieval skates. Skates from this period have also been found in Austria, Belgium, Bulgaria, Denmark, Estonia, Finland, France, Ireland, Latvia, Lithuania, Poland, Russia, Serbia, and Slovakia. In contrast, few medieval bone skates have been found in the region west of Iceland. None have been found at Scandinavian settlements in Greenland or Vinland, and the bone skates known from Iceland are modern. This may be because fewer lakes and large animals (horses and cattle) were available in the west, but other factors may also have contributed. Climate variability is one possibility.

The ability of Scandinavians to travel so far may have been due to the unusually warm weather that earned the period from 800 to 1200 its name, the Medieval Warm Period. This makes a sharp contrast with the evidence from bone skates because skates require

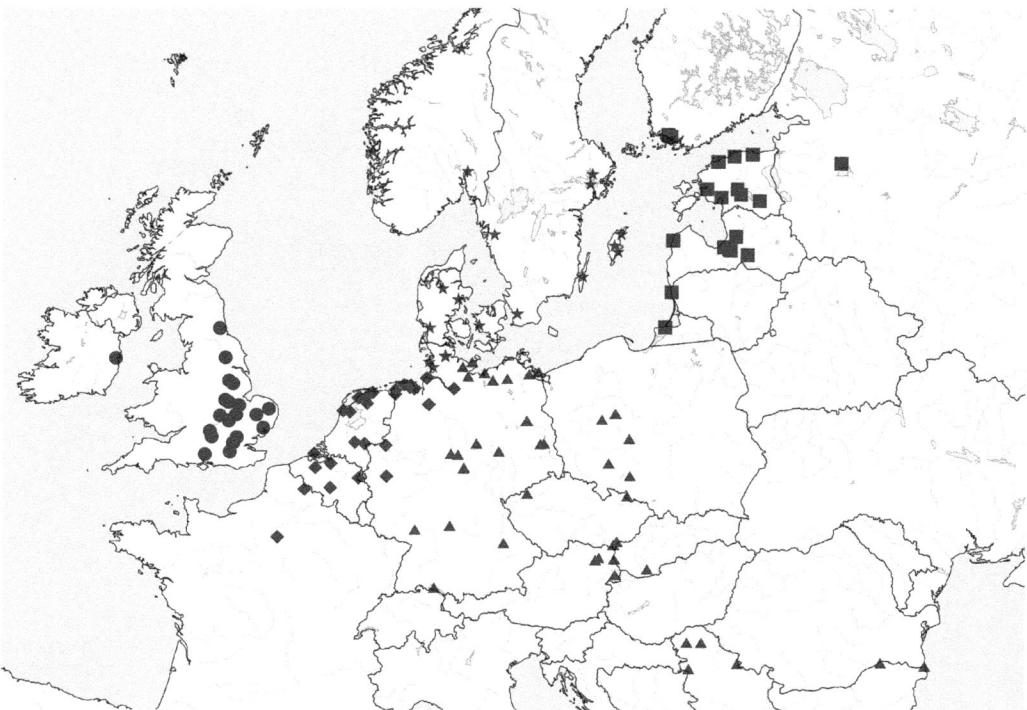

Figure 46: Sites where bone skates dating to the Middle Ages have been found in the five regions: Central (triangles), Great Britain (circles), Northeast (squares), Northwest (diamonds), and Scandinavia (stars), created with QGIS.

ice, which means cold weather. The Medieval Warm Period sounds like the worst time to skate, so why are there so many skates from it? As climate scientists learn more about the Medieval Warm Period, it appears increasingly less likely to have been the global phenomenon it first appeared to be.[4] The warmth may have been confined to the summer months or to particular regions. On the basis of data from tree rings and Greenland ice-sheet isotopes, Lamb[5] concludes that the area in which skate finds are most common experienced strong seasonal temperature variations with little precipitation, which would have been perfect for skating, and indeed, he cites the numerous bone skates found in York as corroborating evidence.

Additional evidence for temperature differences comes from ice cores. Dansgaard et al.[6] show the difference in temperature between England and Greenland between 800 and 1900 CE. The Greenland data are from an ice core collected Godthaab in western Greenland, where the concentration of oxygen-18 corresponds to temperature changes,[7] and the England data are primarily from historical sources. The curve for England follows the curve for Greenland, but lags behind it by approximately 250 years. According to these data, during the time in which bone skates were most popular (approximately 850–1100), England was in a cold phase. In contrast to the situation in England, Greenland seems to have experienced a temperature peak shortly before 1000, which is when Erik the Red discovered it. Dansgaard et al.[8] suggest that at that time Greenland really was green, and the later comment by the anonymous author of *Grœnlendinga saga (The Saga of the Greenlanders)* that Erik called it Greenland to make it seem more attractive was one way of reconciling the fact that in the author's time, the climate was colder and Greenland was no longer green.

The data assembled by Ahmed et al.[9] generally support the idea that Great Britain was cold when bone skates were at their most popular. They found a broadly warm period from 830 to 1100 CE in the northern hemisphere, but it was not without significant regional variation. A careful evaluation of their figure 2 shows that shortly after 1100, Europe, including Great Britain, was very cold. It is quite possible that there were forces at work that produced both cold winters in northern Europe (for skating) and warm northern seas (for summer exploration). One simplistic explanation of this discrepancy involves the North Atlantic Oscillation. This helps explain how it was possible for Scandinavians to establish far-off colonies and skate at the same time. The basic idea is that when there is low pressure over Stykkishólmur, Iceland, and high pressure in the south, usually near the coast of Spain, westerly winds result. This is because fluid (in this case, the atmosphere) flows from high pressure to low pressure, but in the northern hemisphere, the Coriolis force causes air to flow to the right of the direction it would normally flow in.

The North Atlantic Oscillation Index is a measure of the pressure difference. It is positive when the pressure difference is especially high and negative when the pressure system is weaker than normal. When this index is positive, strong westerly winds blow warm ocean air over Europe and cold air from the north comes down over Greenland. When it is negative, this effect is lessened: Europe, including Great Britain, is colder and Greenland is warmer.[10] Sweden and Norway experience the same temperature fluctuations as Greenland and Iceland.[11] This is consistent with the evidence from bone skates: continental Europe and Great Britain have contributed many skates, whereas Greenland has contributed none to date. The broad distribution of sites along the northern coast of Germany and nearby regions shows that skates were used at more sites in regions that probably had cold winters due to the North Atlantic Oscillation. There are relatively few sites with bone skates in Sweden and Norway shown on the map in figure 46, but those sites that have contributed skates have contributed many. Perhaps this reflects local conditions that were particularly conducive to skating despite the generally warm climate of Scandinavia. Overall, the North Atlantic Oscillation helps explain why numerous medieval bone skates have been found in England and Scandinavia and along the northern coast of the Continent, while none have been found in Iceland or Greenland. Other factors were surely involved as well; there is substantial variation from site to site, which may reflect local factors such as the availability of frozen waterways and horses.

The archaeological evidence shows that medieval Scandinavians were avid skaters. They could have taken skates with them throughout their travels and made new skates when they settled in new places. In this way, they could have been a vector for the spread of skating across Europe. However, they were not the only skaters. The other candidate for the most enthusiastic skating culture of the Middle Ages is the Slavs. Slavic groups begin to appear in written records in the sixth century; they are featured in the accounts of Procopius and Jordanes. At that time, the Slavs lived in the area previously occupied by the Goths. In subsequent years, they began to spread and by 900, they occupied a significant part of central Europe.[12] Today, Slavic languages are spoken in much of eastern and central Europe. They include Bulgarian, Czech, Polish, Russian, and Slovak.

The rest of this chapter is based on a statistical analysis of the medieval bone skates described in the literature. The data are drawn from Küchelmann's *Bone Skates Database*[13] and numerous academic articles.[14] A total of 2753 skates are known to date to the Middle Ages. Many of these skates are Scandinavian in origin, including the 471 skates found in Norway and the 679 Swedish skates described by Edberg and Karlsson.[15] I classified the 100+ locations[16] at which these skates were found into the following regions:

1. Britain: This group consists of all the sites in Great Britain (at least 269 skates from 19 locations)[17] and Ireland (three skates from Dublin).
2. Central: This group is made up of sites in the area mainly (though not exclusively) occupied by Slavs. Some of these sites were subjected to raiding or settlement by Scandinavians. It includes Austria, Bulgaria, the Czech Republic, Poland, Serbia, and Slovakia, plus some locations in Germany.[18] It has contributed 634 skates.
3. Northeast: This is the group of sites in the region of Europe primarily occupied by Balts along the eastern coast of the Baltic Sea, where they stood a good chance of being raided or settled by Scandinavians. It includes locations in Estonia, Finland, Latvia, Lithuania, and Russia, as well as some sites in Germany.[19] This region has contributed 125 skates.
4. Northwest: These are the sites in the region of Europe primarily occupied by Germanic groups. Many were subjected to Scandinavian raiding and settlement. Among them are all the sites in Belgium, France, and the Netherlands, plus some locations in Germany.[20] This region has contributed 258 skates.
5. Scandinavia: This group represents sites in the Scandinavian homeland. It includes all sites in Denmark, Norway, and Sweden, plus some in what is now Germany.[21] This is the region that has contributed the most skates, 1464.

Tables 5 through 7 summarize the bone skates from each region. The distribution of bone type is shown in table 5, and the numbers and types of features exhibited by the skates are shown in tables 6 and 7, respectively. The first thing that jumps out from these tables is the large number of skates without full documentation. These are included under "Other/Unknown" bone type. A few of these bones fall under "Other," but the vast majority are unidentified. In the Northeast and Central regions, only around 40 percent of the bone skates have been fully documented. The results from Scandinavia and the Northwest are better—the bone types of around 60 percent of the skates are documented—but still show a need for more detailed work or better dissemination of results. Most of the skates from Great Britain have been fully documented, thanks mainly to Arthur MacGregor's pioneering work.[22]

In tables 6 and 7, only the skates with documented features on the standard list.[23] Bone type is ignored; skates with features based on unidentified bones are included. The decorations on the three skates mentioned in the previous chapter are not included. This is 83 percent of the skates from Britain, 21 percent of the skates from the Central region, 22 percent of the skates from the Northeast, 54 percent of the skates from Scandinavia, and 44 percent of the skates from the Northwest. The skates that are not included may not have features at all—after all, very simple skates were usable—or may not have been fully documented.[24] Table 6 shows how many features each skate has without specifying what they are. This is a way of evaluating how elaborately worked the skates are, which is important for placing them on the manufacturing continuum described in Chapter 5. More extensively worked skates with more features are more likely to have had some sort of social or economic importance to their users. Differences in how extensively worked the skates in the different regions are may suggest that they were used for different purposes.

Table 5

Types of bone used for skates in the five regions. The percentages in the rows under "total identified" are the percentages of identified bones. The skates in the "Other/Unknown" row are mostly unidentified; bones other than horse and cattle radii and metapodia were quite uncommon. The percentages do not always add up to 100 due to rounding.

Region	Great Britain	Central	Northeast	Scandinavia	Northwest
Total	272	634	125	1464	258
Other/Unknown	21 (8%)	354 (56%)	80 (64%)	542 (37%)	109 (42%)
Total identified	251	280	45	922	149
Horse radius	7 (3%)	28 (10%)	3 (7%)	37 (4%)	16 (11%)
Horse metapodium	181 (72%)	160 (57%)	30 (67%)	389 (42%)	87 (58%)
Cattle radius	9 (4%)	27 (10%)	3 (7%)	31 (3%)	13 (9%)
Cattle metapodium	54 (22%)	65 (23%)	9 (20%)	465 (50%)	33 (22%)

The numbers for Scandinavian skates listed in the tables can be used as a baseline for figuring out how medieval skates were used and for identifying the most avid skaters in each region. With Edberg and Karlsson's[25] conclusion that bone skates were primarily used by children and teenagers for recreation in medieval Scandinavia in the background, variations in the type of bone and the number of features provide clues as to the skaters' size and level of competency in boneworking. Scandinavia is the only region where cattle metapodia outnumber horse metapodia. Because cattle metapodia are likely to generally have been smaller, this suggests that children started skating at younger ages in Scandinavia than in other regions. The relative simplicity of the Scandinavian skates is also consistent with the idea of children skating: slightly over half the Scandinavian skates with documented features exhibit only one or two, and only 12 percent exhibit five or more.

Table 6

The numbers of features exhibited by the skates with features from each region. Only features that have been securely identified are included; those listed with question marks or parentheses in the database are excluded. The percentages do not always add up to 100 due to rounding.

Region	Great Britain	Central	Northeast	Scandinavia	Northwest
N	225	130	28	785	114
1	51 (23%)	37 (28%)	11 (39%)	222 (28%)	21 (18%)
2	51 (23%)	51 (39%)	9 (32%)	203 (26%)	29 (25%)
3	71 (32%)	11 (8%)	3 (11%)	162 (21%)	24 (21%)
4	40 (18%)	27 (21%)	1 (4%)	108 (14%)	26 (23%)
5	8 (4%)	3 (2%)	3 (11%)	57 (7%)	8 (7%)
6+	4 (2%)	1 (1%)	1 (4%)	33 (4%)	6 (5%)

Table 7

Skate features by region. Only features that have been securely identified are included; those listed with question marks or parentheses in the database are excluded. Skates without features listed here are also excluded.

Region	Great Britain	Central	Northeast	Scandinavia	Northwest
N	225	130	28	785	114
Axial hole	135 (60%)	5 (4%)	1 (4%)	54 (7%)	8 (7%)
Other binding	73 (32%)	14 (11%)	14 (50%)	183 (23%)	47 (41%)
Chopped	24 (11%)	46 (35%)	12 (43%)	367 (47%)	68 (60%)
Flattened	8 (4%)	8 (6%)	2 (7%)	2 (0%)	3 (3%)
Pointed toe	163 (72%)	44 (34%)	1 (4%)	33 (4%)	17 (15%)
Roughened	7 (3%)	6 (5%)	1 (4%)	18 (2%)	3 (3%)
Split	—	—	2 (7%)	161 (21%)	—
Upswept end	157 (70%)	93 (72%)	12 (43%)	722 (92%)	82 (72%)

Table 7 shows the features exhibited by the skates. Each feature is considered independently; for example, that 72 percent of the skates from Britain were pointed and 70 percent were upswept does not say anything about how many skates exhibited both of these features, though clearly some must have. This table is the key to identifying regional skate styles, which are discussed in the subsequent sections. Carefully examining it shows that the characteristic Scandinavian features are an upswept end with neither a pointed toe nor an axial hole—if a binding apparatus is present, it is of another type. Upsweeping at the proximal end is more common among Scandinavian skates than skates from any other region, and pointed toes and axial holes are rare in Scandinavia. Split bones also seem to be characteristically Scandinavian; they are reasonably common in Scandinavia and virtually nonexistent elsewhere.

Identifying a Scandinavian type of skate is important for studying the bone skates in the Central and Northwest regions. In these areas, people were using bone skates before Scandinavian raiders and settlers arrived. If their bone skates can be distinguished from Scandinavian bone skates, it becomes possible to use bone skates to understand the cultural mixing that went on in these areas. The picture is more clear-cut in Great Britain and the Northeast, where bone skates were not in general use until the Middle Ages. Scandinavian settlers could have introduced bone skates to these two regions, and then, unique regional styles could have evolved.

The goal of the following sections is to show that bone skates contribute to the understanding of the Scandinavian expansion. In the following section, Great Britain is used as a case study. This case is clear-cut because the skates are well documented and there was not an earlier skating tradition. The results fit in well with those of studies based other evidence. The Northeast is treated only briefly because the number of well-documented skates is so small, but it may be possible to make a parallel argument for this region. The pre-existing skating traditions in the Northeast and Central regions make it more difficult to analyze their skates, but a preliminary attempt based on a few sites with well-documented skate finds is made in the third section. The final section is a call to action: it is time to fully document all skate finds because they are important to understanding the Scandinavian expansion.

7.2. *Bone Skates as Scandinavian Artifacts in Great Britain*

The idea that the Danes brought bone skates to England goes back over a century, to Eekhoff's[26] article in the *Nieuwe friesche volks-almanak (New Frisian Folk Almanac)*. More recently, Hall[27] also proposed that the Danes were responsible for bringing bone skates to England. The evidence presented in this section supports this hypothesis and enables bone skates to be brought into the discussion of the Scandinavian settlement of England. What they add is evidence for the presence of adolescents among the settlers. This result supports the results obtained from analyses of other types of evidence. The evidence from language and place-names has been analyzed,[28] as have the descriptions in the Anglo-Saxon Chronicle and other texts. Genetic evidence has also begun to make an appearance.[29] Archaeology has made some contributions,[30] and metalwork has been yielding interesting results: Kershaw[31] describes the implications of finds of brooches in a study that has some parallels to this study of bone skates.

According to Bede, the Angles, Saxons, and Jutes arrived in England in about 450 CE.[32] It is possible that the Saxons used bone skates on the Continent before they moved to England. Six sites that are connected to the Anglo-Saxons have contributed Migration Period skates: three in the Netherlands, one in Belgium, and one in Germany. The sites in the Netherlands are Wijnaldum-Tjitsma (two skates from the fifth or sixth centuries), Englum (two skates from the third–fifth centuries), Dongjum (one skate from the fifth or sixth century), and Leeuwarden (one skate from the fifth to eighth century). The skate from Leeuwarden is slightly later than the other skates from the Netherlands and the earliest of several skates from the site. It provides a transition from the Migration Period to the Early Middle Ages, when skates became very common in this region. There is also a single skate from Vlissegem (near de Haan), Belgium that dates to the fifth through eighth century and is attributed to the Franks in Küchelmann's database. The German site, Feddersen Wierde, contributed three skates dating to the first through fourth century and one early medieval skate.

As shown on the map, Wijnaldum-Tjitsma and Dongjum are located quite close together on the northwestern edge of Friesland, near the coast, while Englum is in western Groningen and a bit farther from the coast. Leeuwarden is located between Englum and Dongjum. This area is important because pottery in the Anglo-Saxon style appeared here during the Migration Period. Nieuwhof[33] takes this as evidence not of migration but of inter-regional contact because this region of the Netherlands was occupied continuously. Bone skates began to appear at Wijnaldum-Tjitsma during the Migration Period and became more numerous during the following centuries as Wijnaldum-Tjitsma made its way into international trade routes, including those that ran through England, Scandinavia, and the region surrounding the Rhine in the sixth and seventh centuries.[34]

With Englum, which is in Groningen rather than Friesland, the chronology is a bit more difficult. The site was abandoned in the third century and reoccupied in the fifth century.³⁵ Bone skates first appeared at Englum during the Migration Period, in the range Prummel, Halici, and Verbaas³⁶ narrow to 425 to 550 CE. This makes the skates the property of the second wave of settlers. Nieuwhof³⁷ tentatively identifies these later inhabitants as Anglo-Saxons because the artifacts attributed to them are similar to those of the people living along Germany's northwestern coast and to early Anglo-Saxon artifacts found in England.

A closer link can perhaps be made through Feddersen Wierde, a site on the northern edge of Germany, an area that is a good candidate for the Saxon homeland.³⁸ Feddersen Wierde was occupied continuously from slightly before 100 CE until about 450 CE.³⁹ Four bone skates have been found there, three from the Roman Iron Age and one from the early Middle Ages. Archaeologists are particularly interested in this site because it may be possible to follow the occupants' travels through the traces of their pottery. Very similar pottery has been found at Mucking, a Germanic site on the Thames estuary dating to shortly after 400 CE, which could be a sign that people from Feddersen Wierde moved to England. Todd⁴⁰ notes that these people may have somehow been connected with Roman authority because the remains of Roman military uniforms were also found at the site. The Romans are connected to bone skates by a single skate from this period that was found with Samian pottery, which is normally a sign of a Roman cultural context, in London.⁴¹ Another London skate may have been found with Roman sandals.⁴² Feddersen Wierde could have been the original home of the hopeful skater who brought the first bone skate or skates to London. It is also quite possible that these skates are both dated incorrectly.

This possible connection to the Anglo-Saxons is of interest because it was during the fifth century that the Anglo-Saxons moved from the Continent to England. However, if they did bring bone skates with them, skating did not catch on. This may have been due to the local climate: Great Britain was probably too warm for skating. In the latter part of the first century, Emperor Domitian banned vineyards in England and Germany, but this law was revoked in about 280 CE, after which the Romans began to grow grapes north of the Alps. Lamb⁴³ suggests that England stopped importing wine in about 300 CE because people living there were able to make their own. This points to winters that were generally not very cold, which does not bode well for skating.

Several centuries passed before the next appearance of skates in England. The bone skates dating to the Middle Ages (no earlier than the eighth century) signal the start of a new era of skating in Great Britain. This era was defined by the presence of Scandinavians, especially Danes. The beginning of this period is usually defined by the attack on Lindisfarne in 793, even though that was not the first viking raid on England. The Danes settled the Danelaw, a region of northeastern England, starting in the ninth century.⁴⁴ This area was defined by a treaty between Alfred, King of the Anglo-Saxons, and Guthrum, leader of a viking army, in the 880s. In the treaty, they defined the Alfred-Guthrun boundary (see figure 47), which separates the Danelaw—the area settled by Danes—from the kingdom of the Anglo-Saxons.⁴⁵

The arrival of Scandinavian settlers marked the beginning of a substantial increase in the number of bone skates found in Great Britain. The list of bone skates I assembled contains 269 bone skates from England and three from Ireland. Most of these are described in MacGregor's⁴⁶ review article, which is based on 168 bone skates in the collections of the British Museum, the Museum of London, the National Museum of Antiquities of Scot-

7.2. Bone Skates as Scandinavian Artifacts in Great Britain

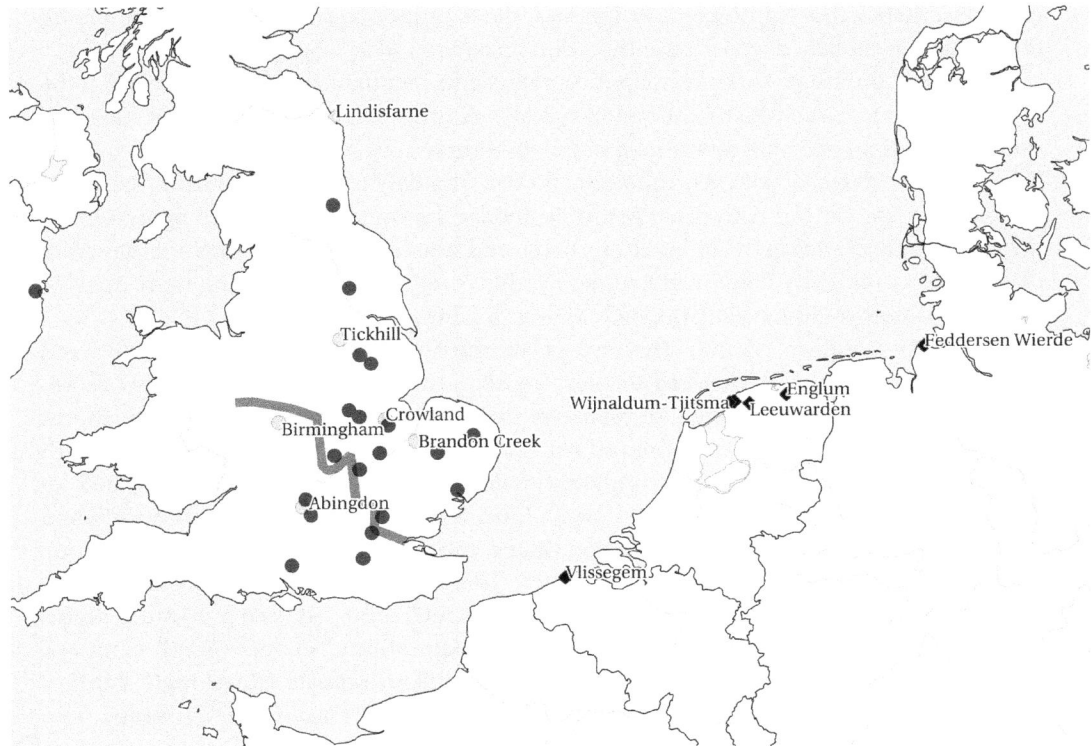

Figure 47: Sites in Great Britain where medieval (dark circles) and later or undated skates have been found and where ethnographic references place bone skates (light circles) and other relevant sites (diamonds). The dark gray line approximates the Alfred-Guthrum boundary (created with QGIS).

land in Edinburgh, the Pitt-Rivers Museum in Oxford, the Scarborough Museum, the Sheffield City Museum, and the Yorkshire Museum. I added skates that have been found in excavations since then. A brief summary of the evidence follows.

York has contributed the most well-stratified examples (60) due to the extensive archaeological investigations conducted there.[47] Because York is in some ways the quintessential Danish city in England, the large number of skates found there makes a strong connection between skating and the Danes. Radley[48] notes that the York skates show significant wear and are highly polished, which he sees as evidence for great skating conditions: long, cold winters and a frozen river to skate on. York may have provided particularly good skating conditions because of its low elevation, which made it prone to floods. MacGregor[49] agrees that the skates show that York must have been a great place to skate but is skeptical about Radley's reasoning because of the dearth of supporting evidence.

The York skates are mostly from Coppergate (42), but a few skates have been found in Piccadilly (3), Fishergate (2), the Bedern Foundry (2), Hungate (1), and unidentified locations (10). They all date to between the eighth century and the sixteenth century. Twelve of the Coppergate skates date to near 975, which is shortly after redevelopment of the city began after the Danes arrived.[50] The earliest date is set by a single skate that is associated with a cross dating to the eighth century and coins from the reign of Eadberht of Northumbria, which lasted from 737 to 758 CE.[51] The context suggests that this skate could date to as early as the eighth century, but later dates cannot be ruled out. The later end of this range is established by a sixteenth-century skate found at Coppergate.[52]

While substantial, York's contribution represents less than a quarter of medieval the bone skates from Great Britain. London has also contributed many skates, but most are not dated. A total of 37 are mentioned in the literature,[53] but most are undated. Only nine have medieval dates, one is early modern, and two may date to the Roman Iron Age. Since London is right on the Alfred-Guthrum boundary, its bone skates could be extremely helpful in understanding the interactions between Danes and Anglo-Saxons—if they were closely and confidently dated. Unfortunately, this is not the case. Nothing more specific than that Danes could have introduced skating to London can be said.

With 19 medieval skates, Thetford is another important contributor.[54] Between them, York, London, and Thetford account for 88 of the 272 medieval skates plus 28 additional skates. Smaller numbers of medieval skates have been found at other locations. MacGregor[55] lists skates from Dublin (three skates from the eleventh through thirteenth centuries), Durham (twelfth century), Huntingdon (probably earlier than the middle of the eleventh century), Ipswich (three skates), Lincoln (two skates, eleventh to twelfth and thirteenth centuries), Northampton (nine skates, including six from 850–1000 and one from 1150–1350), Norwich (twelfth century), Torksey, and Waltham Abbey (c. 1177–1250). A few years later, in another publication, MacGregor[56] adds one or more skates from Bedford[57] and others from Reigate (early medieval) and Oxford (ninth century). Other researchers have contributed medieval skates from Lincoln (three tenth-century skates), Empingham, (two skates made from cattle tibiae),[58] Flaxengate,[59] Thorney (two skates),[60] Winchester (five skates dating from the early eleventh to the fifteenth century),[61] Wallingford (one skate),[62] and Whissendine (one skate).[63]

This accounts for 130 of the 272 medieval skates. In his review article, MacGregor does not associate any of the other 142 medieval skates discussed in this section with a location more specific than "England" or "Great Britain," which is rather unfortunate. Additional undated skates have been found in Aberdeen, Abingdon, Durham, and Ipswich, and a possibly Tudor skate has been found in Crowland.[64]

Nearly all these sites are within the Danelaw. The few skates found south of the Alfred-Guthrum boundary generally date to the eleventh century or later—after Scandinavians began to move into the southern part of England. The southernmost skating site, Winchester, which is deep in southern England, functioned as the Danish headquarters after a second wave of settlement in the latter part of the tenth century.[65] The oldest skate found there dates to the early or mid eleventh century, which means skating could easily have been introduced to Winchester by the Danes. Dublin, the single skating site in Ireland, was founded by Scandinavians in the tenth century.[66]

The skates that are hardest to fit into this picture are the three from Oxford. These need not have belonged to Scandinavians but could have. Blair[67] suggests that many of the Scandinavians living in Oxford in the tenth century were part of a group that had come to the region through nearby areas gradually over time. These early settlers could have introduced bone skates before the Danish community in Oxford began to flourish.

If the conclusions of Edberg and Karlsson are accepted, the presence of bone skates can be interpreted as evidence for Scandinavian children and teenagers.[68] The distribution of bone types in Great Britain (see table 5) is similar to the distribution in Sigtuna; about two thirds (66 percent) of the medieval skates are made from horse metapodia,

7.2. Bone Skates as Scandinavian Artifacts in Great Britain

which are long enough for older children and teenagers. Cattle metapodia, which make up about a fifth (21 percent) of the skates, are generally shorter and therefore, more suitable for younger children. Only 7 percent of the skates are made from radii (both horse and cattle), and the rest are made from other bones or have not been identified. The other bones are an immature cattle tibia from York, a horse tibia from Thetford, and a red deer metatarsus from Reigate.

Edberg and Karlsson[69] conclude that skating was popular among children at both Birka and Sigtuna, but that the greater popularity of horse bones at Sigtuna suggests that children were more likely to continue skating into their teenage years at that site. Since the distribution of bones used for skates in Britain is similar, the skaters were probably similar in size and age. This idea is also supported by the amount of work put into the bone skates from Britain (table 6). Of the skates with features, 66 percent exhibit three or more, which is more than any other region. For comparison, in Scandinavia overall, 46 percent of the skates with features exhibit three or more. However, Edberg and Karlsson[70] note that at Birka and Sigtuna, longer skates tend to be more elaborately worked because they were prepared by older children who were also better skaters.

The idea of adolescents skating is also supported by William fitz Stephen's famous twelfth-century description of skaters in London, which describes them as *juvenes*. Douglas and Greenaway[71] translate this word as "young men," but it could also refer to young women. Latham, Howlett, and Ashdowne[72] note that *juvenis* can refer specifically to people between the ages of 22 and 42 years. These people are old enough that their feet would probably have reached their adult size, which contradicts the Edberg and Karlsson's conclusion. Perhaps fitz Stephen uses this word in a more general sense, which could include teenagers. According to the Norwegian Frostathing law code, "[e]very man is a minor till he is fifteen winters old."[73] This seems to put the transition between childhood and manhood closer to 15 than to 22 and implies that young men could still be growing. Older teenagers (aged 15–22 years) could have had feet that were the right size for horse metapodia. As in Sigtuna, the use of cattle metapodia for skates suggests that younger children also skated, although they were probably not the most enthusiastic skaters in Great Britain.

The age range of skaters may have varied from site to site, in parallel to the difference Edberg and Karlsson[74] note between Birka and Sigtuna. In general, the York skates are smaller than the English skates are overall. Horse metapodia comprise 58 percent of the York skates (35 of 60 skates), and cattle metapodia comprise 25 percent (15 of the 60 skates), while radii account for only 6 percent (10 of the 60 skates). This shows a very slight tilt toward shorter skates (cattle metapodia) compared with Britain overall. In particular, one of the cattle metacarpus skates found at Piccadilly in York is only 15.26 cm long,[75] making it suitable for a child barely beyond the toddler stage. The York skates also tend to be a bit simpler, with 53 percent of the worked skates exhibiting only one or two features, compared with 43 percent of the British skates overall. This implies that life in this quintessentially Scandinavian city may have been somewhat more conducive to younger children skating than country life.

These results suggest that settlers from Scandinavia introduced ice skating to Great Britain primarily as a fun winter activity for children and young people. This implies that there were families in Great Britain who raised their children according to Scandinavians customs—playing Scandinavian games and presumably doing other Scandinavian things rather than assimilating into Anglo-Saxon culture, at least in roughly the ninth or tenth through eleventh or twelfth centuries. Skating on bones continued into at least the thirteenth century in Dublin and London and the sixteenth century in York, according to the archaeological evidence. The ethnographic evidence quoted in the Appendix shows that bone skates endured into the nineteenth and possibly twentieth centuries in Birmingham, Norfolk, and Yorkshire.

Looking at the features of surviving skates reveals that a characteristic skate type emerged in this region. Of the 225 skates with documented features, 135 (60 percent) have an axial hole at the proximal end, presumably for inserting a peg to help attach the skates to the skaters' feet. Two of the York skates were found with wooden pegs in their axial holes.[76] This contrasts sharply with the results for in other regions, where axial holes are found in less than 10 percent of the skates with features. Pointing the toe was also much more common in Great Britain than elsewhere. These two features comprise the characteristic British style. That a special style developed here suggests that the Danish settlement in Britain may have been independent rather than in close contact with the Danish homeland.

The appearance of bone skates in England shortly after the arrival of the Danes supports the hypothesis that the Danes were responsible for introducing them. Even though bone skates could have been brought to England earlier, they seem not have been, or if they were, they did not catch on, perhaps because the climate was unsuitable. Based on Edberg and Karlsson's conclusion that bone skates were primarily used by children and young people in medieval Scandinavia, the bone skates in Anglo-Scandinavian England suggest that young Scandinavians were there to use them. Support for the hypothesis that children were being raised in a Scandinavian cultural setting in England is readily available in studies of other types of evidence. The results from studies of material remains (specifically metalwork) and language suggest that this was the case. Metalwork provides evidence for Scandinavian women living in England, and linguistic evidence suggests that the society was bilingual. Genetic studies also seem promising but have produced a wide range of results to date.

The evidence from metalwork is similar to with the evidence from bone skates. One challenge with using both types of evidence is the numerous objects that lack clear provenances. However, an important difference is that metalwork can be attributed to a particular culture (or seen as a blend of several) based on its style. Kershaw[77] argues that many women wore distinctly Scandinavian clothing in the Danelaw in the ninth and tenth centuries. These women could easily have been mothers, and their clothing is consistent with the idea that they may have raised their children in accordance with Scandinavian customs. In another paper, Kershaw[78] describes the Scandinavian custom of using bullion rather than coin in the Danelaw, which also points to the continuation of Scandinavian culture in that region.

There is one point of disagreement between the results of studying metalwork and the evidence from bone skates. Kershaw[79] reports that most Scandinavian-style metalwork has been found in places outside major cities that were known to be home to Scandinavians, especially in Norfolk and Lincolnshire. The cities with the most bone skates, York and Thetford, have contributed very few Scandinavian metal items, and in general, cities seem to have been less strongly Scandinavian in terms of their culture, which Kershaw attributes to stronger integration in these areas.[80] York and Thetford have also contributed bullion in smaller quantities than other cities, which makes them appear culturally less Scandinavian.[81] In contrast, London has contributed both skates and metalwork, including certain styles of the latter that were very popular in England around the turn of the eleventh century. Kershaw[82] suggests that the King Cnut's Scandinavian court was responsible for this popularity. It is not clear why bone skates have been found in certain areas that are typically considered very Scandinavian in character when Scandinavian metalwork has not, but perhaps women living in less thoroughly Scandinavian locations felt more of a need to display their heritage using their clothing.

The picture of Scandinavian families living in England fits well with the linguistic evidence. Townend[83] describes late Anglo-Saxon England as a place where two languages, Old English, and Old Norse, were used in discrete speech communities. The Scandinavian settlers appear to have stayed together and continued speaking their own language instead of fully integrating into the Anglo-Saxon speech community. The maintenance of a speech community suggests that new native speakers were being raised, i.e., that parents spoke Old Norse with their children. Certainly the first-generation settlers would have been familiar with Scandinavian games as well, and such games could have been passed down. The image of children playing Scandinavian games with mothers wearing Scandinavian clothing connects quite well with the idea that new native speakers of Old Norse were growing up in England.

Genetic evidence also seems like a promising avenue of inquiry, but the results to date are unclear and contradictory. Leslie et al.[84] found no evidence of Scandinavian genes being passed down to modern English people. They interpret this as showing that Scandinavians did not last long in England; it could also show that Scandinavians kept to themselves by intermarrying and raising purely Scandinavian families, then leaving rather than passing their genes down through the centuries. However, one of the genetic markers they used could be either Anglo-Saxon or Danish. This is why Kershaw and Røyrvik[85] encourage further analysis of their data. In a different study, Sykes[86] reports that both matrilinear and patrilinear DNA of modern people living in the Danelaw is somewhat different from that of people living south of it. He suggests that these differences reflect Scandinavian settlers and therefore, that both men and women from Scandinavia must have lived in that region, which is consistent with the idea of Scandinavian families. It seems that both these studies can be used to support the notion of Scandinavian families living in England, the former by showing that Danish settlers kept to themselves and eventually left (unless a re-analysis suggests a different result) and the latter by showing that Scandinavian families were present.

Overall, the evidence from bone skates is consistent with other results that support the idea of Scandinavian families living in England, continuing their culture and remaining relatively separate from the Anglo-Saxons until after the Norman Conquest. Bone skates add evidence for the presence of young people. It is hard to say whether Scandinavian and Anglo-Saxon children played together, but it seems unlikely that this would have been the case at first. Skates have generally been found in places with a strong Scandinavian presence and did not spread widely, which suggests little interaction. The development of a regional type of skate that differs from the Scandinavian type is evidence for an independent Scandinavian community in Great Britain.

A similar analysis could be performed for the Northeast if the skates were better documented. With only 36 percent of the bone types known, little can be said about the region in general, but it is interesting that the bone type distributions in Great Britain and the Northeast are similar, with around 70 percent horse metapodia and 20 percent cattle metapodia in both regions. Comparing the maps in figures 34 and 46 shows that no bone skates earlier than the Middle Ages have been documented there. Because this region is known to have been home to Scandinavian settlers during the Middle Ages, the hypothesis I have developed for Great Britain may apply here: Scandinavians may have introduced bone skates to the Northeast. Because numerous skates from the Middle Ages are known, this seems more plausible than the alternative hypothesis that the results of excavations of earlier sites simply have not been disseminated broadly enough. Evidence for Scandinavians having brought skates to the Northeast includes the fact that this is the only region, other than Scandinavia, where split bones have been found. Two skates with the palmar side removed to expose the medullary cavity and, presumably, make a nice place for the skater's foot to rest: one at Riga, Latvia, (13th–16th century) and one at Turku, Finland, (14th–15th century). Closer examination of the undocumented skates may reveal more interesting patterns.

7.3. Bone Skates on the Continent

In contrast to the situation in Great Britain and the Northeast, there were long-standing skating traditions in the Central and Northwest regions. The map in figure 34 shows that bone skates were already used in both these areas by the Migration Period, which predates the Viking Age. The northern coast of Germany and much of the Northwest region became home to Scandinavian settlers during the Viking Age, but the former was more commonly visited by merchants, whereas the latter is generally thought to have been plagued by raiders. This state of affairs is reflected in figure 45. As new research deepens the understanding of the Scandinavian expansion, it will be necessary to revise this map, perhaps substantially. Bone skates can be used to better understand the situation, but it is necessary to separate the skates that belonged to the people living there before the Scandinavians arrived from those brought by Scandinavians.

Different problems are associated with the skates from these two regions, but they share a major one: poor documentation. Only 45 percent of the bones used for skates in the Central region have been identified; for the Northeast, the number is somewhat better, at 58 percent. Only 130 skates from the Central region and 114 from the Northwest have documented features, which makes it difficult to identify regional types. Still, the first steps can be taken by focusing on the distribution of skates throughout these regions and examining the data from sites with large numbers of well-documented skates.

Scandinavians were busily trading with people along the coast of the North Sea from 700 CE onward,[87] but trading is not the same as settlement. Willemsen[88] explains that the notion that Scandinavians actually settled in this region is relatively new and a response to new finds of hoards containing jewelry and silver that appear Scandinavian on Wieringen. Before those finds were made, D. P. Block's conclusion that few Danes had settled in this area was widely accepted.[89] According to Willemsen,[90] there were Scandinavian settlements in this region during winter, but these settlements were not permanent. Because they were only useful in winter, bone skates could reflect winter rather than permanent settlements.

7.3. Bone Skates on the Continent

Roughly half (136) of the skates from the Northwest are from the Netherlands, and the rest are about equally divided between Germany (62 medieval skates) and the combination of France (22 medieval skates) and Belgium (38 medieval skates). Comparing the overall data for the Northwest in tables 5–7 with that for Scandinavia shows that the distributions were rather different. In the Northwest, horse metapodia were slightly more common at 58 percent of the identified bones compared with 42 percent in Scandinavia, and cattle metapodia were less than half as popular at 22 percent and 50 percent, respectively. Skates in the Northwest also tended to be more extensively worked than in Scandinavia, with 35 percent and 25 percent, respectively, exhibiting four or more features. Upswept ends, which are characteristic of Scandinavian skates, are less common in the Northeast, but chopping is more common.

Because the skates tend to be longer (based on the bone type) and more extensively worked, these results suggest that adolescents who were old enough to go on trading voyages were more likely than younger children with smaller feet to skate at winter settlements in the Northwest. A more detailed analysis of the skates is sorely needed to draw more firm conclusions, and even then, it may be difficult to be sure about the ages of the children living in an area. In Sigtuna, a Scandinavian town where children of all ages must have lived, 61 percent of the bone skates analyzed by Edberg and Karlsson[91] were made from horse bones—all metapodia except for one radius.

Evaluating the skates from the Northwest on a site-by-site basis may be useful for identifying local interactions with Scandinavians. The sites contributing the most skates are Dorestad (70 medieval skates) and Oost-Souburg (42 skates dating to 900–975), which are both in the Netherlands. In third place is the biggest contributor from Germany, Bremen, where 30 skates of various medieval dates have been found, and in fourth place is Saint-Denis, France, which has contributed 25 skates of general medieval date.

Dorestad, located on a branch of the Rhine in the southeastern part of the Utrecht province in the Netherlands near the modern city Wijk bij Duurstede, fits in with the narratives of Great Britain and the Northeast. An international trading city in the eighth and ninth centuries, Dorestad was destroyed by vikings repeatedly in the latter half of that period.[92] At its height, it hosted merchants from Birka and other places.[93] In the mid-ninth century, it was ruled by Roric, a Scandinavian whose rule extended over the surrounding region as far as Frisia to cover approximately the same region as the Netherlands cover today,[94] and remained a major trading port. Scandinavian weapons and jewelry have both been found in Dorestad,[95] making it unique among sites in the Netherlands. Because the Dorestad skates are not closely dated, it is impossible to say much more than that skating and Scandinavians coincided in this town. More detailed study of these skates may shed light on how many Scandinavians lived there and how they lived.

The other cities that have contributed many bone skates were also affected by Scandinavians during the Middle Ages. Oost-Souburg was built toward the end of the ninth century as a way for the people living nearby to defend themselves against viking raiders.[96] Bremen was also attacked in the ninth century, but recovered and became important with the success of the Hanseatic League.[97] Finally, not even Saint-Denis was free of the vikings' touch: they captured the abbot in 858 and held him for ransom.[98]

Examining the types of bone used can help explain some of the goings-on in these areas. However, most of the bones have not been identified. Of the top four sites, only the bones from Oost-Souburg and Bremen have been identified in the literature. The skates from both sites are predominantly (69 percent and 53 percent, respectively) made from horse metapodia. The skates from both sites feature chopping (80 percent at Oost-Souburg and 92 percent at Bremen), but the Oost-Souburg skates tend to be upswept (93 percent) and have some type of binding apparatus (54 percent), whereas the skates from Bremen do not (32 percent are upswept and 24 percent have bindings). This suggests that the skaters at Oost-Souburg may have been older and more skilled than the Bremen skaters. They certainly took more time with their skates; 60 percent of the skates from Oost-Souburg exhibit four or more features, while only 36 percent of the skates from Bremen do. The skates from Oost-Souburg are closer to the Scandinavian style (remember that upswept ends were popular in Scandinavia), which may suggest that that site was more strongly influenced by Scandinavian culture than Bremen was.

As far as it goes, the evidence from bone skates is consistent with the evidence from hoards. Both suggest that Scandinavians actually settled in this area rather than merely visiting to raid or trade. More detailed investigations of the bone skates in combination with other evidence may prove helpful to the study of Scandinavian settlement patterns in the Northwest. Bone skates are a source of information that has not yet been exhausted by researchers in this field. Figuring out exactly when and where bone skates were used may help locate Scandinavian cultural areas. The evidence from bone skates could be used to support other types of evidence and fill out the details of what Scandinavians did in those regions.

It may be easier to separate a local style in the Central region, because it includes many sites that were definitely not influenced by Scandinavians. Comparing the maps in figures 34 and 46 shows a cluster of new skating sites along the northern coast of Germany, which was strongly affected by Scandinavian incursions. Zimmermann and Jöns[99] call these sites "a result of interaction between communities of Scandinavian origin and Slavic people." Barford[100] notes that the Slavs of northern Europe were interested in Scandinavian styles and sometimes adopted them; in particular, some Scandinavian jewelry (brooches and pins) has been found in graves in this region. Based on grave types, Zimmermann and Jöns[101] conclude that a number of different cultural groups came together in a "more or less symbiotic community" at Groß Strömkendorf, a site that contributed two seventh-century skates. Because of their early date, these skates are attributable to Slavs.

Because Scandinavians did not move into the entire region occupied by the Slavs, it may be possible to identify "Slavic" and "Scandinavian" skate types in this region and use them to learn more about the extent of Scandinavian occupation. Since Mikulčice, Czech Republic, is in central Europe, far from the effects of Scandinavian raids or settlement, it is a prime location for identifying a uniquely Slavic style of bone skate, if one exists. This site has contributed 144 bone skates: 4 horse radii, 87 horse metapodia, 1 horse tibia, 22 cattle radii, 23 cattle metapodia, 2 ass metapodia, 2 red deer metacarpi, and 3 unidentified bones.[102] This distribution (60 percent horse metapodia and 16 percent cattle metapodia) is quite different from the distribution in Scandinavia (42 percent horse metapodia and 50 percent cattle metapodia). The number of radii, 18 percent of the total, is also quite different from Scandinavia, where radii comprise only 7 percent of the identified bones used for skates. This difference may be due to some radius-based artifacts being misidentified as skates (as discussed in section 3.3) or may reflect a different skating tradition in central Europe. The numbers for Mikulčice are representative of the numbers for the Central region as a whole, in large part because slightly over half the Central skates with identified bone types are from this one site.

Menzlin, a trading city near the mouth of the Oder, is an example site where Scandinavians and Slavs are known to have interacted. It was a prosperous town with trade contacts around the Baltic.[103] Seven skates dating to the ninth or tenth century have been found there: two cattle metapodia, one horse radius, and four horse metapodia. These skates are identified as Slavic in Küchelmann's database. It is also possible that Scandinavian merchants brought bone skates with them when they moved in. One piece of evidence that suggests these skates were Scandinavian rather than Slavic is that none of them are pointed. This feature was not at all common in Scandinavia, but it was not uncommon in the Central region, where it occurs in about a third of the skates with documented features. The majority of these (all but three) are from sites away from the northern coast of Germany. Furthermore, six of the seven skates from Menzlin feature upswept distal ends, and one also has the proximal end upswept. The other is featureless. Upswept distal ends were common in Scandinavia (90 percent of the skates with features exhibit them), but not unusual in Central skates (71 percent). Overall, they may be considered slightly more Scandinavian than Slavic in type, but a closer examination of all the extant skates is necessary to fully develop descriptions of the two types.

Overall, this analysis shows that bone skates have the potential to contribute to the understanding of the Scandinavian expansion across Europe. Scandinavian settlers are likely to have played a role in the widespread popularity of skating. It is sometimes possible to discern regional skate styles, which suggests that individual areas developed independently. In particular, the popularity of axial holes in the proximal ends of skates in Great Britain and of split bones in Scandinavia show that those areas were autonomous enough to develop their own skate styles. This adds support to the idea that Scandinavians settled permanently enough to raise families across their sphere of influence. It may be possible to find out more about local styles by digging more deeply into the skates in these regions and individual sites, as Edberg and Karlsson did for Birka and Sigtuna in Sweden.

7.4. Directions for Future Research

This chapter shows that careful, detailed study of bone skate finds can be helpful to scholars studying the Scandinavian expansion across Europe during the Middle Ages. Although many medieval skates have been found along the northern coast of Europe, they have not yet been studied systematically. This chapter takes the first steps toward such a study using the data currently available and suggests directions that it might go in. It is limited because the details of many skates have not been published.

The case study on Great Britain provides an example of the results a systematic study could produce. Arthur MacGregor's work on the bone skates from Great Britain provides a set of data that forms the basis for the analysis, and the extensive work on Scandinavian settlement in England completed by Hadley, Kershaw, Sykes, Townend, and others provides a framework for testing the results. The evidence from bone skates, which seem to have been brought to Great Britain by Scandinavians, fits into this framework well. They are consistent with the other evidence and add evidence for the presence of children. They need not be dismissed simply because they were toys.

Because this type of analysis worked for Great Britain, it is reasonable to try it in other regions. The Northeast seems likely to produce similar results because there is no evidence for bone skates before the arrival of Scandinavian settlers in that region. The Central and Northwest regions are more difficult to study because of the pre-existing skating traditions in those regions. There is evidence for Scandinavian incursions in both areas, but the extent of the settlement is not yet clear, especially in the Northwest.

A more detailed study of the bone types and features of bone skates may make it possible to separate those attributable to Scandinavians from those attributable to local traditions. Because skates were made individually, perhaps by children, it is unlikely that identifying discrete regional schools will be possible. It is reasonable to suppose that there are local styles because children play together and learn from their peers. At some sites, it may even be possible to identify playgroups and to say something about how Scandinavians interacted with people living in the areas they settled in.

8

The End of the Bone Age

8.1. *The Emergence of Metal-Bladed Skates*

Today, showing up at a public skating session on bone skates is a good way to pique other skaters' interest—every time I do it, people ask about my unusual skates. Bone skates are now curiosities, but this state of affairs does not go back very far. The transition from bone to metal-bladed skates was long and gradual. Bone skates lasted for centuries after metal-bladed skates were invented, and the first metal-bladed skates were not actually an improvement over bone skates. The shift was not "merely the substitution of a more suitable material" that Munro[1] thought it was; it was a long and complex process that was influenced by many factors. This chapter describes the end of the Bone Age of ice skating through the rise of metal-bladed skates.

Küchelmann lists at least 56 bone skates from the early modern and modern periods in his database, including a few that could be from the later Middle Ages. These skates are from Austria, Estonia, Finland, Germany, Great Britain, Hungary, Iceland, Latvia, and Poland. There also are records of observations of bone skates in use in the nineteenth and twentieth centuries; some of these have already been discussed, and others are listed in the Appendix. They come from a list of countries that overlaps with the archaeological evidence: Estonia, Finland, Germany, Great Britain Hungary, Iceland, Poland, Romania, Russia, and, most especially Sweden. A map of the modern evidence for bone skates is shown in figure 48.

The most recently-used skates on the list are a pair at the National Museum of Iceland that was still in use in 1972; Thordur Thomasson showed them to Hagberg.[2] This pair of skates consists of a cattle metacarpus and a cattle metatarsus with transverse binding holes at both ends and the sides flattened.[3] The most recent recorded observations were made by Thorsteinn Einarsson,[4] who saw bone skates being used in Iceland in 1950, and János Makkay,[5] who noted that bone skates were in use in eastern Hungary after World War II. It is possible that bone skates survived even longer in isolated pockets in Europe.

Based on the archaeological and artistic evidence, it is possible to define three stages in the transition from bone to metal-bladed skates: first, metal-bladed skates were invented based on bone skates no later than 1250; second, people began to push with their feet instead of with poles no later than about 1350; and third, people began to push with the sides of their blades, like modern skaters, no later than 1498. Then, metal-bladed skates and foot-pushing began to spread across Europe and to the rest of the world.

The earliest evidence for metal-bladed skates found to date is from the first half of the thirteenth century. Blauw[6] describes skates from Amsterdam and Dordrecht, the Nether-

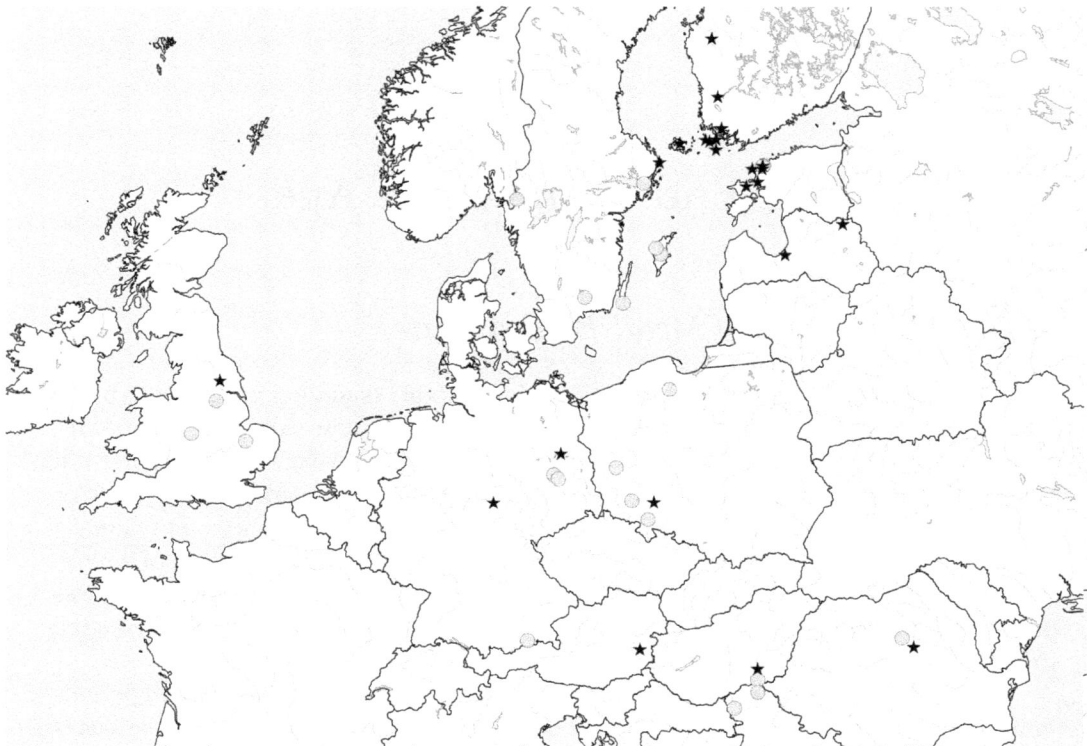

Figure 48: Archaeological (stars) and ethnographic (circles) evidence for early modern and modern bone skates (created with QGIS).

lands, that date to this period. Each skate consists of a wooden platform with an iron bar approximately 7 or 8 mm wide and 12 mm high wrapped around the bottom (see figures 49 and 50). Based on the form of these skates, Mulder[7] supposes that metal-bladed skates were invented no later than the twelfth century. All the earliest metal-bladed skates and references to skating on them that have been found so far are from the Netherlands. This points to the Dutch as the inventors of metal-bladed skates in the absence of a better candidate.

Goubitz[8] thinks the first metal-bladed skates were based on pattens, overshoes worn to protect the wearer's shoes from the environment that consisted of a sole made of wood or leather with a cork filling and a leather strap. A few are shown in figure 51. They were used in the Netherlands during the Roman period, then disappeared until the thirteenth century. Once this fashion had been revived, pattens became common enough to cause trouble: In the early sixteenth century, a Dutch law specified that

> None wearing pattens shall walk among the pilgrims who go barefoot, since they may injure the pilgrims with them.

The fear of injury was increased by the custom of putting pieces of iron on the bottoms of pattens to help people walk on ice without slipping. The extra danger of such pattens is highlighted by another law: Hitting someone with "a patten that has irons on the front or on the bottom" earned the hitter a fine.[9]

Going from rough iron on the bottom of shoes to ice skates is a rather large con-

8.1. The Emergence of Metal-Bladed Skates

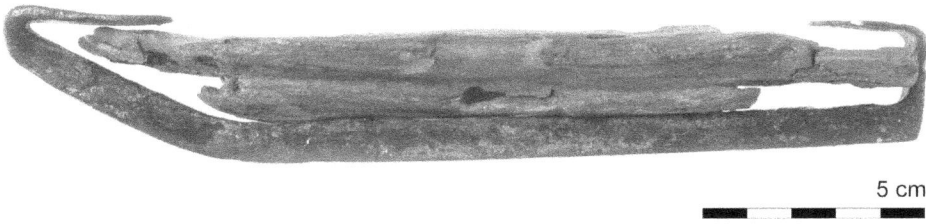

Figure 49: An early metal-bladed skate from Amsterdam. This skate dates to c. 1225–1250 CE (courtesy Amsterdam Museum, the Netherlands).

Figure 50: Another view of the early metal-bladed skate from Amsterdam that dates to c. 1225–1250 CE (courtesy Amsterdam Museum, the Netherlands).

Figure 51: Pattens dating to 1465 from Walraversijde, Belgium, near Oostende (JoJan, courtesy of Wikimedia Commons. Licensed under the Creative Commons Attribution Share Alike 3.0 Unported License, which is available at https://creativecommons.org/licenses/bysa/3.0/deed.en).

ceptual leap. The former are meant to arrest sliding, the latter to assist it. Nonetheless, Goubitz[10] suggests that this leap was made, and there is some early evidence for it: a patten shod with iron and an entry in a late-fourteenth-century account book. The patten was found at same the Amsterdam smithy as one of the earliest skates.[11] This in itself does not mean much for the development of skates but does show that skates and pattens were worked on by the same people. The connection between them is strengthened by the account book entry, made in 1392–93:

> Item mijn heer Van Oestervant screef den scult enen briif dat hi hem zende zoude III paer scouerding ende III paer scaetzen, die Dirc leuerde her Peter vanden Zande, costen XL s.[12]
>
> (Item. My lord van Oestervant wrote to the bailiff a letter that he should send him 3 pairs of *scouerding* and three pairs of *scaetzen*, which Dirk produced for Peter vanden Zande, they cost 40 shillings.)

Mulder[13] interprets this as a request for three pairs of ice skates, each made in two parts: a *scouerding* or metal blade and a *scaetz*, now *schaats*, which was the wooden platform the blade attached to—a patten. The Dutch word *schaats* is from the Old French word *escace, escache*, which means "stilt" and is cognate with French *échasse*. It goes back to the reconstructed Frankish form *skatja* and did not acquire its current meaning of "skate" until the sixteenth century.[14] How exactly it came to mean *skate* remains a mystery, but it seems to be related to the connection between pattens and skates. At first, *schaats* referred only to the wooden foot-stock, which was rather like a stilt; Goubitz[15] calls the foot-bed of a patten "a wooden sole on stilts" due to the two risers at the ends of the foot-bed. As skates became more common, the word could have begun to refer to the combination of a patten and a metal blade.

This illustrates one major difference between bone skates and metal-bladed skates: metal-bladed skates were based on shoes. The underlying concept was walking, not sliding, which may have been the kernel for the idea of foot-pushing. But that came later. It seems likely, for reasons discussed below, that the first metal-bladed skates were used with poles, like bone skates, which suggests that bone skates also played a role in their development. This idea goes back at least a century; Fowler[16] proposed it in 1897. The first metal-bladed skates appear in an area where bone skates are known to have been used, and the broad strip of metal on the bottom of an early metal-bladed skate is similar to the broad sliding surface of a bone skate. Roes[17] calls the earliest metal-bladed skates "nothing but an imitation of the bone skate," and they do indeed look like a combination of pattens and bone skates.

There is also a linguistic connection between bone skates and metal-bladed skates. In their dictionary, Dale and Kruyskamp[18] list six meanings of the modern Dutch word *schenkel*, including a lower leg bone and the iron runners on skates or skis. This provides a conceptual link between bone and metal-bladed skates, which could point to a connection as they were developed or one made later based on their similar functions. The unusual skate shown in figure 2 is a physical example of this connection. This strange skate is very similar to the earliest metal-bladed skates, but the runner is made from bone instead of metal. Herman[19] saw a skate like this in a Berlin museum, and Virchow[20] mentions that such skates were used in Pomerania during the nineteenth century. Balfour[21] notes this similarity and remarks that it may be important because such skates may represent an intermediate form between bone and metal-bladed skates. A similar type of skate is the highly desirable one described by Thunig,[22] which consists of a bone with a board attached to the top. This type of skate is basically a small sled.

The connection between bone and metal-bladed skates is strengthened by their sizes. The early metal-bladed skates that have survived are about the same size as bone skates. Goubitz[23] reports that the two thirteenth-century skates from Dordrecht are 22.5 and 22.8 cm long. Mulder describes the skate from Amsterdam as being 20.3 cm long[24] and a slightly later skate from the Hague as being 21.4 cm long.[25] These numbers are close to the most common size for bone skates at Birka and Sigtuna[26] and suggest that, like bone skates, the earliest metal-bladed skates were the right size for older children, or perhaps small adults.

These small sizes are hard to explain. They imply that these skates may have been primarily used for recreation, but unlike bone skates, metal-bladed skates were probably very expensive. Pattens seem to have been expensive in general: based on the number of leather pattens (the type used indoors) excavated at Dordrecht, Goubitz[27] thinks that only about seven percent of the people living there could have afforded a pair. It is not clear how the costs of leather and wooden pattens would have compared; wooden pattens may have been cheaper, but adding a metal blade could have increased the cost substantially.

In addition to being more expensive than bone skates, metal-bladed skates worked less well. Olaus Magnus claims that people used pieces of flat, polished iron alongside bone skates, but that these metal-bladed skates were poor alternatives to bones because greasing them does not work:

> ...ferro plano, & polito, siue planis ossibus ceruinis, vel bouinis, scilicet tibiis naturalem lubricitatem ob innatam pinguendinem habentibus, pedali longitudine sub plantis affixis, in sola glacie lubrica cursum intendit velocissimum, quemque in glaciali equalitate semper currendo continuat.... Cæteris brauium lucraturi currendo preueniunt, qui ceruinas tibias latè limatas plantis affigunt, porcina axungia peructas, quia gelidas aquæ guttis velut per poros glaciei in vehementi frigore surgentibus, tibiæ sic vnctæ impediri, aut constringi non possunt, sicuti ferrum quantumcunque politum, aut unctum. Nulla nempe vnctio ferro conformior est, quàm ossium tibiarum ceruinarum, aut taurinarum, connaturalem habentium lubricitatem.[28]

> ([Men] attach to the soles of their feet a piece of flat, polished iron, a foot long, or the flat bones of deer or oxen, the shin bones, that is. These are slippery by nature because they have an inherent greasiness and achieve a very great speed, though only on smooth ice, and continue shooting forward without pause as long as the ice remains level. ...The rest are outrun by those competitors in the race who attach to the soles of their feet the shin-bones of deer thoroughly smoothed and greased with pork fat, since, when the cold drops of water rise as it were through the pores of the ice during fierce cold, the bones smeared in this way cannot be hampered or kept in check, as iron can however much it is polished or greased. For no greasing suits iron as much as it does the shinbones of deer or bullocks, which have an innate slipperiness of their own.[29])

These results are supported by recent experiments by Formenti and Minetti,[30] who measured the coefficients of friction for bone and early metal-bladed skates and found values of 0.0103±0.0022 and 0.0147±0.0011, respectively. A lower coefficient of friction means the ice slows the skate down less. This implies that bone skates slide better than the first metal-bladed skates.

Their inferiority makes it hard to explain why metal-bladed skates were not simply abandoned as a bad idea as soon as people started trying them. They may have been more durable than bone skates,[31] but bone skates were easy to replace. The extra durability may or may not have been enough to justify the extra expense. Perhaps there was some notion of prestige attached to the earliest metal-bladed skates. Mulder[32] notes that, unlike bone skates, metal-bladed skates would have been too expensive for workers in land reclamation projects to own and therefore, must have belonged to their landlords. He proposes that the earliest metal-bladed skates were invented to help landlords travel along the narrow canals of the reclaimed land in the western Netherlands.[33]

Class distinctions may also have affected the availability of animal bones. Availability had long been a factor in which bones were selected for making bone skates. It may have been more difficult for landlords to obtain horse and cattle metapodia because they did not work directly with the animals. This distinction became sharper over time: children in the countryside skated on bones until the twentieth century, even after metal-bladed skates had become common in urban areas.

Although they probably made a much wider range of maneuvers possible, including important ones like stopping and turning, the first people to skate on metal-bladed skates had only skated on bones and therefore, probably did not know how to take advantage of this. The revolution in skating came about with the discovery of foot-pushing, which was not normally used with bone skates. Fowler[34] and some of the children at the Lödöse Museum's skating parties had some success with foot-pushing on bone skates,[35] but it requires the skates to be attached to the skater's feet, which was often not done. It seems unlikely that the foot-pushing was developed on bone skates or even as soon as metal-bladed skates were invented. If metal-bladed skates had been used in a substantially different manner than bone skates, it would have been more difficult for Olaus Magnus to conflate them and to explain why bone skates were superior.

Foot-pushing, the type of propulsion used in modern skating, is much more efficient. Formenti and Minetti[36] found an energetic cost of 345 Joules per meter for skating on bones using the pole-pushing method and a cost of 185 Joules per meter for foot-pushing with replicas of thirteenth-century skates. This decrease of 160 Joules per meter occurs despite the higher coefficient of friction of the metal-bladed skates and shows that foot-pushing is much more efficient. Skaters who pushed with their feet would have been able to go faster with less effort and would have won races. This would have made metal-bladed skates much more desirable. The ability to stop and turn would have added to their desirability.

The transition from pole-pushing to foot-pushing can be broken down into three steps. The first step is pole-pushing, as with bone skates: skaters used a pole with a pointed metal tip to push themselves along. Next, this point moved from the end of the pole to the front of the skate. Third, skaters found that the point was unnecessary because they could get a better thrust by pushing their feet outward to press the edge of the blade against the ice. Pole-pushing has already been described, and the fact that the earliest skates imitated bone skates supports the idea that they were used in the same way.

The point moved from the tip of the pole to the toe of the skate by the early fourteenth century. Perhaps the earliest depiction of metal-bladed skates is the February page of the calendar of St. Pierre of Blandigny near Ghent (Oxford, Bodleian Library, MS. Douce 5, folio 1a verso), which dates to the first half of the fourteenth century.[37] At the lower right of the page, a person—perhaps a child—is standing on one foot with the other raised forward, as if to take a big step. The person's hands are raised and empty—no pole is visible. Examining this person's feet reveals skates consisting of two layers: a brown one next to the foot and a white one at the bottom. The brown layer must represent the wood part of the skate, and the white part is the metal blade. Looking at the toes of these skates reveals definite spikes on the toes.

Similar skates have been found in archaeological excavations; an example is shown in figure 52. Mulder[38] and Blauw[39] describe skates with spikes at the toe, one (mentioned by both authors) from the Hague and another (mentioned by Blauw only) from Rotterdam, that date to between 1350 and 1500. In another article, Mulder[40] suggests that the spikes could have been used for pushing. This is consistent with the manuscript image:

8.1. The Emergence of Metal-Bladed Skates

Figure 52: An early metal-bladed skate with a spike at the toe found in the Hague. This skate dates to c. 1350–1450 CE (courtesy of the Department of Archaeology, the Hague).

the skater's free foot is extended in front. Pushing from the side of the blade would be very awkward in this position, but pushing from the toe, as in walking, would work. That these spikes were added at least a century after metal-bladed skates were invented is evidence for the transition from pole-pushing to foot-pushing: skaters may have thought they needed a spike to push from, but also knew they could push with their feet. Therefore, they moved the spike from the pole to the skate.

The modern method of pushing from the side of the blade was probably discovered

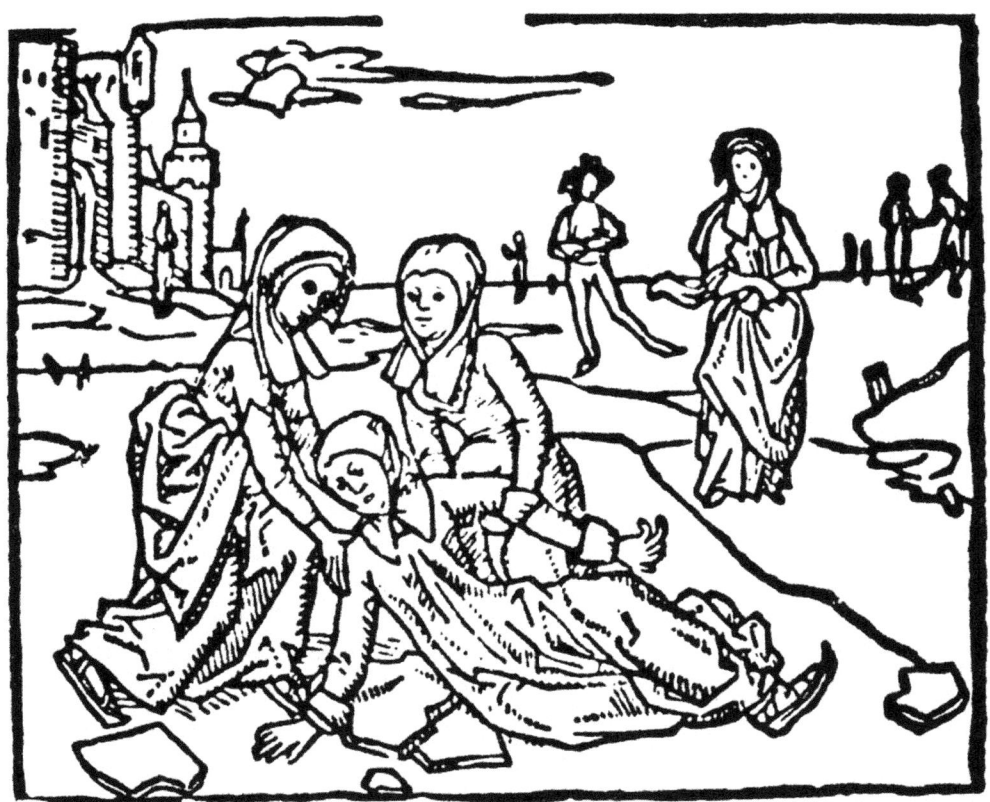

Figure 53: The scene of St. Lydwina's accident. Carefully examining the position of the skater in the background just to the right of center and the skates of one of the skaters in the foreground yields important information about the evolution of skating (Johannes Brugman, *Vita alme virginis Lijdwine*).

not long after this and certainly no later than 1498 CE: the famous woodcut of St. Lydwina's accident from Johannes Brugman's *Vita alme virginis Liidwine (Life of the Blessed Virgin Lidwine)* shows it in use. The skater just to the right of the center in the background of this woodcut (figure 53) has one leg extended to the side, as if he had just pushed off the edge of that skate. Additional evidence from about the same time is provided by Hieronymus Bosch's *Temptation of St. Anthony* from 1500–1510 (figure 54). In this image, a figure uses long metal-bladed skates without spikes on the toes. The length of these blades would have made it difficult to push off with the toe.

This evidence narrows the introduction of modern-style pushing—with the side of the blade rather than the toe—to between 1350 and 1500. The woodcut from Brugmann shows that this may have been a relatively recent innovation, because skates with spiked toes seem not to have disappeared by the time the woodcut was created. The person kneeling by Lydwina, who is lying on the ice, is wearing a skate that may have a small point at the front, as Mulder[41] observes. Instead of curving up to a nicely pointed toe, the blade of this skate bends sharply, leaving a corner. It looks a lot like the spikes on the skate in the Flemish manuscript and the skate from the Hague. At this point, skating was essentially modern, and had only to spread across Europe. After 1498, it took some time for metal-bladed skates to spread out of this area and centuries for them to completely replace bone skates.

Figure 54: Detail from Hieronymus Bosch's *Temptation of St. Anthony* triptych (c. 1501) showing a bird on long wooden skates with metal runners (Wikimedia Commons).

8.2. The Spread of the New Style

The diffusion of metal-bladed skates across Europe may have been facilitated by the climate. Based on a considerable amount of evidence and numerous expert opinions, Soon and Baliunas[42] support the idea of "an objectively discernible climatic anomaly" around the world between 1300 and 1900 CE. This anomaly is usually called the Little Ice Age. Fagan[43] describes the changing climate in Europe and the rest of the world during this period as well as its effect on people in detail; he concludes that in the Netherlands, people were able to take advantage of the colder weather. Inventing metal-bladed skates may have been one way to turn extended cold weather into something good. It may be no coincidence that metal-bladed skates were invented near the beginning of the Little Ice Age.

However, not everyone believes in the Little Ice Age. Kelly and Ó Gráda[44] argue that all the events typically attributed to the onset of the Little Ice Age—such as the failure of the Scandinavian colonies in Greenland[45]—can be explained by other causes. They suggest that the Little Ice Age is merely a statistical artifact resulting from the normal process of smoothing the data. In direct responses, White[46] disagrees and Büntgen and Hellmann[47] report additional evidence from several places around the world and support the idea that the lower temperatures prevailed around the world during this period. These responses both rely on smoothed data, which is what Kelly and Ó Gráda object to. As they note in their response to the responses,[48] nobody has directly addressed their criticism of the statistical methods used to identify the Little Ice Age. It is possible that a local decrease in temperature provided an impetus for the invention of metal-bladed skates. If it existed, the Little Ice Age was the perfect time for a new type of skate.

Although foot-pushing on metal-bladed skates seems to have been invented in the Netherlands by 1350 and to have reached its modern form by 1498, it probably had not reached Sweden by 1555—if it had, Olaus Magnus would presumably have said nicer things about metal-bladed skates. Additionally, the accounts of travelers from central Europe show that this new style had not reached that region in the sixteenth and seventeenth centuries. Other accounts show that metal-bladed skates and foot-pushing had arrived in Sweden and England by the middle of the seventeenth century. These all provide evidence that the spread of the new style was quite slow.

The first piece of evidence is from a description of skating in Brussels in the fifteenth century. From 1456 to 1467, the Bohemian Baron Leo of Rožmitál and Blatná was in Brussels. His squire and armor-carrier, Václav Šašek of Bířkov, wrote a description of the trip that focused on local customs. This account has not survived, but a 1577 Latin translation of it has.[49] This Latin version includes the following description:

> Ea die, qual Dominus Duci valedicebat, mirabile spectaculum conspeximus.
> Vivarium est Bruxellae, arci contiguum, in eoque piscina, cujus summum tum glacie obductum fuit. Id vivarium Dux ministros suos aliquot ingredi jussit, et super piscinam glacie concretam depugnare. Illi—duodetriginta erant—tanta agilitate pedites inter se depugnabant, ut me nunquam tam agiles homines, vel vidisse, vel audivisse affirmare possim. Praecipue unus eorum tanta agilitate praestabat, ut saepissime viginti duos adversus se pugnantes solus sustinuerit. Tantaque erat eorum in cursu velocitas, et in convertendo celeritas, ut nullius equi cursus eis aequiparari possit. Cupidus eram explorandi, quidnam pedibus supponerent, quod se tam celeriter in glacie convertere possent. Nam magnum esset miraculum si supra terram tanta in convertendo celeritate uti quirent, quod quidem facile assecculus fuissem, sed a Domino, qui ex arce cum Duce ea spectabat, discedere not potui.[50]

(On the day on which my lord took leave of the Duke we saw a marvelous spectacle. There is a park adjoining the castle with a lake which was then frozen over. The Duke ordered certain of his courtiers to go out to this park and to run a course on the frozen lake. They—there were twenty-eight of them—fought on foot with such agility that I can declare that never have I seen or heard of such agile men. One in particular was so skillful that he resisted alone the assault of twenty-one men. Such was their speed in running and turning that no horse could have kept up with them. I was curious to see what it was that they had on their feet which enabled them to move so swiftly on the ice. I could easily have done this, but I could not leave my lord who was looking on with the Duke.[51])

It seems likely that these men were using skates because of how quickly they crossed the ice. Their ability to turn quickly shows that these were not bone skates, because such agility would have been beyond bone technology. It is unfortunate that Šašeck did not get a closer look at the men's footwear. His curiosity suggests that he had never seen skating with foot-pushing before, as Letts[52] notes, though he could have been familiar with or even used bone skates. If the translator was faithful to the original, this shows that foot-pushing was used in Brussels, but not Bohemia, by the middle of the fifteenth century.

Corroborating evidence is provided by Márton Szepsi Csombor's description of skating in the Netherlands between 1616 and 1619. Szepsi Csombor was a Hungarian writer who is best remembered for his travelogue, *Europica Varietas*. In it, he directly compares Dutch skaters with the Hungarian skaters he is familiar with:

> Estendőként egyszer, télben harmadnap alatt az egész ország törvénye s szokása szerint pályafutásnak okáért bizonyos jutalom alatt (melyet az magisztrátus teszen le) a leányok az ő idejük szerint való hozzá hasonló ifjú legényeket az tengerre kihiják, az leány az férfiúnak az férfiú az leánynak lábára felköti a csontból vagy csak fából csinált csuszkondót, de ez nem igy vagyon, mint itt Magyarországban láttam, hogy egy nagy vassal megszegezett bot volt kezében az jégen járóknak, hanem szintén csak lábainak mesterségesen való félretaszitása által, felhányása és mozgatása által az egész nép láttára elindulnak, és oly sebességgel mennek, hogy ég alatt nincs oly ló, ki velük elérkeznék.[53]

> (Once a year, during the winter, girls and lads of matching ages are called to the sea within three days according to the law and tradition of the country in order to compete on a course for a prize set by the magistrate. Skates made from bone or simple wood are tied to the feet of men by the maids and on the feet of maids by men. However, it is different here, from what I have seen in Hungary, where ice walkers carry a long rod with an iron bit nailed to it in their hands. People here start in front of a whole crowd of people by simply siding their feet to one side then flinging themselves ahead and moving forward. They attain such a speed that no horse would be able to catch up with them.[54])

This suggests that when he wrote, in 1620, foot-pushing had not reached Hungary. It is noteworthy that Szepsi Csombor describes "skates made from bone or simple wood," which are unlikely to have actually worked well with foot-pushing. He may have been thinking of the types of skates available in his homeland, bone skates and perhaps wooden skates like the later one from Transylvania shown in figure 55 and described by Balfour[55]:

> Each is cut from a single piece, and consists of a straight bar, 11 ins. long, almost square in transverse section (1 by 1 ins.); at the back, or heel end, it is cut off square; in front, the wood is cut so as to curve over the toe of the boot in a long spur-like projection. There are two holes for attachment, bored from side to side, an oblong hole towards the front end, evidently for a strap, and a small circular hole towards the heel end for a cord.... These simple wooden skates are little, if at all, superior to those of bone, and, apart from the material, differ from them chiefly in their squareness, which furnishes them with edges, and in the refinement of a recurved prow.

8.2. The Spread of the New Style

Figure 55: A wooden skate from Transylvania (Henry Balfour, "Sledges with Bone Runners in Modern Use," 253).

Balfour's claim that these skates were no better than bone skates sounds reasonable because wood lacks edges capable of digging into ice the way the edges of metal blades do. It is possible that such skates were what Szepsi Csombor observed, but because the skaters did not use poles, it seems far more likely that he, like Šašek, was unable to get a close enough look at the skates to see the iron runners attached to the wood and instead, described the skates he was familiar with.

By the middle of the seventeenth century, evidence for foot-pushing began to appear outside the Low Countries: first in Sweden, then in England. Swedish acquired a new word for the new type of skate, and two Englishmen described English skaters using metal-bladed skates in 1662. The new Swedish word, *skridskor*, is still used for ice skates today. The English diarists, John Evelyn and Samuel Pepys, were amazed by the skaters.

According to the Swedish Academy, *skridsko* is the word for a modern metal-bladed skate, but it can also be used to refer to bone skates and skates made from blocks of wood with metal blades attached. Its first two recorded uses are in letters written by Johan Ekeblad in 1650 and 1652. In 1650, he wrote that someone "brukar sigh reda med skrittskoer"[56] (busies himself with skating) and in 1652, he wrote to his brother that he himself had "försökte till att löpa på skritsko"[57] (tried to run on skates). This was about a century after Olaus Magnus mentioned people using metal-bladed skates alongside bone skates.

The Swedish Academy suggests that *skridsko* may have been a loan from Middle Low German with the prefix rebuilt to match the Swedish verb *skrida*, the descendant of the Old Norse word *skríða*.[58] In the eighteenth century, *skridskor* coexisted with *isläggar*. Abraham Sahlstedt explains both in his 1773 dictionary: *skridskor* refers to metal-bladed skates, "[c]alceus ligneas, acuato munitus ferro, quo in glacie cursus"[59] (wooden shoes, sharpened with iron, that run on ice), whereas *isläggar* refers to bone skates, "[o]ssa curta, quibus insistens in glacie quis procurrit"[60] (shortened bones, standing on which one runs on ice). The latter, while familiar from its Old Norse predecessor *ísleggr*, which is discussed in Chapter 6 along with *skríða*, is listed in the current edition of the Swedish Academy's dictionary in its singular form, *islägg*, with a brief description of bone skates and their use.

Over the years, *skridskor* has both become the standard Swedish word for ice skates and acquired a general meaning that includes both bone and metal-bladed skates. People referring to bone skates specifically use *islägg* or describe skating as "att åka skridskor på djurben" (to go skating on animal bones). In its description of the winter holiday program for children, the Lödöse Museum in Sweden invites people to "prova på att åka skridskor på djurben"[61] (try skating on animal bones). The word *isläggar* may no longer be familiar to Swedes who lack substantial knowledge of cultural history or archaeology. The decline in the use of *islägg* and the generalization of *skridsko* reflects the modern dominance of metal-bladed skates and rarity of bone skates.[62]

In England, two independent eyewitness accounts show that metal-bladed skates arrived no later than the middle of the seventeenth century. The earliest possible date for their arrival is harder to pin down. Legend has it that metal-bladed skates arrived in England shortly after the Norman Conquest, but this is not supported by the evidence. One story comes from W. H. Barrett, who heard it "from an old fenman" in 1901.[63] According to this story, "men engaged in building Ely Cathedral came stealthily into the deep recesses of the zealously-guarded fens, in winter, to catch wildfowl for food ... they eluded capture by pursuing fenmen who noticed, from a distance, that the marauders wore on their feet objects which enabled them to travel along the frozen streams far more swiftly than they themselves could do on their bone skates. One day the fenmen laid an ambush for the intruders, captured them and removed from their boots the wooden soles which were strapped to them and which had fastened to the bottom of each an iron blade. Trying these on their own boots they found that they were indeed far superior to bones."

The story continues with the fenmen meeting a monk, who made a pair of metal-bladed skates and called them "pattens." It is a nice story, but there is a problem with the chronology: Ely Cathedral was finished in 1083, over a century before the first evidence for metal-bladed skates anywhere. Enid Porter, who collected this story, does not believe it. The first reliable evidence for the new style of skating appears much later. The first use of the word "skate" in English that Foster[64] found is in Henry Hexham's *Copious English and Netherduytch Dictionarie* (1648), and its appearance in his 1648 *Groot Woorden-boeck (Large Dictionary)* is listed as the first use in the *Oxford English Dictionary*.[65]

The next two pieces of evidence follow closely on the heels of Hexham's dictionary: both Samuel Pepys and John Evelyn observed and described skating on December 1, 1662. John Evelyn describes the scene as follows:

> Having seene the strange, and wonderfull dexterity of the sliders on the new Canall in St. James's park, perform'd by divers Gent: & others with Scheets, after the manner of the Hollanders, with what pernicitie & swiftnesse they passe, how sudainly they stop in full carriere upon the Ice, before their Majesties: I went home by Water but not without exceeding difficultie, the Thames being frozen, greate flakes of yce incompassing our boat.[66]

Samuel Pepys describes his trip "over the Parke, where I first in my life, it being a great frost, did see people sliding with their skeates, which is a very pretty art."[67] On the eighth, he returns to the Park "to see them slide with their skeates, which is very pretty."[68] And on the fifteenth, he "followed [the Duke] into the Parke, where, though the ice was broken and dangerous, yet he would go slide upon his scates, which I did not like, but he slides very well."[69]

These diary entries show that metal-bladed skates had reached England by 1662, when they were used by members of the upper class, which is consistent with Mulder's comments on how expensive they must have been. Metal-bladed skates did not reach children in rural areas until much later; as recently as the nineteenth century, it was possible for people to grow up without knowing about metal-bladed skates. Brückner,[70] for example, was an avid skater but did not see metal-bladed skates until he left his childhood home. Bone skates did not go out of use in the Netherlands either; Berg[71] notes that Dutch peasant children learned to skate on bone skates (specifically, rib bones from oxen) before making the transition to metal-bladed skates and adds that this was probably also the case

8.2. The Spread of the New Style 131

in Westphalia. Balfour[72] provides corroborative evidence in the form of a collection of references to skating on ox-ribs in the Netherlands, Iceland, and Westphalia. Küchelmann's database includes a pair of Polish skates dating to about 1880 that are made from cattle ribs.

The rest of the archaeological evidence supports and extends the written descriptions of modern bone skates. Küchelmann's database includes eight nineteenth-century bone skates from Estonia and eight more from Finland, though the dates of six of the latter are questionable. The remaining two skates date to around 1850 and are housed in the Turku Provincial Museum. Hungary contributes 15 nineteenth-century bone skates made from horse metapodia and described by Herman.[73] In addition to the pair of skates from 1972, Iceland contributes two pairs from the nineteenth century, one dating to 1878.

The continuing use of bone skates in rural areas may have been for economic reasons. In a presentation to the Berliner Gesellschaft für Anthropologie, Ethnologie und Urgeschichte, Treichel noted that rural parents did not give their children any money for toys. With no money to buy skates, they had to make their own. Bone skates were much easier to make than metal-bladed skates, but some children were up to the challenge. Treichel describes one type of metal-bladed skate children were able to make. It uses a knife for a blade:

> Ein anderes Instrument zu gleichem Zwecke erwähnte mir Hr. R. G. B. Paschke, das er selbst um 1850 in Gebrauch genommen (Triebel in der Nieder-Lausitz). In ein nach der Form des Fusses [sic] ausgeschnittenes Stück Holz wird ein Messer eingeschlagen, durch ein eingebohrtes Loch ein Bindfaden durchgezogen und oberhalb des Fusses [sic] wie ein Knebel befestigt. Es war dies der sogenannte Hund. Die Stelle des Knochens vertrat hier also, wie bei den modernen Schlittschuhen, die Messerklinge von Eisen.[74]

> (Mr. R. G. B. Paschke mentioned another instrument for the same purpose to me, which he himself used in 1850 (Triebel in lower Lusatia). In a piece of wood shaped like a foot a knife was set, a cord was attached through a bored hole and [the whole thing was] attached to the foot like a knob. This was the so-called dog. In place of the bone was, as in modern skates, a knife blade of iron.)

In Iceland, people were making skates consisting of a wooden platform and a metal blade by the early nineteenth century.[75] According to Säve, homemade metal-bladed skates replaced bone skates in Sweden by the end of the nineteenth century:

> ...när det var glansk-is tog man fisk på söjn-ejs (syn-klar-is) då man lätt såg honom genom den klara träsk-isen. Fordom brukade man så antigen alltid låjpa ejs-läggar eller begagnade man fotbruddar, men numera går man ofta på skridskor, som man sjelf smidt...[76]

> (When there was smooth ice one took fish on the *söjn-ejs* (syn. clear-ice) then it was easy to see them through the clear swamp ice. For that one always ran on ice-legs or one wore crampons in antiquity, but now one often uses [metal-bladed] skates, that one made oneself...)

These homemade metal-bladed skates contrast with the earliest ones, which were probably made by professionals, since one was found in a blacksmith's shop. It seems that by this point, metal objects had become sufficiently widely available to be repurposed for skates. As it became easier to make metal-bladed skates, more people started to use them as there was no reason to use bone skates instead. Despite this gradual shift, as the nineteenth century turned into the twentieth, skating on bones remained a fun diversion for children and young people in rural areas. Their bone skates seem to be essentially the same as the ones used in the Middle Ages. One difference in modern bone skates has been noted by Allen[77]: the bone skates he purchased from a boy in Iceland were "made from the metacarpal bones of the tiny Icelandic ox, not from those of the horse, as is most usually the case with the prehistoric examples." Although he does not mention the age of the boy, the small size of these skates may indicate that he was quite young. The selection of cattle bones instead of horse bones may simply mean that horse bones were less readily available in that place at that time.

According to the written references, modern bone skates were normally attached to the skater's feet. Of the references cited in the Appendix, only three[78] note that they were not attached. Three others[79] describe attachment as optional, and the other references do not mention whether the skates were attached.[80] The archaeological evidence collected in Küchelmann's database tells a slightly different story. Just over half (22, or 56 percent) of the modern bone skates with modifications listed in the database feature some sort of binding apparatus. It is possible that the ethnographers slanted their accounts to meet the expectations of their listeners. People unfamiliar with bone skates find it difficult to believe that skaters did not have to attach their skates to their feet. This encouraged the people describing bone skates to explain how the skates were attached, even though it was unnecessary.

That bone skates lasted for around 800 years after the invention of metal-bladed skates shows that the takeover was not a simple process. The new skates were not an improvement until skaters learned to push with their feet. As skaters were figuring this out, metal-bladed skates began to gradually spread outward from the Low Countries. They came into use in cities first and did not fully replace bone skates in rural areas until quite recently. The disappearance of bone skates may not have been due only to competition from metal-bladed skates. In some areas, skating conditions worsened. Kuckuck[81] notes that between 1845 and 1871, people had to stop using bone skates in the village he grew up in because of local drainage projects. The river disappeared, and the meadows could no longer be flooded, making it impossible to skate. Similar progress could have prevailed at other skating locations across Europe.

The invention of metal-bladed skates, changes in skating conditions, and changes in material availability are all likely to have played roles in the disappearance of bone skates. The first metal-bladed skates did not become serious competition for bone skates until skaters learned to push with their feet. Even after that, bone skates continued to enter the archaeological record, and eventually, ethnographers began describing the bone skates they recalled from their childhoods. From the sixteenth century onward, skating on bones was mainly a pastime for rural children as members of the upper class discovered the joys of metal-bladed skates. Eventually, metal-bladed skates made their way out of cities and into the countryside as they gradually replaced bone skates. Today, metal-bladed skates have taken over, and it is difficult to obtain bones for skating in cities, but it is possible that there are places where children still skate on bones.

8.3. Where to Go from Here

This book is a history of bone skates based on the evidence currently available. The emphasis is on the last two words: one thing I hope to have made clear is that there is a lot more to do. First, there are bone skates that I have been unable to write about. Many languish in museum collections. They have been excavated, but no descriptions of them have been published. Even those that have been mentioned in publications often have not been described in detail. For some, no context is available. Nobody knows where they were found or when they were used. I have included loosely dated skates with only general locations (like the 142 medieval skates from England that Arthur MacGregor[82] was unable to provide dates for) into the analyses of bone types and features. I am hopeful that the type of statistical analysis I present in Chapter 7, which is inspired by the one performed by Edberg and Karlsson,[83] will make it possible to determine where skates are likely to have been made based on their features.

Completely undated skates are not included in these analyses, and their number is not insignificant. In his database alone, Küchelmann lists 676 skates with no date. I have been able to find dates for many of these skates in the literature, bringing the number down to 121, assuming the 300 from Lake Mälaren in Sweden mentioned by Herman[84] are included among the skates analyzed by Edberg and Karlsson.[85] But I have also found more undated skates. The biggest contributor is Jacobi, who, in *De Nederlandse Glissen*, analyses 160 bone skates from the Netherlands, but provides dates for none. Radiocarbon dating and other scientific techniques may help with dates, but even without them, some of these skates are quite interesting. A noteworthy example is the one made from a human femur found at Slánská Hora, Czech Republic, and held at the National Museum in Prague.[86]

Some of the details of the skates residing in museum collections are missing because they have been there since the nineteenth century, when archaeological context was not considered as important as it is now. Archaeologists were not always as careful to record find locations as they are now. One thing that remains difficult is accurately and confidently identifying artifacts as skates. Wear patterns are helpful, but, as the confusion about the mysterious radius-based artifacts shows, they are not always sufficient. Some of the ongoing research in metrology seems promising; Stemp, Watson, and Evans[87] have suggested some ways new scientific methods can be applied to bone tools. In particular, it would be useful to understand how the wear patterns produced by friction against ice, plants, and hides differ. Furthermore, the rate of wear of bone skates has not been investigated at all yet. A study of how bone wears down against ice could yield insight into diagnostic wear patterns while also allowing researchers to better understand the climate conditions of their archaeological sites. The number of hours a bone skate was used for skating says something about how people lived and how cold their environment was.

In the absence of such a study, simply measuring these skates and their details, such as the distance between binding holes and the ends of the bones, will create a quantitative dataset that can be used to develop statistical arguments. Edberg and Karlsson have shown how useful such data can be. Bone skates have the potential to be powerful tools for historians and archaeologists. The Scandinavian expansion is one particular application. This is a good place to start mining the data from bone skates because so many have been found in areas visited, raided, and occupied by medieval Scandinavians. Careful descriptions of these thousands of skates will turn them into data points for statistical analyses that can support or refute arguments developed by historians. I started doing this in Chapter 7, but there is much more to be done. The Northeast is particularly ripe for this type of analysis. Because there does not seem to have been an earlier skating tradition in Estonia, Latvia, and nearby areas, parallels to the situation in England can be drawn.

The bone skates from this region can be combined with other types of evidence to better understand Scandinavian settlement patterns. In central Europe and the Northwest, the situation is more complex due to the earlier skating tradition, but with sufficiently detailed data, it may be possible to separate Scandinavian skates and learn about trade and settlement patterns from them.

More work on the origin of ice skating is also needed. New results from the Eurasian steppes—new artifacts, broadly disseminated information, and details of known artifacts—have the potential to add to the understanding of the origin of ice skating. In Chapter 4, I outline the types of evidence that could support the hypothesis that bone skates were an adaptation of skis to the conditions and resources of the steppes. If this is true, it provides another layer of detail about the motion of prehistoric people through the steppes. Even a few skate finds could yield important clues about the interactions that led steppe peoples to adopt skis and how they adapted them to steppe conditions.

In this book, I have identified ways in which bone skates can be used to glean important information about certain parts of European prehistory and history. They yield clues about climate conditions, settlement patterns, and interactions among peoples. Because they are most strongly associated with children and young people, they may be useful in studies of family life. Bone skates have the potential to become important tools in the study of the past. They are much more than a footnote to the history of ice skating.

Appendix: Modern Descriptions

This appendix quotes the modern eyewitness descriptions of bone skates I used while writing this book in full. They are grouped by general area and placed in context.

A.1. Germany and Poland

The references in the first group are from Germany and Poland, mainly the area surrounding Berlin. Most of these are parts of a long conversation about bone skates had by members of the Berliner Gesellschaft für Anthropologie, Ethnologie und Urgeschichte. In this conversation, discussions of archaeological finds and whether they were actually skates, tools in the textile industry, or other things alternated with members' recollections of skating on bones as children. Page 19 of Foster's *Bibliography of Skating* lists numerous references to bone skates in the proceedings of this society.

Ethnographic contributions to the conversation began on June 24, 1871, when Rudolf Virchow read a letter from a Mr. Kuckuck to the Berliner Gesellschaft für Anthropologie, Ethnologie und Urgeschichte. It tells the story of Kuckuck's experience using bone skates as a child in Sulechow, Poland, which is fairly close to Berlin.

> Vielleicht interessirt [sic] Sie die Mittheilung, dass ich mich ganz bestimmt erinnere, dass in der Umgegend meiner Heimath (Züllichau in der Mark) vor ungefähr 25 Jahren von Bauernknaben Knochen zum Schlittschuhlaufen ziemlich häufig gebraucht wurden; in welcher Weise dieselbe aber befestigt waren, ist mir nicht mehr erinnerlich. Dass dieselben jetzt noch daselbst zu diesem Zwecke in Gebrauch sind, bezweifle ich jedoch, denn die seitdem vorgenommenen Entwässerungen haben die Gelegenheit zum Schlittschuhlaufen ausserordentlich beeinträchtigt, da durch dieselben eine Ueberschwemmung von grossen Wiesenflächen verhindert ist und bei dem Fehlen eines Flusses das Schlittschuhlaufen nicht mehr so allgemein betrieben werden kann wie damals.[1]

> (Perhaps you are interested in the communication, that I remember very clearly, that in the neighborhood of my homeland (Züllichau in der Mark) about 25 years ago, bones were used for skating by farmer boys quite often; I no longer recall the manner in which they were fastened. That they are still in use for this purpose I rather doubt, because the drainage done since then has seriously damaged the area for skating in that the flooding of large meadows is prevented and due to the lack of a river, skating can no longer be done as frequently common as it was then.)

At the December 9, 1871, meeting, Herm. Grimm gave a very brief presentation on the use of bone skates in Jüterbog, a town in Brandenburg located approximately 65 km southwest of Berlin.

Es sind dies Metatarsalknochen vom Pferde, wie sie noch gegenwärtig von der Jugend beim Schlittschuhlaufen auf dem dortigen See gebraucht werden. Sie stimmen in der Art, wie die eine Seite abgeglättet ist und die Knochenvorsprünge an den durchbohrten Gelenkenden abgetragen sind, ganz mit den aus alten Fundstätten früher vorgelegten überein. Eingeschlagene Nägel dienten offenbar zur Befestigung der Schnüre.[2]

(These are metatarsal bones of horses, which are still used by young people for skating on the lake there. They are of the same type; one side is smoothed and the bony projections are removed from the drilled-through joints, just like the previously presented ones from the old find sites. Inserted nails obviously served for fastening the strings.)

The most commonly cited contribution to this conversation is Brückner's presentation to the Berliner Gesellschaft für Anthropologie, Ethnologie und Urgeschichte on January 13, 1872. It is one of the most detailed accounts of the modern use of bone skates. As a child in Legnica, Poland, Brückner grew up using bone skates with his friends.

Brückner gives distances in German miles. Before the metric system was adopted in Germany, units of measurement were not universal standards. In particular, a mile was not a fixed distance but varied from place to place. Küchelmann and Zidarov[3] give the conversion factor 7420.44 meters per mile. In Prussia, one mile was 7532.50 meters, according to Staatsbibliothek zu Berlin.[4] The miles have been converted using the Prussian value in translation below, but inches are left in the original units because the size of an inch did not vary as much. The values should be taken as approximate rather than exact.

Brückner's presentation provoked numerous responses from other members of the Society who also remembered using bone skates while growing up. The most relevant of those responses are listed below.

Bis zu meinem 15. Jahre—mit diesem Alter verliess ich das elterliche Haus—habe ich keine Stahlschlittschuhe zu Gesicht bekommen. Die gesammte männliche Jugend meines Heimatsortes Gross-Läswitz, eines bis in die Fünfziger Jahre auffallend isolirten [sic], 3 Meilen nordöstlich von Liegnitz gelegenen Dorfes, bezog ihren Bedarf an Schlittschuhen vom Schindanger. Im Herbste wurde derselbe mit vielem Fleisse von uns durchwühlt; es galt hierbei entweder die eigenen, im vorhergegangenen Winter abgelaufenen "Knochen" durch neue zu ersetzen oder den in der thönernen Sparbüche deponirten [sic] Schatz um ein paar "Böhmen" zu vermehren. Die von uns gesuchten Knochen waren etwa 10–13" lang und dem Fusse des Pferdes (ob Voder- oder Hinterfuss, weiss ich nicht anzugeben) entnommen. Sie wurden mit dem Taschenmesser von Fleisch und Haut sorgfältig gereinigt und dann übernahm ich es als der Sohn des Müllers, die untere Seite des Knochens auf dem Mühlsteine eben zu schleifen; hiernach wurden sie in Brauch genommen. Ein Befestigen der Knochen fand nicht statt; man stellte sich aufrecht auf sie und stiess sich mit einem unten mit einem Stachel versehenen Stabe, den man mit beiden Händen anfasste und vor sich in der Mitte in das Eis einhieb, fort. Die Geübteren unter uns fuhren in 15 Minuten eine starke halbe Meile. Wenn das Eis vollkommen eben war und die "Knochen" durch langen Gebrauch eine spiegelartige Glätte erhalten hatten, konnten wir uns durch Ausspannen unserer Jacken vom Winde treiben lassen. Die "Knochen" wurden öfter an dem einen Ende durch bohrt, aber nur, um durch Zusammenbinden derselben ihr Tragen nach entfernteren Eisflächen bequemer und sicherer zu machen.[5]

(Until my fifteenth year—at that age I left my parents' house—I had never seen steel skates. All the male youths of my hometown Gross-Läswitz, one of the villages that was strikingly isolated until the fifties, 22.60 km [14.04 modern miles] northeast of Liegnitz, met their need for skates in the knacker's yard. In fall they were made by us with great diligence; in this case it was either to replace the worn-out "*bones*" from the previous winter with new ones or to increase the treasure deposited in the clay money box by a pair of "*Böhmen*." The bones sought by us were about 10–13" long and taken from the foot of a horse (whether fore or hind foot, I do not know specifically). They were carefully cleaned of flesh and hair with a pocket knife and then I, as the son of the miller, undertook to grind the bottom side of the bone flat on the mill stone; then they were put into use. There was no fastening system for the bones; one stood upright on them and propelled oneself with a staff with a spike underneath, which one held with both hands and put into the ice directly in front of oneself. The practiced among us went a good 3.77 km [2.34 modern miles] in 15 minutes. When the ice was completely even and the "bones" had gained a mirror-like smoothness through long use, we could allow ourselves to be driven by the wind by opening our jackets. The "bones" were often bored through at one end, but only to make it easier and more secure to carry them to far-away ice by tying them together.)

Mr. Koenig commented on Brückner's remarks during the discussion period. He offered to follow up, but if he did, he never presented his results to the Society. His contribution simply reads:

> Herr Koenig bemerkt, dass er vor 50 Jahren als Knabe bei Jüterbogk nur Knochenschlittschuhe gekannt und benutzt, diese jedoch mittelst Fäden an die Füsse befestigt habe, welche durch in die Knochen eingebohrte Löcher gezogen wurden. Derselbe vermuthet, dass ein ähnlicher Gebrauch noch jetzt daselbst herrsche, wird aber noch nähere Erkundigungen darüber einziehen.[6]

> (Mr. Koenig remarks that 50 years ago, as a young boy near Jüterbog, he knew and used only bone skates, however, these were normally fastened to the feet with cords, which were threaded through holes bored in the bones. He supposes that a similar use still prevails there but will make further inquiries about it.)

Nearly a year later, at the December 14, 1872, meeting of the Berliner Gesellschaft für Anthropologie, Ethnologie und Urgeschichte, Mr. Thunig wrote in response to Brückner's presentation.

> Auch ich habe in meinen Kinderjahren in Schlesien die Schlittknochen zum Eislaufen vielfach benutzt. Ich sowohl wie die anderen Jungen bezogen Rindsochen [*sic*] vom Fleischer, brannten in der Schmiede Löcher durch und befestigten sie mittelst Stranglitzen (Provinzialismus für starken Bindfaden, während der schwache Spuckat heisst) an unsere Füsse. Söhne von Zimmerleuten oder Tischlern befestigten auf diese Knochen Klötze resp. Bretter, so dass ein schlittschuhähnlicher Gegenstand zum Vorschein kam. Ich kann mich noch sehr gut daran erinnern, dass mir meine Mutter zu Weihnachten 1820 oder 21 ein Paar dergleichen Schlittknochen mit Brettern schenkte. Auch mein 70 Jahre alter Administrator Fordan, in Nieder-Schlesien geboren und erzogen, während ich aus Peilau bei Reichenbach stamme, kann sich ebensowenig, als ich, entsinnen, jemals undurchbohrte Schlittknochen resp. derartige Knochen ohne Bänderbefestigung im Gebrauch gesehen zu haben.[7]

> (In my childhood in Silesia, I also used bone skates for ice skating many times. The other boys and I obtained cattle bones from the butcher, burned holes through them in the smithy, and fastened them with *Stranglitzen* (a provincial term for strong binding cords, while the weak ones are called *Spuckat*) to our feet. The sons of carpenters or joiners fastened to these bones pads or boards, so that a skate-like object came into view. I still can remember very well, that for Christmas 1820 or 21 my mother gave me a pair of such bone skates with boards. My 70-year-old administrator Fordan, born and raised in Lower Silesia, while I come from Peilau by Reichenbach, can recall just as little as I having ever seen skates that had not been bored through or such bones without attachments in use.)

After this, some years passed before ethnographic discussions of bone skates resumed. At the October 17, 1885, meeting of the Berliner Gesellschaft für Anthropologie, Ethnologie und Urgeschichte, the following letter from Herm. Grimm was read.

> Auf Umfrage habe ich den Gebrauch von Schlittknochen bis in die neueste Zeit noch häufig angetroffen.
> Als Knochen nahm man dazu solche aus den Gerippen vom Hammel oder Pferd. Indem man einen Stock dazu nahm, fuhr man damit wie auf einem Piekschlitten. So machten's die Alten und

so machen's die Jungen in Bauerndörfern, wenn ihre Eltern ihnen natürlich kein Geld geben zum Ankaufe von Schlittschuhen. Der Knochen wurde übrigens an seinen dünneren Stellen oder am Gelenke durch Bindfaden an dem Fusse [sic] befestigt.[8]

(In a survey, I quite have frequently run across the use of bone skates up to the present time.
People took bones for this purpose from the carcasses of sheep or horses. One used a stick to drive oneself as on a pick-sled. So did the old do and so do the young in farming villages, where naturally their parents give them no money for purchasing skates. The bone was frequently fastened to the foot with string on its narrower part or at the joint.)

Outside the Berliner Gesellschaft für Anthropologie, Ethnologie und Urgeschichte, others were discussing bone skates in this area. Herman quotes Ernst Friedel's description of bone skates in use. Friedel's location is not specified, but because he published in *Brandenburgia*, it seems likely that he was describing the situation in Brandenburg, Germany, like several other members of the Berliner Gesellschaft. Friedel was well aware of the Berliner Gesellschaft's work and presented on other topics at some of the meetings.[9]

Die Schlittschuhknochen sind entweder undurchbohrt oder zum Hindurchziehen von Riemen oder Bindfaden durchbohrt. Die undurchbohrten sind natürlich die primitiveren.
Bei undurchbohrten Schlittknochen, bei denen selbstredend die Köpfe der Rinder- oder Pferde-Röhrknochen abegeschlagen sind, stand der Eisläufer einfach auf dem Knochen aufrecht, wozu lediglich Vorübung und Gewandtheit gehörhte, und trieb sich mit einem Stachelstock, besser mit zweien—in jeder Hand mit einem, vorwärts. Auch wurde zwischen den beiden Stöcken wohl ein Tuch als Segel befestigt. Setzte sich der Wind dort hinein, so war man imstande, außerordentlich schnell auf glatter Eisbahn vorwärts zu kommen. So haben noch vor wenigen Jahren uns alte Leute berichtet.[10]

(The bone skates are either undrilled or drilled through for attachment with straps or cords. The skates that have not been drilled through are naturally the more primitive. With the skates that were not drilled through, after the heads of the long bones of horses or cattle had been cut off, the skater simply stood upright on the bones, for which only preliminary practice and dexterity were necessary, and pushed forward with a prick-stick, or better with two, one in each hand. Also, a cloth was fixed between the two poles as a sail. When the wind blew into it, one could go forwards exceedingly fast on a smooth ice surface. So the elders informed us a few years ago.)

A.2. *Central Europe*

The conversation among the members of the Berliner Gesellschaft für Anthropologie, Ethnologie und Urgeschichte was not limited to the local area. In 1887, A. Treichel passed on some information about the use of bone skates in Bavaria that he received from Dr. K. von Maurer.

In Bezug auf Schlittknochen in Bayern versichert Herr Prof. Dr. K. von Maurer in München, dass selbige in dortigen Bergen, zumal bei Knaben, noch immer in Gebrauch sind, z.B. auf dem Schliersee, der leicht zufriert, sei es ein beliebter Spass, auf Knochen zu laufen.[11]

(With respect to bone skates in Bavaria, Professor K. von Maurer in Munich reports that they remain in use by young men in the mountains there, for example, on the Schliersee, which freezes easily; it is said to be popular fun to skate on bones.)

The next report is a departure from the Berliner Gesellschaft to a similar society in Austria, the Anthropologischen Gesellschaft in Wien. F. von Luschan gave a presentation on bone skates there in 1876. In it, he responded to the discussion of an archaeological find at an earlier meeting of the Society. He related it to what his friend Dr. Heinrich Kraus told him about using bone skates as a child in Transylvania. The report is longer than the others quoted here, but beautifully written in German. The figures from this report are reproduced in this book. The original figure numbers are left in the German, but the translation has been updated to reflect their new locations.

> Auf Seite 247 des ersten Bandes unserer Zeitschrift beschreibt Jeitteles einen 209 Mm. langen Metacarpus vom Pferd, dessen dorsale Fläche einen spiegelglatt abgeschliffenen Streifen trägt, während die palmare an beiden Enden 20 Millim. grosse unregelmässige Löcher besitzt; er hielt ihn damals für einen Schlittschuh eines Knaben, ohne auf die bedeutung der beiden Löcher einzugehen.
>   Um dieselbe Zeit erhielt ich von meinem Freunde, Dr. Heinrich Kraus, der gegenwärtig in seiner Siebenbürgischen Heimat eine sehr segensreiche chirurgische Thätigkeit entfaltet, mehrere Metatarsusknochen vom Pferd, deren dorsale Fläche in derselben Weise abgeschliffen war, wie beim Olmützer Knochen. Ueber ihre Auffassung als Schlittschuhe konnte kein Zweifel bestehen, mein Freund war in seiner Kindheit auf ihnen gelaufen, und konnte mir auch über ihre Herstellung und ihren Gebrauch sehr werthvollen Aufschluss geben. Die Schuljungen sammeln für solchen Zweck brauchbare Knochen—also in erster Linie Metacarpus und Metatarsus von Pferd und Rind—und lassen ein ausgewähltes Paar vom Fleischhauer des Ortes oder sonst einem kundigen Mann mit scharfer Axt so zuhauen, dass die dorsalen Flächen geebnet und auf von den palmaren etwa zu sehr vorspringende Leisten entfernt werden; die weitere Ausarbeitung, den "höheren Schliff" besorgen sie sich dann selbst in der Weise, dass sie ihre ärmeren oder physisch und intelektuell zurückstehenden Mitschüler durch Geld oder Schläge veranlassen, dieselben durch unermüdlichen "Gebrauch" zu glätten. Ist das einmal erreicht, dann ist der Schlittschuh auch bereits fertig; ohne weitere Befestigung wird er nun so unter die Sohle gelegt, dass die Ferse oder der Absatz des Stiefels frei bleibt, dann wird ein langer Stock, der mit einem eisernen Stachel versehen ist, mit beiden Händen ergriffen und zwischen die Beine gesetzt—und in halber Kniebeuge beginnt der Lauf, der an Schnelle und Annehmlichkeit dem auf unseren besten Halifax nichts nachgeben soll, wenn er ihm auch kaum in aesthetischer Beziehung zu vergleichen sein dürfte.
>   Fig. 3. unserer I. Tafel [Figure 56] gibt eine Darstellung eines solchen recenten Siebenbürgischen Schlittschuhes und in Fig. 4. [Figure 17] habe ich als höchst interessantes Beispiel deutscher Hausindustrie den oberen Theil eines 130 Cm. langen Stockes abbilden lassen, den ich gleichfalls der Güte meines Freundes Kraus verdanke und der vorläufig noch meiner Sammlung angehört. Schnitzmesser und ein hohler Meisel sind die einzigen Werkzeuge gewesen, die zu der zierlichen und geschmackvollen Arbeit gedient haben. Die Ornamente selbst und besonders das runenhafte Zeichen in ihrer Mitte, das wohl die Marke des Künstlers oder Eigenthümers sein wird, machen jenen eminent prähistorischen Eindruck, den wir von den Producten der Hausindustrie abgelegener, von den Wogen der Cultur noch wenig bespülter Länder zu empfangen gewohnt sind. Das untere Ende des Stockes, der die Zeichen Jahrzehnte langen Gebrauches aufweist, ist mit einem eisernen Stachel und einem schmalen eisernen Bande versehen, das in fünf Spiraltouren schützend um das selbe herumgeführt ist.
>   Löcher, ähnlich denen des Olmützer, fehlen den recenten Siebenbürger Schlittknochen gänzlich, und es ist auch schwer einzusehen, wie derartige Löcher zur Befestigung des Knochens am Fuss gedient haben mögen; es schien mir schon von Anfang an wahrscheinlicher, dass der Knochen von Olmütz eher als Beschlag eines Schlittens aufzufassen sei, eine Ansicht, deren Richtigkeit mir sofort klar wurde, als mein College Dr. Soltan mir gelegentlich mittheilte, dass speciell [sic] in der Nähe von Olmütz noch heute kleine Schlitten unter dem Namen "Palakrlatan" in Gebrauch seien, die nicht auf eisenbeschlagenen Leisten, sondern auf einem Brette stünden, dem Röhrenknochen von Pferden und Rindern aufgenagelt sind.[12]

(On page 247 of the first volume of our journal, Jeitteles describes a 209-mm-long horse metacarpus whose dorsal surface bears a mirror-smooth, striated strip, while the palmar side exhibits irregular holes 20 mm wide at both ends; he thinks it is a skate belonging to a young man, without getting into the meaning of the two holes.
  At the same time, I received from my friend, Dr. Heinrich Kraus, who is currently employed

in very beneficial surgical activity in his Transylvanian homeland, some horse metatarsus bones whose dorsal surfaces were ground in the same way as the bone from Olomouc. As to their identification as skates there could be no doubt; my friend skated on them in his childhood and was able to give me very important information about their production and their use. The schoolboys collect for such a purpose usable bones—the first choice being metacarpi and metatarsi from horses and cattle—and have a selected pair trimmed by the butcher of the area or by a skillful man with a sharp axe so that the dorsal surfaces are even and any bits that protrude too far are removed from the palmar [side]; the further preparation, the "higher polish," they procure themselves by inducing their poorer or physically and intellectually inferior classmates by means of money or violence to smooth through untiring "use." Once that is complete, the skate is finished; without additional fastening it is laid under the sole [of the foot], so that the heel or the heel of the boot remains free, then a long pole with an iron spike is held with both hands and placed between the legs—and in a half squat the run begins, to which in speed and pleasantness our best Halifax gives nothing up, because there is hardly anything to compare in the aesthetic realm.

Figure 56 depicts a recent Transylvanian skate and in figure 17 I have reproduced, as a very interesting example of German home-industry, the upper part of a 130 cm long staff, for which I thank the kindness of my friend Kraus; it temporarily belongs to my collection. A carving knife and a concave chisel were the only tools that served in the dainty and tasteful work. The ornament itself and especially the runic sign in the middle, which is probably the mark of the artist or the owner, make the eminently prehistoric impression, that we are accustomed to receiving from the products of home industry of remote lands even less touched by the waves of culture. The lower end of the staff, which shows signs of decades-long use, is held by an iron spike and a narrow iron band, which is wrapped around in five protective spiral turns.

Holes like those on the Olomouc skate are completely missing from the recent Transylvanian skates, and it is hard to determine how the holes may have served to fasten the bones to the feet; it has seemed to me from the beginning more likely, that the bone from Olomouc was previously a part of a sled. This idea's correctness became clear to me when my colleague Dr. Soltan incidentally told me that, especially near Olomouc, even today, little sleds called "Palakrlatan" are in use, which stand not on iron-shod runners but rather on a board to which long-bones of horses and cattle are nailed.)

The next few recollections move out of the learned societies of Germany and Austria and into Hungary and Romania. In 1885, Otto Herman, author of one of the founding papers on bone skates, published a paper in Hungarian. In it, he reported that bone skates were still used among poor children in Hungary at that time.

"Bone skates exist even today. The children of the poor in Szeged, Zenta, Doroszló and other places merrily slide on them. These specimens usually show no sign of drilling...."[13]

The next two anonymous descriptions of bone skates in Romania were collected by Sandór Petényi. The first informer describes the popularity of bone skates among Székelys, most of whom lived in part of Romania. The second is a child's description of making bone skates; its details make it one of the more important first-hand accounts. This child is from Gyergyóújfalu (Suseni).

Figure 56: A nineteenth-century bone skate from Transylvania (F. von Luschan, "Mitteilungen aus dem Museum der Gesellschaft," 142).

A gilics leginkább ló vagy (ritkábban) marha első lábcsontjából alkalmatosan készitett csontdarab, melyet korcsolya gyanánt használ s igen kedves játékszernek ismer a gyergyói székelyfiú. Csákly-abottal hajtja magát, amely legtöbb esetben fenyőfából készül: alsó végéből 4–5 cm hosszú jégseg áll ki. Arra szolgál, hogy a lábak között a jégbe ütve az erős rugaszkodáshoz támasztékul szolgáljon.

[...]

Mikor megszáradt a két essö lábszárcsont, hát vettem a fészit osztán elöl-hátul már itt a talpánál lecsaptam, hogy legyen srégen s osztán itt elöl s hátul, ahova a lábujkóm s a sarkam jő—hát itt ahol egy kicsit kiemelkedik, mind egy gerinc, átalfúrtam, mer osztan lássa, ebbe a két lyukba van fűzve az ispárga na s ... lássa az úrfi, amik ispárga elöl van azt a csidma vaj már a bocskor orrára húzom fel s amik hátul van, evval a bokámnál kötöm keresztül...—de amik becsületes gilics, hogy egy kicsit görbül felfelé azt nem is kell erősen megkötni.[14]

(Runners were a worked piece of bone made mostly from the front metapodials of horse or sometimes (less frequently) cattle, which were used as skates and greatly liked by Sekler boys. He pushes himself forward with a grappling rod which is made from pinewood in most cases. A 4 to 5 cm long ice nail is mounted on the bottom of such a rod in order to provide a firm hold when stabbed into the ice between the two legs.

[...]

When the two frontal leg bones dried, I took an axe and hacked it in front and back to form a sole, then I hacked it obliquely in front and back where my toes and heel would fit—and here, where it is a bit elevated like a backbone, I drilled the bone through, because you see, this is where the string is threaded through, well ... you see, young mister, the string up front is pulled over my boots or moccasin while the ones in the back are tied across my ankles ... however, when the runner is really good and bent a little bit upwards, it does not even have to be tied.[15])

This section ends with a tantalizing hint that bone skates were still in use after World War II. This is one of the most recent pieces of evidence for bone skates, and the most recent for central Europe.

Children in Eastern Hungary in the 1940s were using such improvised single skates even after World War II.[16]

A.3. Great Britain

I have found three accounts of bone skates being used in England around the end of the nineteenth century: one from Birmingham, one from Tickhill in South Yorkshire, and one from Brandon Creek in Norfolk.

I am informed by Professor H. A. Miers that bone-skates were in common use in Birmingham about the year 1881, when he saw them, and may be so still. They were tied on to the feet.[17]

A Mr. Hartshorn of Tickhill, South Yorkshire, recalled using skates made from rib bones in his youth, about the turn of the century.[18]

When the rivers were frozen a 90-year-old man from Soham Fen used to take the opportunity to visit his out-lying relations. One day he arrived at his niece's—she lived next door to us—having skated from Soham Lode into the Ouse, then on to Denver Sluice and back through Littleport. After a huge meal with his niece and a short rest he called to borrow from my father a file to smooth down a chip in one of his skates. These skates, which the old man boasted were two hundred years old, were simply a pair of sheep bones, beautifully shaped and polished. Holes had been bored through them for the leather thongs which he passed crisscross style over his boots and up his legs.[19]

A.4. The Northeast

The four accounts of bone skates in the Northeast are short but descriptive. First is the discovery Henry Balfour made during a trip to Finland:

> In Finland, too, precisely similar bone runner-skates (bein läggar) were till very recently in use, especially near Åbo. Specimens from Korpo and Houtskär are in the Ethnographical Museum at Helsingfors.... They also were used with the assistance of an iron-pointed staff.[20]

The next account is a graphic description of what could go wrong on bone skates provided by Antti Juvel of Finland in 1915. At the time, he was 75 years old.

> ...useasti oli maa silti mustana silmissä, kun niilä kaatui ... jos oli sula edessä, ei auttanut muuta kuin mennä siihen, vauhdista oli vaarallista jäällekään pitkälleen heittäytyä.[21]
>
> (...if there was open water ahead, you had no choice but going into it, for it would be too dangerous to fall on the ice in such a great speed and turning was impossible.[22])

Mechanical details are given in the following description of skating in the 1920s and 1930s in Stora and Lilla Rågö, Estonia.

> På vintern åker man kälke så gott det går i de små backarna, och när isen ligger blank över det grunda sundet, komma is-läggia fram, och skridskoåkningen börjar. De göras av *äik-knokar*, dvs hästens mellanfotsben, "underbenet." Den ena "sidan" avhugges för att erhålla såbred yta som möjligt för foten, den motsatta sidan slipas plan och glatt. I bägge ändarna borras hål och genom dem trädes de snören, med vilka isläggen bindes vid foten. Då man nu ska *skrill*, går det naturligtvis ej att förfara så, som då man har stålskridskor. Man använder sig istället av en järnskodd pik, *is-kölva*. Denna håller man med båda händerna, järndubben hugges i isen och pressas bakåt mellan benen. Det kan bli en riktigt god fart, beroende på hur kvick man är med iskolven. Att vända går sämre, då måste farten bromsas in och piken tas till hjälp. Några isprinsessor lära aldrig kunna utbildas påläggarna, men barnen ha roligt, och det är i det här fallet huvudsaken.[23]
>
> (In the winter, one goes on low slopes as best as one can, and when the ice lies smoothly on the shallow strait, the bone skates come out, and the ice skating begins. They are made from *äik-knokar*, that is, horse metapodia or "lower legs." One "side" is shaped to provide a place for the foot that is as broad as possible, the opposite side is sanded flat and smooth. At both ends, holes are drilled and the cords that bind the bone skates to the foot are stretched through them. Then when one would *skrill*, moving forward naturally does not work as it does when one has steel skates. Instead, one pushes oneself with an iron-shod pick, an *is-kölva*. One holds this with both hands, the iron end cuts into the ice and is pressed back between the legs. It can be a very good run, depending on how quick one is with the *is-kölva*. To reverse does not work as well; then the run must be stopped and the staff used for help. No ice princesses can ever be made on bone skates, but the children have fun, and that is, in this case, the main thing.)

This section ends with Alexander von Middendorff's trip to Siberia in the first half of the nineteenth century. While there, he observed that

> Krustet der Schnee im Frühjahre so bindet der Tunguse ein paar Knochen-Plättchen unter seinen Schneeschuh, damit dieser besser gleite und doch nicht seitlich ausweiche.[24]
>
> (When the snow crusts in the spring, a Tungus binds a pair of small bone plates under his ski, so that it glides better and does not slide sideways.)

A.5. Scandinavia

Most of the nineteenth-century Scandinavian descriptions of bone skates are from Sweden. This reflects the state of scholarship on the Scandinavian bone skates: the largest

A.5. Scandinavia

and most detailed study of bone skates to date, the one completed by Edberg and Karlsson, is based on skates from Sweden. In this study, they also collect modern references to bone skates. The first three spent much of a century languishing in Swedish archives before seeing publication. In the oldest of them, Nils Johan Petersson (born in Folaboda in 1833) and Matilda Petersson (born in Väckelsång in 1844) of Folaboda, Urshults parish, Småland, Sweden, described old-fashioned skates to their daughter, Christina Marcusson, in 1929.

> ...i gamla tider, innan skridskorna kom i bruk, fanns en sorts andra 'skridskor,' om man så får kalla dem. De hetade 'isläggar' och voro gjorda av ben, och det var så ställt att de kunde bindas fast under foten. De voro gjorda av det längsta benet, som är i nötkreaturens ben. Dessa isläggar voro mycket hala så att det var värst att stanna om så skulle behövas. De, som hade sådana, kunde ju ej taga några skär utan gagnade en broddakäpp att föra sig framåt med. En broddakäpp var en stark käpp, skodd med en brodd i nedra ändan. Bland de unga männen var skridskosporten mycket älskad förr. Var det vacker skridskois färdades såväl män som unga flickor helst över isen, dåde skulle till kyrkan. Och rätt hastigt gick det också, så att somliga av de unga männen, som tränat sig lite mera nog med framgång kunnat tävla i en hastighetsåkning.[25]

> (In the old days, before [metal-bladed] skates came into use, there was another type of "skate," if you can call them that. They were called "*isläggar*" and were made from bone, and were such that they could be bound fast under the feet. They were made from the longest bones, which are cattle bones. These ice-legs were very slippery, so that it was worst to stand on them, should it be needed. Those who had them could not take a stroke but rather used a *broddakäpp* to move themselves forward. A *broddakäpp* was a strong staff shod with a point at the bottom. Among the young men, skating was very popular. When there was good skating ice, both men and young girls traveled across the ice when they went to church. And it also went very fast, so that some of the young men, the ones who practiced a little more, could successfully compete in a race.)

The next is from a 1931 interview with Petter Andersson (born in 1843) of Salberga, Gräsgård parish, Sweden.

> Ja då var jag väl inte mer än en åtta äller nio år, när jag började med isläggar.—Det var framläggra av ett nöt. Och så ispik. Det skulle smeden göra. Vi brukade vara väster om i hagarna fram till kyrkan. Om söndagsäftermiddagen var vi på isen.
> (Intervjuaren: Var det några barn i byn som inte fick gåpåisen?)
> Å nä det vet jag inte, de var med allihop.[26]

> (Yes, I was not more than eight or nine years old when I began with bone skates. They were the front legs of a cow. And an ice-pick. The smith would make that. We used to be west of the church in the pastures. On Sunday afternoons we were on the ice.
> Interviewer: Were there any children in the village who did not go on the ice?
> I do not know [of any], they were for everyone.)

The third of the newly-published archival texts was recorded in 1932 by Atle Pettersson (born in 1853) of Sneckered, Nössemark, Dalsland, Sweden.

> I forna tider gick de på isen med "isläggar." De gjordes av hästens smalben. Det sades att en person, som gick ned sig på isen och drunknade med ett par "isläggar" på fötterna inte fick "körgål." Han fick inte kristlig begravning.[27]

> (In the old days, they went across the ice with "*isläggar*." They were made from the shinbones of a horse. It is said that a person who went down on the ice and drowned with a pair of "*isläggar*" on the feet did not receive a "*körgål*." He did not receive a Christian burial.)

The next few descriptions were published alongside archaeological artifacts and in collections of reminiscences of the old days on Gotland, an island off the coast of Sweden. The first is by Fredrik Nordin (born in 1852), an archaeologist who was able to comment knowledgably on a bone skate found in an excavation on Gotland because he had used bone skates in his youth, which was helpful to Klindt-Jensen.[28]

> Sådana begagnas ännu på ön, åtminstone har jag som barn sett dem användas och sjelf haft ett par dylika. De göras af skenbenet af en häst på det enkla sätt, att benet genomborras med tvenne hål. Genom dessa trädas de snören, med hvilka man fastbinder isläggen vid foten. Med dylika fortskaffningsmedel och med tillhjelp af en isstaf kan man åka med stor hastighet på glatt is.[29]
>
> (They are still used on the island; at least, as a child I saw them in use and had a pair of them. They are easily made from the shin-bones of a horse; the bone is bored with two holes. A string is put through these, with which the bone skates are tied to the feet. With such enhancement and with the help of an ice-staff, one can go very quickly on glassy ice.)

The details are filled in by the longer description of bone skates being used on Gotland in modern times that P. A. Säve (1811–1887) included in his collection of traditional games, published in 1948.

> **101.** *Att ro is-laggar.*
>
> Detta är deremot allmänt på landet, och öfvas på glansis på tillfrusna *bryor* och myrar.—Man står då på tvenne häst-läggar, som man på undra sidan på slip-sten gjort alldeles glatta, och på öftra sidan något utgräft, så att fötterna kunna der stå något fast: men dessa *is-läggar* äro ej på något sätt bundna eller fästade vid fötterna, utan dessa stå helt lösa på dem. Fart tar man derigenom, att man med en stake med brodd uti ändan och som kallas *is-käpp*, *is-spett* eller *is-pigg* hugger i isen mellan fötterna. När isen är rätt glansk, går det i en brinnande fart; det farligaste för en ovan är att alldeles spräcka opp sig, ty isläggarna ränna gerna åt sidorna, och det att man i farten omöjligt kan vika åt sidan eller göra en lof, om man ock såge den djupaste vak för sig. Det enda medel i detta sednare fall är sätta iskäppen mellan benen, sätta sig vackert ned på den och låta den rispa uti isen och att sålunda något hejda farten; men då skall man ej vara för nära faran.
>
> [I marg:en står:] Arsboar (karlar och qvinnor) gingo öfver Bästeträsk på is-läggar till Fleringe kyrka på 11 minuter, och gömde sedan is-läggarna i enes-buskarne till återfärden.[30]
>
> **(101.** *To go on ice-legs.*
>
> This, however, is common in the country, and practiced on the smooth surfaces of frozen springs and marshes. One stands on two horse-legs, the underside of which has been thoroughly smoothed with a grindstone, and the upper side often somewhat carved, so that the feet can stand fast there. These ice-legs are not bound or fastened to the feet but rather stay loose on them. One builds speed by placing a stake with a point at the end that is called an *is-kapp*, *is-spett*, or *is-pigg* in the ice between the feet. When the ice is very glassy, it goes at a burning speed; the most dangerous part of the above is splitting apart: the ice-legs easily run to the sides. And when sliding, one cannot turn aside or change course, if one sees a deep hole in front. The only thing to do in this case is to set the pole between the legs, lean back on it, and let it scratch the ice to slightly arrest the motion; but then, one should not go too close to the danger.
>
> [Marginal note:] People living in *Ars* (men and women) went across Lake Bästetrask on ice-legs to the Flering church in 11 minutes, and then hid the ice-legs in the bushes for the return trip.)

Further evidence for the modern use of bone skates on Gotland is provided by Munro's account of his visit to a museum there.

> I also saw some bone skates in the public museum at Visby, in the island of Gotland, in regard to which the curator remarked that he himself in his earlier years had actually used similar skates.[31]

Despite its distance from Berlin, Scandinavia was not neglected by the Berliner Gesellschaft für Anthropologie, Ethnologie und Urgeschichte. On June 10, 1871, A. Bastian read excerpts from a letter from Dr. Hans Hildebrand of Stockholm that included what his father remembered about bone skates.

Mein Vater hat in seiner Jugendzeit oft von Schlittknochen gehört, sie aber nie gesehen. Sie waren an den Enden durchbohrt und wurden, sagte man, durch Schnüre am Fusse befestigt. Sie sollten ausgezeichnet sein, wenn man schnell laufen wollte; mit ihnen auszuweichen war unmöglich. Wir haben im Museum einige Knochen, die offenbar in angegebener Weise benutzt sind. Sie sind in Süd-Schweden gefunden.[32]

(My father often heard about bone skates in his youth, but never saw them. They were drilled through at the ends and were, it was said, fastened to the feet with cords. They were supposed to be excellent, when one wanted to go quickly; with them it was not possible to turn. In the museum, we have some bones that were probably used in this manner. They were found in southern Sweden.)

In the final reference from Sweden, Gunnar Hyltén-Cavallius (1818–1899) describes split bones, the special Scandinavian bone skates with the palmar side completely removed so that the medullary cavity is exposed.

Om julen, när det var glansk is, rände man härvid på is-läggor eller tvänne kluvna och inunder glättade läggben på vilka man stående sköt sig fram med stålskodd staf, kallad is-brodd eller brodda-käpp. Is-läggor hafva i Värend varit brukade från äldsta tid till för en mans-ålder tillbaka.[33]

(At Christmas, when there was glassy ice, people ran on bone skates or two legbones that have been split and smoothed on the bottom on which they, standing, shot themselves forward with a steel-shod staff called an *is-brodd* or *brodda-käpp*. Bone skates were used in Värend from the earliest times until a generation ago.)

The next two descriptions are by Englishmen visiting Iceland. The figure numbers have been updated to reflect their locations in this book. First, Mr. A. Heneage Cocks describes the bone skates he bought from a boy at what he calls Kalderhohdi Farm on the Log River in southwest Iceland. This is probably Kaldárhöfði Farm in Grímsnes- og Grafningshreppur, Iceland. Frederick W. W. Howell visited it in about 1900. A photograph he took survives in the collection of Cornell University Library and is reproduced in figure 57.

I noticed the bone skates hanging up in Kalderhohdi Farm, on the Log River, S.W. Iceland, when putting up there in August 1878. I remember carefully concealing my feeling of excitement when I saw what they were, and the mutual satisfaction of the boy to whom they belonged and myself when they changed owners for the consideration of 20 öre (3d.) They are fastened to the foot by strings exactly as were the sandals formerly worn by ladies in England up to thirty or thirty-five years ago. Without having a pattern before me I may very likely make some mistake in detail, but I think the sketch (see [figure 13]) will explain it sufficiently. The string for fastening the skate to the foot is passed through the front hole under the toe and the two ends crossed over the lower part of the instep, and then inserted in opposite directions through the hole at the back under the heel, thence they pass upwards over the higher part of the instep and round the ankle, being tied in a knot in front. These skates are made from the metacarpal bones of the tiny Icelandic ox, not from those of the horse, as is most usually the case with the prehistoric examples.[34]

Second, Henry Balfour compares a modern Icelandic skate in his collection with Cocks's skate.

Not long since (in 1895), through the kindness of Mr. C. E. Peek, I received from Iceland a somewhat similar pair of skates, differing, however, from those obtained by Mr. Cocks, in being made of horse bones (entire *radii*), and also in the method of fixing to the boot, which shows a slight improvement. The runners are about 1 ft. 1 in. long and altogether larger than the usual bones employed for skates. They are bored transversely through from side to side, at some distance from either end [figure 15]. Through each hole is passed a cord, which is prevented from slipping through its hole by stop knots. The hinder cord is on either side of the bone tied in a small loop about 3 ins. from the stop-knot, and the ends of the cord are knotted through a band of leather,

Figure 57: Kaldárhöfði Farm, Iceland, in about 1900 (Frederick W. W. Howell, courtesy Division of Rare and Manuscript Collections, Cornell University Library, Ithaca, New York).

> which forms a heel strap. The foremost cord has its ends quite free. The boot rests on the palmar surface of the bone, and the ends of the foremost cord are crossed over the instep, passed through the loops on the hinder cord, which form an excellent "purchase," and after being drawn tight, are fastened together above the instep [bottom image]. This pair of skates has not been flattened by grinding, and not having been used, shows no ice polishing on the under surface. They were the work of an Icelander who had made and used such skates many a time in his youth, but he stated that they were now quite obsolete in his district.[35]

The section ends with two very recent sightings of bone skates being used in Iceland. Thorsteinn Einarsson actually saw them in use in about 1950, and Thordur Thomasson showed a pair that had been in use as recently as 1972 to Ulf Erik Hagberg.

> In about 1950, I saw teenagers using "ice-legs" and this type, with and without bindings; a boy, on bones without bindings, used a pole, and I saw a girl use a single ice-leg, using her free foot to push with from time to time, in between which she rested it on her other foot or moved it out sideways to keep her balance. I had difficulty grasping the skill involved here, and only appreciated it fully when I saw skate-boarding for the first time in 1979.[36]

> Thordur Thomasson, Lehrer und Museumsvorsteher in Skogar, Südisland, hat mir Schlittknochen gezeigt, die noch im Jahr 1972 im Gebrauch waren und bei denen der Läufer sich mit einem Stock vorwärts bewegte.[37]

> (Thordur Thomasson, teacher and museum docent in Skogar, southern Iceland, showed me bone skates that were in use in the year 1972 and with which the skater moved himself forward with a pole.)

Chapter Notes

Preface

1. J. Romilly Allen, "The Primitive Bone Skate," *The Reliquary and Illustrated Archaeologist* 2 (1896): 35.
2. Henry Balfour, "Notes on the Modern Use of Bone Skates," *The Reliquary and Illustrated Archaeologist* 4 (1898): 32.
3. Asbjørn E. Herteig, *Kongers havn og handels sete: Fra de arkeologiske undersøkelser på Bryggen i Bergen 1955–68* (Oslo: H. Aschehoug, 1969), 198.
4. Unless otherwise noted, all translations are my own.
5. Richard Hall, *Book of Viking Age York*, English Heritage (London: B. T. Batsford, 1994), 104.
6. Robert Munro, "Notes on Ancient Bone Skates," *Proceedings of the Society of Antiquaries of Scotland* 28 (1894): 197.
7. G. Herbert Fowler, *On the Outside Edge: Being Diversions in the History of Skating*, ed. B. A. Thurber (Evanston, IL: Skating History Press, 2018), 30.
8. Arthur MacGregor, "Bone Skates: A Review of the Evidence," *Archaeological Journal* 133 (1976): 67.
9. Examples of broad surveys of the evidence are MacGregor, "Bone Skates: A Review of the Evidence," and Hans Christian Küchelmann and Petar Zidarov, "Let's Skate Together! Skating on Bones in the Past and Today," in *From Hooves to Horns, from Mollusc to Mammoth: Manufacture and Use of Bone Artefacts from Prehistoric Times to the Present, Proceedings of the 4th Meeting of the ICAZ Worked Bone Research Group at Tallinn, 26th–31st of August 2003*, ed. Heidi Luik et al., Muinasaja Teadus 15 (Tallinn: Ajaloo Instituut, 2005), 425–445. Detailed analyses of skates from a particular area include Alice M. Choyke and László Bartosiewicz, "Skating with Horses: Continuity and Parallelism in Prehistoric Hungary," *Revue de Paléobiologie* spéc. 10 (2005): 317–326 and Rune Edberg and Johnny Karlsson, *Isläggar från Birka och Sigtuna. En undersökning av ett vikingatida och medeltida fyndmaterial*, Stockholm Archaeological Reports 43 (Stockholm: Institutionen för arkeologi och antikens kultur, Stockholms universitet, 2015).
10. Küchelmann and Zidarov, "Let's Skate Together! Skating on Bones in the Past and Today"; Hans Christian Küchelmann, *Bone Skates Database*, updated May 28, 2018, https://www.knochenarbeit.de/bone-skates-database/?lang=en.
11. Rune Edberg and Johnny Karlsson, "Bone Skates and Young People in Birka and Sigtuna," *Fornvännen* 111 (2016): 7–16.
12. William fitz Stephen's account of bone skates is part of his *Description of London*. A modern translation can be found in David C. Douglas and George W. Greenaway, eds., *English Historical Documents*, vol. 2: 1042–1189 (Oxford: Oxford University Press, 1968). It is recounted in R. Jones and W. E. Cormack, *A Treatise on Skating*, ed. B. A. Thurber (Evanston, IL: Skating History Press, 2017).
13. J. R. R. Tolkien, *The Lord of the Rings*, 50th anniversary ed., ed. Wayne G. Hammond and Christina Scull (Boston: Houghton Mifflin Harcourt, 2004), 1041.
14. Kevin Crossley-Holland, *Bracelet of Bones*, Viking Sagas 1 (New York: Quercus, 2014), 190–192.
15. Cressida Cowell, *How to Steal a Dragon's Sword: The Heroic Misadventures of Hiccup the Viking*, How to Train Your Dragon 9 (New York: Little, Brown, 2012), 168.
16. Joan Lennon, *Ice Road*, The Wickit Chronicles 3 (La Jolla, CA: Kane/Miller Book Publishers, 2008).
17. The professor's name, "Free-leg Change-of-edge," combines two terms specific to figure skating.
18. Edberg and Karlsson, "Bone Skates and Young People in Birka and Sigtuna."

Chapter 1

1. James Craigie Robertson, ed., *Materials for the History of Thomas Becket, Archbishop of Canterbury*, vol. 3 (London: Longman, 1877), 11–12.
2. Douglas and Greenaway, *English Historical Documents*, 961.
3. Edberg and Karlsson, "Bone Skates and Young People in Birka and Sigtuna."
4. Ulf Erik Hagberg, "Fundort und Fundgebiet der Modeln aus Torslunda," *Frühmittelalterliche Studien* 10 (1976): 330.
5. For experiments, see, for example, MacGregor, "Problems in the Interpretation of Microscopic Wear Patterns: The Evidence from Bone Skates," and Küchelmann and Zidarov, "Let's Skate Together! Skating on Bones in the Past and Today." Two reenactment groups that have used bone skates are Hurswic (see William R. Short, *Viking-Age Ice Skates*, Hurstwic, accessed December 15, 2018, http://www.hurstwic.com/history/articles/daily_living/text/ice_skates.htm) and the Dark Ages Re-Creation Company (see Steve, "Bone Skates or, Vikings on Ice," Dark Ages Re-Creation Company, updated August 26, 2017, http://www.darkcompany.ca/projects/skates/index.

php). Hagberg ("Isläggar och pälsskinn—en frusen kulturhistorisk rapsodi," 81) and Frans van Liere at Calvin College (described in Short, *Viking-Age Ice Skates*) have used bone skates in college courses.

6. Küchelmann and Zidarov, "Let's Skate Together! Skating on Bones in the Past and Today," 433.

7. Edberg and Karlsson, *Isläggar från Birka och Sigtuna. En undersökning av ett vikingatida och medeltida fyndmaterial*, 16.

8. Marie Jonasson Schmidt, email message to author, August 29, 2018.

9. Alice Marriott, "Indians on Horseback," in *Saynday's People* (New York: University of Nebraska Press, 1963), 102.

10. Stewart Culin, *Games of the North American Indians* (New York: AMS Press, 1907), 616–647.

11. Patricia D. Sutherland and Robert McGhee, *Vergessene Welten unter Schnee und Eis: Die Vorläufer der Eskimos vor 4000 Jahren. Begleitheft zur Ausstellung: Vergessene Welten unter Schnee und Eis, Übersee-Museum, 29 November 1998 bis 14. März 1999* (Bremen: Bremen Übersee-Museum, 1998), 8, cited in Küchelmann, *Bone Skates Database*.

12. MacGregor, "Bone Skates: A Review of the Evidence," 58.

13. Henry Balfour, "Sledges with Bone Runners in Modern Use," *The Reliquary and Illustrated Archaeologist* 4 (1898): 253–254.

14. Mary Mapes Dodge, *Hans Brinker or the Silver Skates: A Story of Life in Holland* (New York: James O'Kane, 1865).

15. Oxford University Press, *OED Online*, 2011, "skate," http://www.oed.com/.

16. Thomas Percy, trans., *Five Pieces of Runic Poetry, Translated from the Islandic Language* (London: R. and J. Dodsley, 1763), 80, 99.

17. MacGregor, "Problems in the Interpretation of Microscopic Wear Patterns: The Evidence from Bone Skates."

18. Küchelmann and Zidarov, "Let's Skate Together! Skating on Bones in the Past and Today"

19. MacGregor, "Bone Skates: A Review of the Evidence."

20. Toby F. Martin, *The Cruciform Brooch and Anglo-Saxon England* (Woodbridge: Boydell Press, 2015).

21. Edberg and Karlsson, *Isläggar från Birka och Sigtuna. En undersökning av ett vikingatida och medeltida fyndmaterial*.

22. Munro, "Notes on Ancient Bone Skates."

23. Otto Herman, "Knochenschlittschuh, Knochenkufe, Knochenkeitel: Ein Beitrag zur näheren Kenntnis der prähistorischen Langknochenfunde," *Mittheilungen der anthropologischen Gesellschaft in Wien* 32 (1902): 217–238.

24. Gösta Berg, "Isläggar och skridskor," *Fataburen*, 1943, 79–90; Gösta Berg, "The Origin and the Development of the Skis throughout the Ages," in *Finds of Skis from Prehistoric Time in Swedish Bogs and Marshes* (Stockholm: Generalstabens litografiska anstalts förlag, 1950), 9–64; Gösta Berg, "Skier und Schlittschuhe: Zwei nordische Fortbewegungsmittel," *Tribus: Jahrbuch des Linden-Museums Stuttgart* 2 (1952): 188–195; Gösta Berg, "Skates and Punt Sleds: Some Scandinavian Notes," in *Vriendenboek voor A. J. Bernet Kempers*, ed. P. J. Meertens and Hermanna W. M. Plettenburg (Arnhem: Nederlands Openluchtmuseum, 1971), 4–13.

25. MacGregor, "Problems in the Interpretation of Microscopic Wear Patterns: The Evidence from Bone Skates"; MacGregor, "Bone Skates: A Review of the Evidence."

26. Arthur MacGregor, *Bone, Antler, Ivory and Horn: The Technology of Skeletal Materials since the Roman Period* (1985; repr., London: Routledge, 2015).

27. Cornelia Becker, "Bemerkungen über Schlittknochen, Knochenkufen und ähnliche Artefakte, unter besonderer Berücksichtigung der Funde aus Berlin Spandau," in *Festschrift für Hans R. Stampfli*, ed. Jörg Schibler, J. Sedlmeier, and H. Spycher, Beiträge zur Archäologie, Anthropologie, Geologie und Paläontologie (Basel: Helbing and Lichtenhahn, 1990), 19–30.

28. Alice M. Choyke and Jörg Schibler, "Prehistoric Bone Tools and the Archaeozoological Perspective: Research in Central Europe," in *Bones as Tools: Current Methods and Interpretations in Worked Bone Studies*, ed. Christian Gates St-Pierre and Renee B. Walker, vol. 1622, British Archaeological Reports International Series (Oxford: Archaeopress, 2007), 51–65; Alice M. Choyke, "Bone Skates: Raw Material, Manufacturing and Use," special issue *Pannonia and Beyond: Studies in Honor of L. Barkoczi*, ed. A. Vaday. *Antaeus: Communicationes ex Instituto Archaeologico Academiae Scientiarum Hungaricae* 24 (1997/1998): 148–156, 651–654; Alice M. Choyke, "Worked Animal Bone at the Sarmatian Site of Gyoma 133," in *Cultural and Landscape Changes in South-East Hungary*, ed. Andrea H. Vaday, vol. 2 (Budapest: Archaeological Institute of the Hungarian Academy of Sciences, 1996), 307–322; Choyke and Bartosiewicz, "Skating with Horses: Continuity and Parallelism in Prehistoric Hungary."

29. Becker, "Bemerkungen über Schlittknochen, Knochenkufen und ähnliche Artefakte, unter besonderer Berücksichtigung der Funde aus Berlin Spandau."

Chapter 2

1. MacGregor, "Problems in the Interpretation of Microscopic Wear Patterns: The Evidence from Bone Skates."

2. Küchelmann and Zidarov, "Let's Skate Together! Skating on Bones in the Past and Today."

3. Hagberg, "Isläggar och pälsskinn—en frusen kulturhistorisk rapsodi."

4. Short, *Viking-Age Ice Skates*.

5. Magnus Magnusson, *Vikings!* (New York: Elsevier-Dutton, 1980), 18–19.

6. Fowler, *On the Outside Edge: Being Diversions in the History of Skating*, 39.

7. Federico Formenti and Alberto E. Minetti, "Human Locomotion on Ice: The Evolution of Ice-Skating Energetics through History," *Journal of Experimental Biology* 210 (2007): 1825–1833, doi:10.1242/jeb.002162.

8. Edberg and Karlsson, *Isläggar från Birka och Sigtuna. En undersökning av ett vikingatida och medeltida fyndmaterial*, 30–31.

9. Choyke and Schibler, "Prehistoric Bone Tools and the Archaeozoological Perspective: Research in Central Europe," 59.

10. The two horse metacarpi found in pit 41 at Gyoma 133 are quite similar, according to Choyke, "Worked Animal Bone at the Sarmatian Site of Gyoma 133," 311. It is possible that they are a pair.

11. A pair of bone skates consisting of two bones from the same horse found at Baumgarten an der March is described in Günther Karl Kunst, "Tierreste aus der frühmittelalterlichen Siedlung von Baumgarten an der March, Niederösterreich," *Mitteilungen der Prähistorischen Kommission* 68 (2009): 198.

12. David Parma et al., "Netradični materiál, neobvyklý předmet Opomijený segment kostěné industrie mladší doby bronzové (Non-Traditional Material and a Non-Traditional Object: A Neglected Sort of the Late Bronze Age Bone Industry)," *Archeologické rozhledy* 43, no. 1 (2011): fig. 5.1–5.2.

13. I. W. Cornwall, *Bones for the Archaeologist* (London: Phoenix House, 1956), 176.

14. MacGregor, *Bone, Antler, Ivory and Horn: The Technology of Skeletal Materials since the Roman Period*, 142.

15. Ibid.

16. Hagberg, "Isläggar och pälsskinn—en frusen kulturhistorisk rapsodi," 81.

17. MacGregor, *Bone, Antler, Ivory and Horn: The Technology of Skeletal Materials since the Roman Period*, 44.

18. Edberg and Karlsson, *Isläggar från Birka och Sigtuna. En undersökning av ett vikingatida och medeltida fyndmaterial*, 38.

19. Choyke, "Bone Skates: Raw Material, Manufacturing and Use," 151.

20. Edberg and Karlsson, "Bone Skates and Young People in Birka and Sigtuna," 14.

21. Brückner, "Über den heutigen Gebrauch von Schlittknochen in Schlesien." In Verhandlungen der Berliner Gesellschaft für Anthropologie, Ethnologie und Urgeschichte, *Zeitschrift für Ethnologie* 4 (1872): 42.

22. This particular skate is a fragment of a horse metacarpus found in pit 41.

23. Küchelmann and Zidarov, "Let's Skate Together! Skating on Bones in the Past and Today," 433.

24. Marie Jonasson Schmidt, email message to author, August 31, 2018.

25. Edberg and Karlsson, *Isläggar från Birka och Sigtuna. En undersökning av ett vikingatida och medeltida fyndmaterial*, 36–37.

26. Brückner, "Über den heutigen Gebrauch von Schlittknochen in Schlesien," 42.

27. Marie Jonasson Schmidt, email message to author, August 31, 2018.

28. Short, *Viking-Age Ice Skates*.

29. Küchelmann and Zidarov, "Let's Skate Together! Skating on Bones in the Past and Today."

30. Short, *Viking-Age Ice Skates*.

31. MacGregor, *Bone, Antler, Ivory and Horn: The Technology of Skeletal Materials since the Roman Period*, 30.

32. Formenti and Minetti, "Human Locomotion on Ice: The Evolution of Ice-Skating Energetics through History."

33. Latin from Olaus Magnus, *Historia de gentibus septentrionalibus* (Rome: J. M. de Viottis, 1555), 41–42, translation from Olaus Magnus, *Description of the Northern Peoples*, ed. P. G. Foote, trans. Peter Fisher and Humphrey Higgens (London: Hakluyt Society, 1996), 58.

34. Magnus, *Historia de gentibus septentrionalibus*, 42.

35. Magnus, *Description of the Northern Peoples*, 58.

36. Choyke and Bartosiewicz, "Skating with Horses: Continuity and Parallelism in Prehistoric Hungary," 322.

37. Magnus Magnusson, *Vikings!*, 18–19.

38. Georg E. Fantner et al., "Influence of the Degradation of the Organic Matrix on the Microscopic Fracture Behavior of Trabecular Bone," *Bone* 35 (2004): 1013–1022, doi:10.1016/j.bone.2004.05.027.

39. Choyke, "Worked Animal Bone at the Sarmatian Site of Gyoma 133," 310.

40. MacGregor, *Bone, Antler, Ivory and Horn: The Technology of Skeletal Materials since the Roman Period*, 64.

41. Sándor Petényi, *Games and Toys in Medieval and Early Modern Hungary*, Studia archaeologica mediae recentisque aevorum Universitatis Scientiarum de Rolando Eötvös nominatae 1 (Krems: Medium Aevum Quotidianum, 1994), 111–112.

42. Küchelmann and Zidarov, "Let's Skate Together! Skating on Bones in the Past and Today," 427.

43. F. von Luschan, "Mitteilungen aus dem Museum der Gesellschaft," *Mitteilungen der anthropologischen Gesellschaft in Wien* 6 (1876): 146.

44. Brückner, "Über den heutigen Gebrauch von Schlittknochen in Schlesien," 42.

45. For skates shaped with a knife, see Nicola S. H. Rogers, *Anglian and Other Finds from Fishergate*, The Archaeology of York 17: The Small Finds, fasc. 9 (York: York Archaeological Trust for Excavation and Research, 1993), 1406 and Karl Hucke, "Frühgeschichtliche Geweihund Knochengeräte von der Insel Olsborg im Großen Plöner See in Holstein," *Zeitschrift für Morphologie und Anthropologie* 44, nos. 1–2 (1952): 113–114. Choyke and Bartosiewicz, "Skating with Horses: Continuity and Parallelism in Prehistoric Hungary," 322 describes skates made with stone tools and axes.

46. MacGregor, "Bone Skates: A Review of the Evidence," 57.

47. Steve, "Bone Skates or, Vikings on Ice."

48. Küchelmann and Zidarov, "Let's Skate Together! Skating on Bones in the Past and Today."

49. H. W. Jacobi, *De Nederlandse Glissen*, student project, University of Amsterdam, June 1976, 11–12.

50. Choyke, "Bone Skates: Raw Material, Manufacturing and Use."

51. Ibid., 151, 652–653.

52. Short, *Viking-Age Ice Skates*.

53. Marie Jonasson Schmidt, email message to author, August 31, 2018.

54. Berg, "Skates and Punt Sleds: Some Scandinavian Notes," 5.

55. Brückner, "Über den heutigen Gebrauch von Schlittknochen in Schlesien," 42.

56. Anna Roes, *Bone and Antler Objects from the Frisian Terp-Mounds* (Haarlem: H. D. Tjeenk Willink and Zoon N.V., 1963), 59.

57. Clifford D. Long, "Excavations in the Medieval City of Trondheim, Norway," *Medieval Archaeology* 19 (1975): 25.

58. Marie Jonasson Schmidt, email message to author, August 31, 2018.

59. J. S. Kovacs, "Hogyan gilicseznek Gyergyóban," *Erdély*, 1908, 18, quoted in Petényi, *Games and Toys in Medieval and Early Modern Hungary*, 111.

60. Ibid., 112.

61. Thunig, "Über Schlittknochen und Gräbeurnen." In Verhandlungen der Berliner Gesellschaft für Anthropologie, Ethnologie und Urgeschichte, *Zeitschrift für Ethnologie* 4 (1872): 280.

62. Roes, *Bone and Antler Objects from the Frisian Terp-Mounds*, 59.
63. *Ibid.*
64. Küchelmann and Zidarov, "Let's Skate Together! Skating on Bones in the Past and Today," 435.
65. Herman, "Knochenschlittschuh, Knochenkufe, Knochenkeitel: Ein Beitrag zur näheren Kenntnis der prähistorischen Langknochenfunde," 221.
66. Edberg and Karlsson, *Isläggar från Birka och Sigtuna. En undersökning av ett vikingatida och medeltida fyndmaterial*, 42.
67. Küchelmann and Zidarov, "Let's Skate Together! Skating on Bones in the Past and Today," 434.
68. *Ibid.*, 435.
69. Jacobi, *De Nederlandse Glissen*, 11.
70. Küchelmann and Zidarov, "Let's Skate Together! Skating on Bones in the Past and Today," 428.
71. Herman, "Knochenschlittschuh, Knochenkufe, Knochenkeitel: Ein Beitrag zur näheren Kenntnis der prähistorischen Langknochenfunde," 220, fig. 123, 124.
72. Allen, "The Primitive Bone Skate," 32.
73. Robertson, *Materials for the History of Thomas Becket, Archbishop of Canterbury*, 11.
74. Douglas and Greenaway, *English Historical Documents*, 961.
75. Küchelmann and Zidarov, "Let's Skate Together! Skating on Bones in the Past and Today," 436.
76. Marie Jonasson Schmidt, email message to author, August 31, 2018.
77. Short, *Viking-Age Ice Skates*.
78. Fowler, *On the Outside Edge: Being Diversions in the History of Skating*, fig. 2.
79. Quoted in Short, *Viking-Age Ice Skates*.
80. Kovacs, "Hogyan gilicseznek Gyergyóban," 18, quoted in Petényi, *Games and Toys in Medieval and Early Modern Hungary*, 111.
81. *Ibid.*, 112.
82. Formenti and Minetti, "Human Locomotion on Ice: The Evolution of Ice-Skating Energetics through History," 1827.
83. Ole Klindt-Jensen, "Economic and Daily Life at Vallhagar," in *Vallhagar: A Migration Period Settlement on Gotland, Sweden*, ed. Mårten Stenberger and Ole Klindt-Jensen, vol. 2 (Copenhagen: Einar Munksgaards Forlag, 1955), 858.
84. *Ibid.*, 857.
85. Formenti and Minetti, "Human Locomotion on Ice: The Evolution of Ice-Skating Energetics through History," 1827.
86. MacGregor, "Bone Skates: A Review of the Evidence," 59.
87. Küchelmann and Zidarov, "Let's Skate Together! Skating on Bones in the Past and Today," 436.
88. Jacobi, *De Nederlandse Glissen*, 6.
89. Latin from Robertson, *Materials for the History of Thomas Becket, Archbishop of Canterbury*, 11; translation from Douglas and Greenaway, *English Historical Documents*, 961.
90. See, for example, Lauwerier and Van Heeringen, "Skates and Prickers from the Circular Fortressof Oost-Souburg, the Netherlands (AD 900–975)" and Becker, "Bone Points: No Longer a Mystery? Evidence from the Slavic Urban Fortification of Berlin-Spandau."
91. Berg, "Skates and Punt Sleds: Some Scandinavian Notes," 6.
92. See, for example, Layard, "Bone Skates and Skating Stakes," 74; Library Committee of the Corporation of the City of London *Catalogue of the Collection of London Antiquities in the Guildhall Museum*, 154; and, more recently, Porter, "Fen Skating," 43; Lauwerier and Van Heeringen, "Skates and Prickers from the Circular Fortress of Oost-Souburg, the Netherlands (AD 900–975)," 124–125; and Choyke and Bartosiewicz, "Skating with Horses: Continuity and Parallelism in Prehistoric Hungary," 324.
93. MacGregor, *Bone, Antler, Ivory and Horn: The Technology of Skeletal Materials since the Roman Period*, 174.
94. Küchelmann and Zidarov, "Let's Skate Together! Skating on Bones in the Past and Today," 439.
95. Lauwerier and Van Heeringen, "Skates and Prickers from the Circular Fortress of Oost-Souburg, the Netherlands (AD 900–975)," 124–125.
96. Becker, "Bone Points: No Longer a Mystery? Evidence from the Slavic Urban Fortification of Berlin-Spandau."
97. MacGregor, *Bone, Antler, Ivory and Horn: The Technology of Skeletal Materials since the Roman Period*, 175.
98. Berg, "Skates and Punt Sleds: Some Scandinavian Notes," 6.
99. Von Luschan, "Mitteilungen aus dem Museum der Gesellschaft," 146.
100. *Ibid.*
101. Herman, "Knochenschlittschuh, Knochenkufe, Knochenkeitel: Ein Beitrag zur näheren Kenntnis der prähistorischen Langknochenfunde."
102. P. Söderbäck, *Rågöborna* (Stockholm: P. A. Norstedt, 1940).
103. Josef Markwart, "Ein arabischer Bericht über die arktischen (uralischen) Länder aus dem 10. Jahrhundert," *Hungarische Jahrbücher* 4 (1924): 261–334.
104. V. Minorsky, *Sharaf Al-Zaman Zamān Ṭāhir Marvazī on China, the Turks, and India* (London: The Royal Asiatic Society, 1942), 34.
105. Küchelmann and Zidarov, "Let's Skate Together! Skating on Bones in the Past and Today," 439.
106. *Ibid.*, 437.
107. Herman, "Knochenschlittschuh, Knochenkufe, Knochenkeitel: Ein Beitrag zur näheren Kenntnis der prähistorischen Langknochenfunde," 220–221.
108. Küchelmann and Zidarov, "Let's Skate Together! Skating on Bones in the Past and Today," 440.
109. *Ibid.*, 436.
110. Latin from Magnus, *Historia de gentibus septentrionalibus*, 41–42, translation from Magnus, *Description of the Northern Peoples*, 58.
111. P. A. Säve, *Svenska lekar*, ed. Herbert Gustavson, vol. 1: Gotländska lekar (Uppsala: Almqvist och Wicksells Boktryckeri AB, 1948), 77.
112. Küchelmann and Zidarov, "Let's Skate Together! Skating on Bones in the Past and Today," 439–440.
113. Jacobi, *De Nederlandse Glissen*, 11–12.
114. J. Katajisto, "Turun Kaupunkialuen luuluistimet: tarkasteltuna osteologiselta ja historialliseltakannalta" (Unpublished manuscript, University of Turku, Turku, 2002), 4, translated by Auli Touronen, Turku University, Finland, and quoted in Küchelmann and Zidarov, "Let's Skate Together! Skating on Bones in the Past and Today," 440.
115. Edberg and Karlsson, *Isläggor från Birka och Sigtuna. En undersökning av ett vikingatida och medeltida fyndmaterial*, 16.

116. Marie Jonasson Schmidt, email message to author, August 31, 2018. A video of several skaters performing pirouettes on one foot on bone skates is on the Lödöse Museum's Facebook page. The pirouettes are performed by pushing with the toe of one skate and standing on the other. The skaters are on synthetic ice with their skates firmly bound to their feet. No poles are visible.

117. Berg, "Skates and Punt Sleds: Some Scandinavian Notes," 6.

118. E. Friedel, "Mitteilungen über altertümliche Geräte. d. Schlittknochen," *Brandenburgia, Monatsblatt der Gesellschaft für Heimatskunde u. s. w. zu Berlin* 6 (1898): 318–327, quoted in Herman, "Knochenschlittschuh, Knochenkufe, Knochenkeitel: Ein Beitrag zur näheren Kenntnis der prähistorischen Langknochenfunde," 221.

119. MacGregor, "Bone Skates: A Review of the Evidence," 61.

120. Küchelmann and Zidarov, "Let's Skate Together! Skating on Bones in the Past and Today," 439.

121. Edberg and Karlsson, "Bone Skates and Young People in Birka and Sigtuna," 15.

122. Küchelmann and Zidarov, "Let's Skate Together! Skating on Bones in the Past and Today," 436.

123. Short, *Viking-Age Ice Skates*.

124. Thorsteinn Einarsson, "Winter Sport in Iceland," in *Winter Games Warm Traditions*, ed. Matti Goksøyr, Gerd von der Lippe, and Kristen Mo (Oslo: The Norwegian Society of Sports History / The International Society for the History of Physical Education / Sport, 1994), 59.

125. Heidi Luik, "Luust Uisud Eesti Arheoloogilises Leiumaterjalis," *Eesti Arheoloogia Ajakiri* 4, no. 2 (2000): 149–150, cited in Küchelmann and Zidarov, "Let's Skate Together! Skating on Bones in the Past and Today," 439.

126. Anonymous, *Deutscher Eis-Sport*, 31 March 1898, quoting F. Meyer, cited in Balfour, "Sledges with Bone Runners in Modern Use," 253.

127. Friedel, "Mitteilungen über altertümliche Geräte. d. Schlittknochen," quoted in Herman, "Knochenschlittschuh, Knochenkufe, Knochenkeitel: Ein Beitrag zur näheren Kenntnis der prähistorischen Langknochenfunde," 221–222.

128. Brückner, "Über den heutigen Gebrauch von Schlittknochen in Schlesien." In Verhandlungen der Berliner Gesellschaft für Anthropologie, Ethnologie und Urgeschichte, 42.

129. J. M. Heathcote, "Skating as a Recreation," in *Skating*, Badminton Library of Sports and Pastimes (London: Longmans, Green, 1892), 211–217.

130. Henry A. Buck, "Ice-Sailing," in *Skating*, Badminton Library of Sports and Pastimes (London: Longmans, Green, 1892), 407–428.

131. Liv Arnesen, Ann Bancroft, and Cheryl Dahle, *No Horizon Is So Far: Two Women and their Extraordinary Journey across Antarctica* (Cambridge, MA: Da Capo Press, 2003).

132. Latin from Robertson, *Materials for the History of Thomas Becket, Archbishop of Canterbury*, 11, translation from Douglas and Greenaway (*English Historical Documents*, 961).

133. Latin from Saxo Grammaticus, *Historia Danica*, ed. Petrus Erasmus Müller (Havniae: Libraria Gyldendalianæ, 1839), 131, translation from Grammaticus (*The History of the Danes, Books I–IX*, I.79).

134. Latin from Wilhelmus Rubruquensis, "Itinerarium ad partes orientales," in *Sinica Franciscana*, ed. Anastasius van den Wyngaert, vol. 1: Itinera et relationes Fratrum Minorum saeculi XIII et XIV (Florence: Quaracchi-Florence, 1929), 269, my translation.

135. Latin from Magnus, *Historia de gentibus septentrionalibus*, 122, translation from Magnus (*Description of the Northern Peoples*, 176).

136. Latin from Magnus, *Historia de gentibus septentrionalibus*, 41, translation from Magnus (*Description of the Northern Peoples*, 58).

137. Formenti and Minetti, "Human Locomotion on Ice: The Evolution of Ice-Skating Energetics through History," 1829.

138. Küchelmann and Zidarov, "Let's Skate Together! Skating on Bones in the Past and Today," 439.

139. Brückner, "Über den heutigen Gebrauch von Schlittknochen in Schlesien." In Verhandlungen der Berliner Gesellschaft für Anthropologie, Ethnologie und Urgeschichte, 42.

140. Staatsbibliothek zu Berlin, *Projekt zur Erschliessung historisch wertvoller Altkartenbestände: Anleitung und Sonderregeln mit Beispielen für die Aufnahme*, 2001, §4.3.3, http://ikar.sbb.spk-berlin.de/werkzeugkasten/sonderregeln/index.htm.

141. Säve, *Svenska lekar*, 77.

142. This reference to women using bone skates is unusual. Berg ("Isläggar och skridskor," 82) notes that bone skates were mainly the province of men, while women used crampons in Urshult in Småland. In general, the sources describing bone skates do not mention women specifically.

143. Länsstyrelsen Gotlands Län, *Bästeträsk*, accessed December 15, 2018, https://www.lansstyrelsen.se/gotland/besok-ochupptack/naturreservat/bastetrask.html.

144. Formenti and Minetti, "Human Locomotion on Ice: The Evolution of Ice-Skating Energetics through History," 1826.

145. G. J. van Ingen Schenau, "The Influence of Air Friction in Speed Skating," *Journal of Biomechanics* 15, no. 6 (1982): 449–458, doi:10.1016/0021-9290(82)90081-1.

146. Guillaume Y. Millet et al., "Poling Forces during Roller Skiing: Effects of Technique and Speed," *Medicine and Science in Sports and Exercise* 30, no. 11 (November 1998): 1645–1653, doi:10.1097/00005768-199811000-00014.

147. Formenti and Minetti, "Human Locomotion on Ice: The Evolution of Ice-Skating Energetics through History," 1830.

148. Anton Englert, "Trial Voyages as a Method of Experimental Archaeology: The Aspect of Speed," chap. 6 in *Connected by the Sea: Proceedings of the Tenth International Symposium on Boat and Ship Archaeology, Denmark 2003*, ed. Lucy Blue, Frederick M. Hocker, and Anton Englert (Oxford: Oxbow Books, 2017), n.p.

149. Graeme Davis, *Vikings in America* (Edinburgh: Birlinn Limited, 2009), 16.

150. Küchelmann and Zidarov, "Let's Skate Together! Skating on Bones in the Past and Today," 439.

151. Michael B. Schiffer, *Formation Processes of the Archaeological Record* (Albuquerque: University of New Mexico Press, 1987), 7.

152. Choyke and Bartosiewicz, "Skating with Horses: Continuity and Parallelism in Prehistoric Hungary," 324.

153. Layard, "Bone Skates and Skating Stakes," 187–188.
154. MacGregor, "Bone Skates: A Review of the Evidence," 387.
155. Jacobi, *De Nederlandse Glissen*, 12.
156. Choyke and Bartosiewicz, "Skating with Horses: Continuity and Parallelism in Prehistoric Hungary," 324.
157. Edberg and Karlsson, *Isläggor från Birka och Sigtuna. En undersökning av ett vikingatida och medeltida fyndmaterial*, 22, 24, 32–33.
158. Brückner, "Über den heutigen Gebrauch von Schlittknochen in Schlesien." In Verhandlungen der Berliner Gesellschaft für Anthropologie, Ethnologie und Urgeschichte, 42.
159. Porter, "Fen Skating," 43.
160. Choyke and Bartosiewicz, "Skating with Horses: Continuity and Parallelism in Prehistoric Hungary," 322.
161. Ibid., 320.

Chapter 3

1. Wiebe Blauw, *Van glis tot klapschaats: Schaatsen en schaatsenmakers in Nederland, 1200 tot heden* (Franeker: Van Wijnen, 2001), 45.
2. Fred. W. Foster, *A Bibliography of Skating* (London: B. W. Warhurst, 1898), 18.
3. Charles Roach Smith and Alfred Smee, "Bone Skate Found at Moorfield," *Archæologia, or Miscellaneous Tracts Relating to Antiquity* 29 (1842): 397–399.
4. Brückner, "Über den heutigen Gebrauch von Schlittknochen in Schlesien"; Rudolf Virchow, "Geglättete Knochen zum Gebrauche beim Schlittschuhlaufen und Weben." In Verhandlungen der Berliner Gesellschaft für Anthropologie, Ethnologie und Urgeschichte, *Zeitschrift für Ethnologie* 3 (1870): 19–21; Kuckuck, "Über den Gebrauch von Schlittknochen in neuester Zeit." In Verhandlungen der Berliner Gesellschaft für Anthropologie, Ethnologie und Urgeschichte, *Zeitschrift für Ethnologie* 3 (1871): 104.
5. Von Luschan, "Mitteilungen aus dem Museum der Gesellschaft," 146.
6. British Archaeological Association, "Proceedings of the Association," *Journal of the British Archaeological Association* 28 (1872): 76–77.
7. Munro, "Notes on Ancient Bone Skates."
8. Herman, "Knochenschlittschuh, Knochenkufe, Knochenkeitel: Ein Beitrag zur näheren Kenntnis der prähistorischen Langknochenfunde."
9. Wawn, *The Vikings and the Victorians: Inventing the Old North in Nineteenth-Century Britain*, 22–23.
10. Jones and Cormack, *A Treatise on Skating*.
11. Wawn, *The Vikings and the Victorians: Inventing the Old North in Nineteenth-Century Britain*, 117–118.
12. B. A. Thurber, "A New Interpretation of Frithiof's Steel Shoes," *Scandinavica* 50, no. 2 (2011): 6–30.
13. Fowler, *On the Outside Edge: Being Diversions in the History of Skating*, 26.
14. Guido Weiss, "Ueber den Eissport im sechzehnten Jahrhundert." In Verhandlungen der Berliner Gesellschaft für Anthropologie, Ethnologie und Urgeschichte, *Zeitschrift für Ethnologie* 3 (1871), 60; John Stow, *A Survay of London: Contayning the Originall, Antiquity, Increase, Moderne Estate, and Description of that Citie: Written in the Yeare 1598* (London: John Wolfe, 1598).
15. Weiss, "Ueber den Eissport im sechzehnten Jahrhundert," 60.
16. Vandervell and Witham, *A System of Figure-Skating: Being the Theory and Practice of the Art as Developed in England, with a Glance at Its Origin and History*, 3.
17. Fowler, *On the Outside Edge: Being Diversions in the History of Skating*, 44.
18. Smith and Smee, "Bone Skate Found at Moorfield."
19. Percy, *Five Pieces of Runic Poetry, Translated from the Islandic Language*, 78.
20. Theodore M. Andersson and Kari Ellen Gade, *Morkinskinna: The Earliest Icelandic Chronicle of the Norwegian Kings (1030–1157)*, Islandica 51 (Ithaca: Cornell University Press, 2000), 149.
21. Finnbogi Guðmundsson, ed., *Orkneyinga saga. Legenda de Sancto Magno. Magnúss saga skemmri. Magnúss saga lengri. Helga þáttr ok Úlfs*, Íslenzk fornrit 34 (Reykjavík: Hið íslenzka fornritafélag, 1965), 130.
22. Percy, *Five Pieces of Runic Poetry, Translated from the Islandic Language*, 80–81.
23. Charles Roach Smith, "Ancient Bone Skates," *Collectanea Antiqua* 1 (1848): 168.
24. Nigel Brown, *Ice-Skating: A History* (London: Nicholas Kaye Limited, 1959), 22.
25. Figure skaters and fans of figure skating may recognize Axel Paulsen as the inventor of the popular Axel jump.
26. Guðni Jónsson, ed., *Edda Snorra Sturlusonar, nafnaþulur og skáldatal*, Íslendingasagnaútgáfan (Akureyri: Prentverk Odds Björnssonar, 1954), 70–71.
27. Thomas Percy, *Northern Antiquities: Or a Description of the Manners, Customs, Religion and Laws of the Ancient Danes, Including Those of Our Own Saxon Ancestors*, 2nd ed. (Edinburgh: C. Stewart, 1809), II.92.
28. John Kennedy, *Translating the Sagas: Two Hundred Years of Challenge and Response*, Making the Middle Ages 5 (Turnhout, Belgium: Brepols, 2007), 55.
29. Margaret Clunies Ross, *The Old Norse Poetic Translations of Thomas Percy*, Making the Middle Ages 4 (Turnhout, Belgium: Brepols, 2001), 13.
30. Fowler, *On the Outside Edge: Being Diversions in the History of Skating*, 46.
31. Foster, *A Bibliography of Skating*, 29.
32. *Ibid*.
33. Edwin H. Zeydel, "More Oddities and Novelties for the German Literary Historian," *The German Quarterly* 8, no. 1 (1945): 29. The other two odes are "Der Eislauf" (Ice Skating) and "Braga."
34. Vereeniging ter Beoefening der Geschiedenis van 's-Gravenhage, "'s Gravenhage onder deregering der graven uit de Huizen van Holland, Henegouwen en Beijeren," *Mededeelingenvan de Vereeniging ter Beoefening der Geschiedenis van 's-Gravenhage* 1 (1863): 319.
35. Fritz Pieter van Oostrom, *Court and Culture: Dutch Literature, 1350–1450* (Berkeley: University of California Press, 1992), 20.
36. Antheun Janse, *Ridderschap in Holland: Portret van een adellijke elite in de late Middeleeuwen*, Adelgeschiedenis 1 (Hilversum, Netherlands: Verloren BV, 2001), 346.

37. Niko Mulder, "Ten IJse (4)—Met scaetzen en souerding op de hofgracht," *Kouwe Drukte* 13, no. 36 (October 2009): 40.
38. Foster, *A Bibliography of Skating*, 17.
39. Alessandro Guagnini, "Arma Magni Ducis Moschouiæ," in *Sarmatiae Europeæ descriptio, quae Regnum Poloniæ, Lituaniam, Samogitiam, Russiam, Masouiam, Prussiam, Pomeraniam, Liuoniam, & Moschouiæ, Tartariaeque partem complectitur* (Krakow: Matthia Wirzbietae, 1578), folio 16r.
40. Roelof van Straten, *An Introduction to Iconography: Symbols, Allusions and Meaning in the Visual Arts*, Revised ed., trans. Patricia de Man, Documenting the Image (Yverdon: Gordon and Breach Science Publishers, 1994), 4.
41. Virchow, "Geglättete Knochen zum Gebrauche beim Schlittschuhlaufen und Weben," 20–21.
42. Petrus Tergast, *Die heidnischen Alterthümer Ostfrieslands* (Emden: W. Haynel, 1879), 43, fig. 49.
43. Fowler, *On the Outside Edge: Being Diversions in the History of Skating*, 38.
44. S. T. Kjellberg, "Gnida, mangla och stryka," *Kulturen*, 1940, 68–91, summarized in MacGregor, "Bone Skates: A Review of the Evidence," 57.
45. S. A. Semenov, *Prehistoric Technology: An Experimental Study of the Oldest Tools and Artefacts from Traces of Manufacture and Wear*, trans. M. W. Thompson (Bath: Adams and Dart, 1964), 191–193.
46. Brown, *Ice-Skating: A History*, 18.
47. Allen, "The Primitive Bone Skate," 35.
48. Balfour, "Notes on the Modern Use of Bone Skates," 32.
49. Arthur MacGregor, "Bone, Antler and Horn: An Archaeological Perspective," *Journal of Museum Ethnography* 2 (1991): 29.
50. MacGregor, "Bone Skates: A Review of the Evidence," 57.
51. MacGregor, "Problems in the Interpretation of Microscopic Wear Patterns: The Evidence from Bone Skates."
52. MacGregor, *Bone, Antler, Ivory and Horn: The Technology of Skeletal Materials since the Roman Period*, 141.
53. Hans-Joachim Barthel, "Schlittknochen oder Knochengeräte?" *Alt-Thüringen* 10 (1968–1969): 211–212.
54. Küchelmann and Zidarov, "Let's Skate Together! Skating on Bones in the Past and Today," 427.
55. Becker, "Bemerkungen über Schlittknochen, Knochenkufen und ähnliche Artefakte, unter besonderer Berücksichtigung der Funde aus Berlin Spandau," 23.
56. Genevieve M. LeMoine, "Use Wear on Bone and Antler Tools from the Mackenzie Delta, Northwest Territories," *American Antiquity* 59, no. 2 (April 1994): 322–325, doi:10.2307/281935.
57. Becker, "Bemerkungen über Schlittknochen, Knochenkufen und ähnliche Artefakte, unter besonderer Berücksichtigung der Funde aus Berlin Spandau," 24–25.
58. MacGregor, "Problems in the Interpretation of Microscopic Wear Patterns: The Evidence from Bone Skates," 388–389.
59. W. James Stemp, Adam S. Watson, and Adrian A. Evans, "Surface Analysis of Stone and Bone Tools," *Surface Topography: Metrology and Properties* 4 (2016): 013001.
60. MacGregor, *Bone, Antler, Ivory and Horn: The Technology of Skeletal Materials since the Roman Period*, 141.
61. Rozália Kustár and Beáta Tugya, "'Csontkorcsolyák' a késő bronzkorban (Bone 'Skates' in the Late Bronze Age)," in *Csont és bőr: Az állati eredetű nyersanyagok feldolgozásának törtnéte, régészete és néprajza (Bone and Leather: History, Archaeology and Ethnography of Crafts Utilizing Raw Materials from Animals)*, ed. János Gömöri and Andrea Körösi, Az anyagi kultúra a Kárpát-medencében 4. (Budapest: Magyar Tudományos Akadémia Veszprémi Akadémiai Bizottsága (MTA VEAB), 2010), 79.
62. Edberg and Karlsson, "Bone Skates and Young People in Birka and Sigtuna," 9.
63. MacGregor, *Bone, Antler, Ivory and Horn: The Technology of Skeletal Materials since the Roman Period*, 142.
64. Ewald Schuldt, *Altslawisches Handwerk: Ausstellung zur 800-jahrfeier der Stadt Schwerin: Sonderausstellung 1960*, Bildkataloge des Museums für Ur- und Frühgeschichte Schwerin 2 (Schwerin: Museum für Ur- und Frühgeschichte, 1960).
65. MacGregor, *Bone, Antler, Ivory and Horn: The Technology of Skeletal Materials since the Roman Period*, 142.
66. Thunig, "Über Schlittknochen und Gräbeurnen," 280.
67. Jakov Pertrovic Gerškovič, *Studien zur spätbronzezeitlichen Sabatinovka-Kultur am unteren Dnepr und an der Westküste des Azov'schen Meeres*, Archäologie in Eurasien 7 (Rahden, Westphalia: Marie Leidorf, 1999).
68. Valentin Pankovskiy, *Bone and Antler Industry from the Late Bronze Age Settlement Sabatynivka-I (Ukraine, South Bug)*, Abstract #31 of the Sixth Meeting of the Worked Bone Research Group of the International Council forArchaeozoology, August 27th–31th [sic] 2007, Paris. August 2007.
69. Gerškovič, *Studien zur spätbronzezeitlichen Sabatinovka-Kultur am unteren Dnepr und an der Westküste des Azov'schen Meeres*, 65.
70. *Ibid.*
71. Cornwall, *Bones for the Archaeologist*, 176.
72. MacGregor, "Bone Skates: A Review of the Evidence," 65.
73. H. H. Andersen, P. J. Crabb, and H. J. Madsen, *Århus Søndervold: En byarkaeologisk undersøgelse*, Jyst Arkeologisk Selskabs skrifter 9 (Copenhagen: Nordisk Forlag, 1971), 142.
74. Anna Drzewicz, *Wyroby z kości i poroża z osiedla obronnego ludności kultury łużyckiej w Biskupinie* (Warsaw: Wydawnictwo Naukowe 'Semper,' 2004), 81–82.
75. Choyke and Bartosiewicz, "Skating with Horses: Continuity and Parallelism in Prehistoric Hungary," 318; Kustár and Tugya, "'Csontkorcsolyák' a késő bronzkorban (Bone 'Skates' in the Late Bronze Age)."
76. Alice M. Choyke, "Hidden Agendas: Ancient Raw Material Choice for Worked Osseous Objects in Central Europe and Beyond," in *From These Bare Bones: Raw Materials and the Study of Worked Osseous Objects, Proceedings of the Raw Materials Session at the 11th ICAZ Conference, Paris, 2010*, ed. Alice M. Choyke and Sonia O'Connor (Oxford: Oxbow Books, 2013), 8.
77. Drzewicz, *Wyroby z kości i poroża z osiedla obronnego ludności kultury łużyckiej w Biskupinie*, 81.
78. Václav Furmánek, Ladislav Veliačik, and Jozef

Vladář, *Die Bronzezeit in Slowakischen Raum*, Prähistorische Archäologie in Südosteuropa 15 (Rahden, Westfalia: Marie Leidorf, 1999), 145.

79. Choyke and Bartosiewicz, "Skating with Horses: Continuity and Parallelism in Prehistoric Hungary," 325.

80. The Smuszewie artifacts are from the Lusatian culture during the Hallstatt period; see Dobromir Durczewski, *Gród ludności kultury łużyckiej z okresu halsztackiego w Smuszewie, woj. pilskie*, Biblioteka Fontes archaeologici Posnanienses 6 (Poznań: Muzeum Archeologiczne w Poznaniu, 1985), Table 57:1–4.

81. The Słupcy artifacts are from the Lusatian culture during the Early Iron Age; see Tadeusz Malinowski, "Osadnictwo kultury łużyckiej wczesnej epoki żelaza w Słupcy," *Fontes Archaeologici Posnanienses: Annales Musei Archaeologici Posnaniensis* 8–9 (1957–1958): fig. 37:9–12.

82. The Jankowie artifacts date to the Hallstatt period; see Janusz Ostoja-Zagórski, *Gród halsztacki w Jankowie nad Jeziorem Pakoskim* (Wrocław: Zakład Narodowy imienia Ossolinskich, 1978), fig. 10e, 12a.

83. The Łagiewnikach artifacts date to the Late Bronze or Early Iron Age; see Szamałek, *Kruszwicki Zespół Osadniczy w Młodszej Epoce Brązu i w Początkach Epoki Żelaza*, 92.

84. Choyke and Bartosiewicz, "Skating with Horses: Continuity and Parallelism in Prehistoric Hungary," 325.

85. Furmánek, Veliácik, and Vladář, *Die Bronzezeit in Slowakischen Raum*, 145.

86. Kazimierz Lukasiewicz and Zdzisław Rajewski, "Przedmioti rogowe i kościane z grodu kultury 'łużyckiej' w Biskupinie," in *Gród prasłowiański w Biskupinie w powiecie żnińskim. Sprawozdanie z prac wykopaliskowych w latach 1937 i 1937 z uwględnieniem lat 1934–1935*, ed. Józefa Kostrzewskiego (Poznań: Nakladem Instytutu Prehistorycznego Uniwersytetu Poznańskiego, 1938), 46–47, plate XXXVIII.

87. Z. Rajewski, "Przedmioty z kości i rogu w obróbka obu surowców w grodach 'łużyckich' z wczesnego okresu żelaza," *Sprawozdanie z prac wykopaliskowych w grodzie kultury łużyckiej w Biskupinie w powiecie żnińskim* (Poznań) 3 (1950): 173.

88. Małgorzata Mogielnicka-Urban, *Warsztat ceramiczny w kulturze łużyckiej* (Wrocław: Zakład Narodowy imienia Ossolińskich, 1984), 24.

89. Drzewicz, *Wyroby z kości i poroża z osiedla obronnego ludności kultury łużyckiej w Biskupinie*, 81–82.

90. Ibid., 82.

91. Becker, "Bemerkungen über Schlittknochen, Knochenkufen und ähnliche Artefakte, unter besonderer Berücksichtigung der Funde aus Berlin Spandau," 21, 23.

92. Joanna Sofaer, Lise Bender Jørgensen, and Alice M. Choyke, "Craft Production: Ceramics, Textiles, and Bone," in *The Oxford Handbook of the European Bronze Age*, ed. Harry Fokkens and Anthony Harding (Oxford: Oxford University Press, 2013), 483.

93. Choyke, "Hidden Agendas: Ancient Raw Material Choice for Worked Osseous Objects in Central Europe and Beyond," 8.

94. Choyke and Bartosiewicz, "Skating with Horses: Continuity and Parallelism in Prehistoric Hungary."

95. Joanna Sofaer, "Technology and Craft," in *Organizing Bronze Age Societies: The Mediterranean, Central Europe, and Scandinavia Compared*, ed. Timothy Earle and Kristian Kristiansen (Cambridge: Cambridge University Press, 2010), 198–199.

96. Choyke and Bartosiewicz, "Skating with Horses: Continuity and Parallelism in Prehistoric Hungary," 318–319.

97. Choyke, "Hidden Agendas: Ancient Raw Material Choice for Worked Osseous Objects in Central Europe and Beyond," 8.

98. Choyke and Bartosiewicz, "Skating with Horses: Continuity and Parallelism in Prehistoric Hungary," 319.

99. Alice M. Choyke, email message to author, August 7, 2018.

100. Anton Točík, "Knochen- und Geweihindustrie der Maďarovce-Kultur in der Südwestslovakei," *Studijné Zvesti Archeologicky Ustav Slovenskej Akadémie Vied* 3 (1959): 45–46.

101. O. Sellmann, "Latènezeitliche Grab- und Wohngrubenfunde von der Aue bei Mühlhausen i. Th.," *Jahresschrift für die Vorgeschichte der sächsisch-thüringischen Länder* 10 (1911): 69–70, pl. IX.6–7.

102. Albert Kiekebusch, *Die Ausgrabung des bronzezeitlichen Dorfes Buch bei Berlin*, vol. 1 (Berlin: Dietrich Reimer, 1923), 75.

103. Otto Tschumi, "Die steinzeitlichen Epochen," in *Urgeschichte der Schewiz*, ed. Otto Tschumi, vol. 1 (Frauenfeld: Verlag Huber, 1949), 621.

104. Stemp, Watson, and Evans, "Surface Analysis of Stone and Bone Tools," 12.

Chapter 4

1. MacGregor, "Bone Skates: A Review of the Evidence," 67.

2. Küchelmann and Zidarov, "Let's Skate Together! Skating on Bones in the Past and Today," 430.

3. Choyke and Bartosiewicz, "Skating with Horses: Continuity and Parallelism in Prehistoric Hungary," 325.

4. László Bartosiewicz, "The Archaeology of Domestic Animals," in *Hungarian Archaeology at the Turn of the Millennium*, ed. Zsolt Visy (Budapest: Ministry of National Cultural Heritage, Teleki László Foundation, 2003), 116.

5. Federico Formenti and Alberto E. Minetti, "The First Humans Travelling on Ice: An Energy-Saving Strategy?" *Biological Journal of the Linnean Society* 93 (2008): 1–7, doi:10.1111/j.1095–8312.2007.00991.x.

6. Stefan Lovgren, "Bone Ice Skates Invented by Ancient Finns, Study Says," *National Geographic News*, 2008.

7. Wikipedia contributors, *Ice Skating*, Wikipedia, accessed September 14, 2017, https://en.wikipedia.org/wiki/Ice_skating.

8. Munro, "Notes on Ancient Bone Skates," 197.

9. Choyke and Bartosiewicz, "Skating with Horses: Continuity and Parallelism in Prehistoric Hungary," 318.

10. Točík, "Knochen- und Geweihindustrie der Maďarovce-Kultur in der Südwestslovakei," 45–46; bone type unspecified.

11. Sellmann, "Latènezeitliche Grab- und Wohngrubenfunde von der Aue bei Mühlhausen i. Th.," 69–70; two radii.

12. Kiekebusch, *Die Ausgrabung des bronzezeitlichen Dorfes Buch bei Berlin*, 75.

13. Tschumi, "Die steinzeitlichen Epochen," 621.
14. Gerškovič, *Studien zur spätbronzezeitlichen Sabatinovka-Kultur am unteren Dnepr und an der Westküste des Azov'schen Meeres*, 91.
15. Yakov P. Gershkovich, "Farmers and Pastoralists of the Pontic Lowland during the Late Bronze Age," in *Prehistoric Steppe Adaptation and the Horse*, ed. Marsha Levine, Colin Renfrew, and Katie Boyle (Cambridge: McDonald Institute for Archaeological Research, 2003), 309.
16. Josef Schranil, *Die Vorgeschichte Böhmens und Mährens* (Berlin: W. de Gruyter, 1928), 169; Parma et al., "Netradični materiál, neobvyklý předmet Opomıjený segment kostěné industrie mladšı doby bronzové (Non-Traditional Material and a Non-Traditional Object: A Neglected Sort of the Late Bronze Age Bone Industry)," 149.
17. Furmánek, Veliačik, and Vladář, *Die Bronzezeit in Slowakischen Raum*, 145.
18. Choyke and Bartosiewicz, "Skating with Horses: Continuity and Parallelism in Prehistoric Hungary," 320–21.
19. Marco Bertolini and Ursula Thun Hohenstein, "Evidence of Butchery Marks and Anthropic Modifications on Horse Remains in a Late Bronze Age Site of Northern Italy: The Case of Bovolone," *Journal of Archaeological Science: Reports* 9 (2016): 477, doi:10.1016/j.jasrep.2016.08.031.
20. See Herman, "Knochenschlittschuh, Knochenkufe, Knochenkeitel: Ein Beitrag zur näheren Kenntnis der prähistorischen Langknochenfunde," 222, which refers to the site by its Hungarian name, Verebély; their dating, warns Küchelmann (*Bone Skates Database*), is doubtful.
21. Choyke, "Worked Animal Bone at the Sarmatian Site of Gyoma 133," 310–317.
22. Alice M. Choyke, "The Bone Tool Manufacturing Continuum," *Anthropozoologica* 25–26 (1997): 71.
23. Andrea Vaday, "Introduction," in *Cultural and Landscape Changes in South-East Hungary*, ed. Andrea H. Vaday, vol. 2 (Budapest: Archaeological Institute of the Hungarian Academy of Sciences, 1996), 13.
24. E. Gál, "Animal Remains from the Multi-Period Site of Hajdúnánás-Fürjhalom-Dűlő," *Acta Archaeologica Academiae Scientarum Hungaricae* 61 (2010): 226, doi:10.1556/AArch.61.2010.1.7.
25. László Bartosiewicz, "Animal Exploitation at the Sarmatian Site of Gyoma 133," in *Cultural and Landscape Changes in South-East Hungary*, ed. Andrea H. Vaday, vol. 2 (Budapest: Archaeological Institute of the Hungarian Academy of Sciences, 1996), 372.
26. Magdolna Vicze, "The Prehistoric Settlement at the Site of Gyoma 133," in *Cultural and Landscape Changes in South-East Hungary*, ed. Andrea H. Vaday, vol. 2 (Budapest: Archaeological Institute of the Hungarian Academy of Sciences, 1996), 28.
27. Herodotus, *The History*, trans. David Grene (London: University of Chicago Press, 1987), 4.110, 4.116.
28. David W. Anthony, *The Horse, the Wheel, and Language: How Bronze-Age Riders from the Eurasian Steppes Shaped the Modern World* (Princeton: Princeton University Press, 2007), 18.
29. Küchelmann and Zidarov, "Let's Skate Together! Skating on Bones in the Past and Today," 432.
30. J. G. D. Clark, *Prehistoric Europe: The Economic Basis* (Stanford: Stanford University Press, 1952), 295.
31. P. V. Addyman et al., "Palaeoclimate in Urban Environmental Archaeology at York, England: Problems and Potential," *World Archaeology* 8, no. 2 (1976): 227, doi:10.1080/00438243.1976.9979666.
32. Bartosiewicz, "The Archaeology of Domestic Animals," 62.
33. Panayoti Kelaidis, "Introduction: Principal Steppe Regions," in *Steppes: The Plants and Ecology of the World's Semi-arid Regions* (Portland, OR: Timber Press, 2015), 17–18.
34. K. Bharatdwaj, *Physical Geography: Atmosphere* (New Delhi: Discovery Publishing House, 2006), 340.
35. Konstantin V. Kremenetski, "Steppe and Forest-Steppe Belt of Eurasia: Holocene Environmental History," in *Prehistoric Steppe Adaptation and the Horse*, ed. Marsha Levine, Colin Renfrew, and Katie Boyle (Cambridge: McDonald Institute for Archaeological Research, 2003), 11, 14–15, 25; Anthony, *The Horse, the Wheel, and Language: How Bronze-Age Riders from the Eurasian Steppes Shaped the Modern World*, 300.
36. Bartosiewicz, "The Archaeology of Domestic Animals," 62.
37. Anthony, *The Horse, the Wheel, and Language: How Bronze-Age Riders from the Eurasian Steppes Shaped the Modern World*, 198–199, 201, 204, 220.
38. David W. Anthony, "Bridling Horse Power: The Domestication of the Horse," in *Horses through Time*, ed. Sandra L. Olsen (Boulder, CO: Roberts Rinehart, 1996), 70.
39. Ibid., 63.
40. Anthony, *The Horse, the Wheel, and Language: How Bronze-Age Riders from the Eurasian Steppes Shaped the Modern World*, 300–03.
41. Ibid., 390.
42. Kremenetski, "Steppe and Forest-Steppe Belt of Eurasia: Holocene Environmental History," 23.
43. Constantin V. Kremenetski, Olga A. Chichagova, and Nathalia I. Shishlina, "Palaeoecological Evidence for Holocene Vegetation, Climate and Land-Use Change in the Low Don Basin and Kalmuk Area, Southern Russia," *Vegetation History and Archaeobotany* 8 (1999): 233.
44. Kremenetski, "Steppe and Forest-Steppe Belt of Eurasia: Holocene Environmental History," 11.
45. T. Alekseeva et al., "Late Holocene Climate Reconstructions for the Russian Steppe, Based on Meralogical and Magnetic Properties of Buried Palaeosols," *Palaeogeography, Palaeoclimatology, Palaeoecology* 249 (2007): 122, doi:10.1016/j.palaeo.2007.01.006.
46. Philip L. Kohl, *The Making of Bronze Age Eurasia*, Cambridge World Archaeology (Cambridge: Cambridge University Press, 2007), 144–158.
47. Gershkovich, "Farmers and Pastoralists of the Pontic Lowland during the Late Bronze Age," 310.
48. Anthony, *The Horse, the Wheel, and Language: How Bronze-Age Riders from the Eurasian Steppes Shaped the Modern World*, 402.
49. Laura Dietrich, "Projectile Weapons of the Late Bronze Age in Eastern Europe. The Case of the Noua-Sabatinovka-Coslogeni Cultural Complex," in *Archäologische Wanderungen zwischen Ostund Westeuropa: Studien für Tiberius Bader zum 75. Geburtstag*, vol. 29/1, Studii şi comunicări Satu Mare (Satu Mare, Romania: Editura Muzeului Sătmărean, 2013), 191.
50. Anthony, *The Horse, the Wheel, and Language: How Bronze-Age Riders from the Eurasian Steppes Shaped the Modern World*, 223–224.

51. Choyke and Bartosiewicz, "Skating with Horses: Continuity and Parallelism in Prehistoric Hungary," 321.
52. Markwart, "Ein arabischer Bericht über die arktischen (uralischen) Länder aus dem 10. Jahrhundert," 307.
53. W. Schott, "Über die ächten Kirgisen," *Abhandlungen der Königlichen Akademie der Wissenschaften zu Berlin*, 1864, 448.
54. Grigoriy M. Burov, "Some Mesolithic Wooden Artifacts from the Site of Vis I in the European North East of the U.S.S.R.," in *The Mesolithic in Europe*, ed. Clive Bonsall (Edinburgh: John Donald Publishers, 1985), 391–394.
55. A. L. Mongyat, "XII vek. Puteschestvie v Rossiyu," *Nauka i zhizn'* 1 (1965): 35–36, translated and quoted in Burov, "Some Mesolithic Wooden Artifacts from the Site of Vis I in the European North East of the U.S.S.R.," 394.
56. Burov, "Some Mesolithic Wooden Artifacts from the Site of Vis I in the European North East of the U.S.S.R.," 394.
57. Bo Wang, "Research on Altay, the Original Place of Skiing," in *The Original Place of Skiing—Altai Prefecture of Xingjiang, China*, ed. Zhaojian Shan and Bo Wang (Beijing Shi: Ren min ti yu chu ban she : Xinjiang ren min chu ban she, 2011), 195.
58. E. John B. Allen, *The Culture and Sport of Skiing: From Antiquity to World War II* (Amherst: University of Massachusetts Press, 2007), 24.
59. Nils Larsen, "The Timeless Skiers of the Altay," in *The Original Place of Skiing—Altai Prefecture of Xingjiang, China*, ed. Shan Zhaojian and Wang Bo (Beijing Shi: Ren min ti yu chu ban she : Xinjiang ren min chu ban she, 2011), 298–300.
60. Choyke and Bartosiewicz, "Skating with Horses: Continuity and Parallelism in Prehistoric Hungary."
61. Berg, "The Origin and the Development of the Skis throughout the Ages," 19.
62. John Weinstock, *Skis and Skiing: From the Stone Age to the Birth of the Sport* (Lewiston, NY: Edwin Mellen Press, 2003), 23–33.
63. Daniel Sutherland Davidson, *Snowshoes*, Memoirs of the American Philosophical Society 6 (Philadelphia: American Philosophical Society, 1937).
64. Allen, *The Culture and Sport of Skiing: From Antiquity to World War II*, 10.
65. Markwart, "Ein arabischer Bericht über die arktischen (uralischen) Länder aus dem 10. Jahrhundert," 304–309.
66. Berg, "Skier und Schlittschuhe: Zwei nordische Fortbewegungsmittel," 193.
67. LeRoy J. Dresbeck, "The Ski: Its History and Historiography," *Technology and Culture* 8, no. 4 (October 1967): 473–474.
68. Author's translation of Schott, "Über die ächten Kirgisen," 448.
69. Carlo Cargiolli, ed., *Il Viaggio Settentrionale di Francesco Negri* (Bologna: Nicola Zanichelli, 1883), 65.
70. Paul Lunde and Caroline Stone, eds., *Ibn Fadlān and the Land of Darkness* (New York: Penguin, 2012), xxxii*n*4.
71. Author's translation of the German translation in Markwart, "Ein arabischer Bericht über die arktischen (uralischen) Länder aus dem 10. Jahrhundert," 289.
72. Minorsky's translation of Marwazī first appeared in Minorsky, *Sharaf Al-Zaman Zamān Ṭāhir Marvazī on China, the Turks, and India* and has been reprinted in Lunde and Stone, *Ibn Fadlān and the Land of Darkness*, 184–185, with slight modifications. The new version is reproduced here.
73. Lunde and Stone, *Ibn Fadlān and the Land of Darkness*, 179.
74. Rubruquensis, "Itinerarium ad partes orientales," 269.
75. Alexander Theodor von Middendorff, *Reise in den äussersten Norden und Osten Sibiriens während der Jahre 1843 und 1844*, vol. 4.2 (St. Petersburg: Kaiserlichen Akademie der Wissenschaften, 1875), 1350.
76. Michael Brian Schiffer, *Studying Technological Change: A Behavioral Approach* (Salt Lake City: University of Utah Press, 2011), 63.
77. Anthony, *The Horse, the Wheel, and Language: How Bronze-Age Riders from the Eurasian Steppes Shaped the Modern World*, 307–11.
78. Burov, "Some Mesolithic Wooden Artifacts from the Site of Vis I in the European North East of the U.S.S.R.," 393.
79. Weinstock, *Skis and Skiing: From the Stone Age to the Birth of the Sport*, 20–21.
80. *Ibid.*, 17–20.
81. Burov, "Some Mesolithic Wooden Artifacts from the Site of Vis I in the European North East of the U.S.S.R.," 393, fig. 2.1, fig. 3.
82. Rubruquensis, "Itinerarium ad partes orientales," 269.
83. Middendorff, *Reise in den äussersten Norden und Osten Sibiriens während der Jahre 1843 und 1844*, 1350.
84. Schiffer, *Formation Processes of the Archaeological Record*, 69.
85. Davidson, *Snowshoes*.
86. Choyke, "Hidden Agendas: Ancient Raw Material Choice for Worked Osseous Objects in Central Europe and Beyond," 7.
87. David W. Anthony, "Two IE Phylogenies, Three PIE Migrations, and Four Kinds of Steppe Pastoralism," *Journal of Language Relationship (Moscow)* 9 (2013): 11–12.
88. J. P. Mallory, *In Search of the Indo-Europeans* (New York: Thames and Hudson, 1991), 223.
89. G. R. Isaac, "The Origins of the Celtic Languages: Language Spread from East to West," chap. 1 in *Celtic from the West: Alternative Perspectives from Archaeology, Genetics, Language and Literature*, ed. Barry Cunliffe and John T. Koch, Celtic Studies Publications 15 (Oxford: Oxbow Books, 2012), 153–166.
90. Jean Manco, *Blood of the Celts* (New York: Thames and Hudson, 2015), 61.
91. Anthony, *The Horse, the Wheel, and Language: How Bronze-Age Riders from the Eurasian Steppes Shaped the Modern World*, 305–306.
92. Wolfgang Haak et al., "Massive Migration from the Steppe Was a Source for Indo-European Languages in Europe," *Nature* 522 (2015): 207–211, doi:10.1038/nature14317.
93. J. P. Mallory, "The Indo-Europeanization of Atlantic Europe," chap. 1 in *Celtic from the West 2: Rethinking the Bronze Age and the Arrival of Indo-European in Atlantic Europe*, ed. John T. Koch and Barry Cunliffe (Oxford: Oxbow Books, 2013), 33.
94. Jocelyne Desideri and Marie Besse, "Swiss Bell Beaker Population Dynamics: Eastern or Southern Influences?" *Archaeological and Anthropological Science* 2 (2010): 171, doi:10.1007/s12520-010-0037-9.
95. Choyke and Bartosiewicz, "Skating with

Horses: Continuity and Parallelism in Prehistoric Hungary," 318.

96. Anthony, *The Horse, the Wheel, and Language: How Bronze-Age Riders from the Eurasian Steppes Shaped the Modern World*, 367.

97. Volker Heyd, "Überregionale Verbindungen der süddeutschen Glockenbecherkultur anhand der Siedlungen," in *Siedlungen der Glockenbecherkultur in Süddeutschland und Mitteleuropa*, ed. Volker Heyd, Ludwig Husty, and Ludwig Kreiner, Arbeiten zur Archäologie Süddeutschlands 17 (Büchenbach: Verlag Dr. Faustus, 2004), 194.

98. Norbert Benecke and Angela von den Driesch, "Horse Exploitation in the Kazakh Steppes during the Eneolithic and Bronze Age," in *Prehistoric Steppe Adaptation and the Horse*, ed. Marsha Levine, Colin Renfrew, and Katie Boyle (Cambridge: McDonald Institute for Archaeological Research, 2003), 77; Anthony, *The Horse, the Wheel, and Language: How Bronze-Age Riders from the Eurasian Steppes Shaped the Modern World*, 203–04.

99. Sándor Bökönyi, "The Earliest Waves of Domestic Horses in East Europe," *Journal of Indo-European Studies* 6, nos. 1–2 (1978): 59.

100. The Late Bronze Age skates are those listed in Küchelmann (*Bone Skates Database*) with the Knoviz skate from Přemýšlení moved from the Early Bronze Age to the Late Bronze Age plus skates from Kustár and Tugya, "'Csontkorcsolyák' a késő bronzkorban (Bone 'Skates' in the Late Bronze Age)" and Bertolini and Hohenstein, "Evidence of Butchery Marks and Anthropic Modifications on Horse Remains in a Late Bronze Age Site of Northern Italy: The Case of Bovolone." When specific numbers are not given, the lowest number of skates that can account for the entry is assumed, i.e., only one horse metatarsus skate and only one cattle radius skate are assumed for Sabatinovka because Pankovskiy (*Bone and Antler Industry from the Late Bronze Age Settlement Sabatynivka-I (Ukraine, South Bug)*) notes that skates of these types were found without specifying how many.

101. The list of Iron Age and Migration Period skates is compiled from the data in Küchelmann (*Bone Skates Database*) with additions from Gabriella Petrucci, Giancarla Malerba, and Giacomo Giacobini, "Manufatti in osso dal castelliere di Pozzuolo del Friuli," chap. 2 in *La prima età del ferro nel settore meridionale del castelliere: Le attività produttive e i resti faunistici*, ed. Paolo Càssola Guida et al., vol. 2.2, Pozzuolo del Friuli (Rome: Edizioni Quasar di Severino Tognon s.r.l., 1998), 142–147; Olivier Putelat, "Les relations homme-animal dans le monde des vivants et des morts: Étude archéozoologique des établissements et des regroupements funéraires ruraux de l'Arc jurassien et de la Plaine d'Alsace de la fin de l'Antiquité Tardive au premier Moyen Âge" (PhD diss., University of Paris Panthéon-Sorbonne, 2015), 614–617; Hagberg, "Isläggar och pälsskinn—en frusen kulturhistorisk rapsodi," 80–81; and Hagberg, "Fundort und Fundgebiet der Modeln aus Torslunda," 330. The Leeuwarden skate may be slightly later—Küchelmann (*Bone Skates Database*) lists it as Migration Period or Early Middle Age.

102. Charles Christopher Mierow, trans., *The Gothic History of Jordanes* (Princeton: Princeton University Press, 1915), 53.

103. Tacitus, *Agricola, Germany*, trans. A. R. Birley, Oxford World's Classics (Oxford: Oxford University Press, 1999).

104. D. H. Green, *Language and History in the Early Germanic World* (Cambridge: Cambridge University Press, 1998), 145–146.

105. Ibid., 165.

106. Bruno Genito, "The Archaeological Cultures of the Sarmatians with a Preliminary Note on the Trial Trenches at Gyoma 133: A Sarmatian Settlement in South-Eastern Hungary (Campaign 1985)," *Annali, Istituto Orientale di Napoli* 48, no. 2 (1988): 99.

107. Mierow, *The Gothic History of Jordanes*, 101, 131.

108. Ágnes B. Tóth, "The Gepids," in *Hungarian Archaeology at the Turn of the Millennium*, ed. Zsolt Visy (Budapest: Ministry of National Cultural Heritage, Teleki László Foundation, 2003), 298.

109. Malcolm Todd, *The Early Germans*, The Peoples of Europe (Malden, MA: Blackwell Publishers, 1992), 235–236.

110. Simina Stanc and Luminita Bejenaru, *Bone and Antler Manufacturing in the IVth–VIth AD Centuries, on the Territory of Romania*, presented at the 6th Meeting of the International Council for Archaeology Worked Bone Research Group, August 2007, cited in Küchelmann, *Bone Skates Database*.

111. Green, *Language and History in the Early Germanic World*, 168.

112. Karl H. Marschalleck, "Die ostgermanische Siedlung von Kliestow bei Frankfurt (Oder)," *Praehistorische Zeitschrift* 30–31, nos. 1–2 (1940): 302, 305–306.

113. H. H. Lamb, *Climate, History, and the Modern World*, First (London: Methuen, 1982), 153.

114. R. P. Duncan-Jones, "Economic Change and the Transition to Late Antiquity," chap. 2 in *Approaching Late Antiquity: The Transformation from Early to Late Empire*, ed. Simon Swain and Mark Edwards (Oxford: Oxford University Press, 2004), 50.

115. Ulf Büntgen et al., "2500 Years of European Climate Variability and Human Susceptibility," *Science* 331 (February 2011): 578–582, doi:10.1126/science.1197175.

116. Malcolm Todd, "Germans and Germanic Invasions," in *The Oxford Companion to Archaeology*, ed. Brian Fagan (Oxford: Oxford University Press, 1996), 250–251.

117. Lamb, *Climate, History, and the Modern World*, 158.

118. Dave G. Ferris et al., "South Pole Ice Core Record of Explosive Volcanic Eruptions in the First and Second Millennia A.D. and Evidence of a Large Eruption in the Tropics around 535 A.D.," *Journal of Geophysical Research* 116 (2011): D17308, doi:10.1029/2011JD015916; L. B. Larsen et al., "New Ice Core Evidence for a Volcanic Cause of the A.D. 536 Dust Veil," *Geophysical Research Letters* 35 (2008): L04708, doi:10.1029/2007GL032450.

119. Lamb, *Climate, History, and the Modern World*, 157–158.

120. Ibid., 154.

121. Ibid., 163–64.

122. Bo Gräslund and Neil Price, "Twilight of the Gods? The 'Dust Veil Event' of AD 536 in Critical Perspective," *Antiquity* 86, no. 332 (January 2012): 437, doi:10.1017/S0003598X00062852.

123. Guðni Jónsson, *Edda Snorra Sturlusonar, nafnaþulur og skáldatal*, 86.

Chapter 5

1. Berg, "Skates and Punt Sleds: Some Scandinavian Notes," 6.
2. Olaus Magnus's (*Historia de gentibus septentrionalibus*, 42) 8–12 Italian miles is equal to 12–18 km, according to Fisher and Higgens (Magnus, *Description of the Northern Peoples*, 86n1:25.7). The 5–8 km calculated by Berg ("Isläggar och skridskor," 84) and cited by others appears to involve a mathematical error.
3. Halldór Halldórsson, "Leggir og skautar," in *Nordæla: Afæmliskveðja til Sigurðar Nordals*, ed. Halldór Halldórsson et al. (Reykjavík: Helgafell, 1956), 75–89.
4. W. H. Barrett, quoted in Porter, "Fen Skating," 43.
5. Berg, "Skates and Punt Sleds: Some Scandinavian Notes," 5–6.
6. Säve, *Svenska lekar*, 77.
7. Christina Marcusson, unpublished record, Folkslivarkivet, Lund, No. 2898–69, 1929, quoted in Edberg and Karlsson, *Isläggar från Birka och Sigtuna. En undersökning av ett vikingatida och medeltida fyndmaterial*, 11.
8. Küchelmann and Zidarov, "Let's Skate Together! Skating on Bones in the Past and Today," 441.
9. Formenti and Minetti, "The First Humans Travelling on Ice: An Energy-Saving Strategy?"
10. Rubruquensis, "Itinerarium ad partes orientales," 269.
11. Herman, "Knochenschlittschuh, Knochenkufe, Knochenkeitel: Ein Beitrag zur näheren Kenntnis der prähistorischen Langknochenfunde," 226.
12. Mårten Stenberger, "The Finds and the Dating of the Vallhagar Settlement," in *Vallhagar: A Migration Period Settlement on Gotland, Sweden*, ed. Mårten Stenberger and Ole Klindt-Jensen, vol. 2 (Copenhagen: Einar Munksgaards Forlag, 1955), 1106.
13. Berg, "Isläggar och skridskor," 82–83.
14. P. A. Säve, *Hafvets och fiskarens sagor samt spridda drag ur Gotlands odlingssaga och strandallmogens lif*, 2nd ed. (Visby: Gotlands Allehandas tryckeri, 1892), 87.
15. Edberg and Karlsson, *Isläggar från Birka och Sigtuna. En undersökning av ett vikingatida och medeltida fyndmaterial*, 42–43.
16. Berg, "Skates and Punt Sleds: Some Scandinavian Notes," 6.
17. Edberg and Karlsson, "Bone Skates and Young People in Birka and Sigtuna," 15.
18. Säve, *Hafvets och fiskarens sagor samt spridda drag ur Gotlands odlingssaga och strandallmogens lif*, 87.
19. Berg, "Isläggar och skridskor," 82.
20. Kuckuck, "Über den Gebrauch von Schlittknochen in neuester Zeit," 104.
21. Petényi, *Games and Toys in Medieval and Early Modern Hungary*, 113.
22. Choyke, "Bone Skates: Raw Material, Manufacturing and Use," 152.
23. Von Luschan, "Mitteilungen aus dem Museum der Gesellschaft," 146.
24. Edberg and Karlsson, "Bone Skates and Young People in Birka and Sigtuna," with a more detailed version published in Swedish as Edberg and Karlsson, *Isläggar från Birka och Sigtuna. En undersökning av ett vikingatida och medeltida fyndmaterial*.
25. Edberg and Karlsson, *Isläggar från Birka och Sigtuna. En undersökning av ett vikingatida och medeltida fyndmaterial*, 43–45.
26. Howard V. Meredith, "Human foot length from embryo to adult," *Human Biology* 16, no. 4 (December 1944): 230.
27. Charles B. Davenport, "The Growth of the Human Foot," *American Journal of Physical Anthropology* 17 (1932): 167–211, cited in Meredith, "Human foot length from embryo to adult," 235.
28. Ibid., 259–261.
29. Gustafsson et al. ("Stature and Sexual Stature Dimorphism in Sweden, from the 10th to the End of the 20th Century," 865) report average heights between 170 and 175 cm for men and between 157 and 162 cm for women in tenth century Sweden compared with c. 178 cm and 165 cm, respectively, in 2000.
30. Edberg and Karlsson, *Isläggar från Birka och Sigtuna. En undersökning av ett vikingatida och medeltida fyndmaterial*, 45.
31. Choyke, "The Bone Tool Manufacturing Continuum," 66.
32. Küchelmann and Zidarov, "Let's Skate Together! Skating on Bones in the Past and Today," 435.
33. Choyke, "The Bone Tool Manufacturing Continuum," 71.
34. Küchelmann and Zidarov, "Let's Skate Together! Skating on Bones in the Past and Today," 436.
35. Choyke and Bartosiewicz, "Skating with Horses: Continuity and Parallelism in Prehistoric Hungary," 321.
36. Marsha A. Levine, "Eating Horses: The Evolutionary Significance of Hippophagy," *Antiquity* 72, no. 275 (1998): 98–99, doi:10.1017/S0003598X00086300.
37. Choyke, "Bone Skates: Raw Material, Manufacturing and Use," 149; MacGregor, *Bone, Antler, Ivory and Horn: The Technology of Skeletal Materials since the Roman Period*, 31.
38. Anthony, *The Horse, the Wheel, and Language: How Bronze-Age Riders from the Eurasian Steppes Shaped the Modern World*, 182–184.
39. Quoted in Küchelmann and Zidarov, "Let's Skate Together! Skating on Bones in the Past and Today," 439.
40. Ibid.
41. The database was last updated 5/28/2018. The additions are the three (or more) described in Hagberg, "Isläggar och pälsskinn—en frusen kulturhistorisk rapsodi," 80–81; the two described in Kustár and Tugya "'Csontkorcsolyák' a késő bronzkorban (Bone 'Skates' in the Late Bronze Age)"; the two described in Bertolini and Hohenstein, "Evidence of Butchery Marks and Anthropic Modifications on Horse Remains in a Late Bronze Age Site of Northern Italy: The Case of Bovolone"; the nine described in Petrucci, Malerba, and Giacobini, "Manufatti in osso dal castelliere di Pozzuolo del Friuli," 142–147; and the 20 described in Putelat, "Les relations homme-animal dans le monde des vivants et des morts: Étude archéozoologique des établissements et des regroupements funéraires ruraux de l'Arc jurassien et de la Plaine d'Alsace de la fin de l'Antiquité Tardive au premier Moyen Âge," 614. I added modifications 1, 2, and 8 to the skate from Lébeny-Kaszás domb, feature 100, based on the photograph in figures 38 and 39 and corrected the date of the Knoviz skate from Early Bronze Age in Küchelmann's database to Late Bronze Age following Schranil (*Die Vorgeschichte Böhmens und Mährens*, 169). Based on a comment from an anonymous reviewer, I changed the date of the four skates

from Zwingendorf, Austria, from "Roman Iron Age" to Migration Period and the date of the skate from Prellenkirchen, Austria, from Early Iron Age to Migration Period/Early Middle Age. One skate from Leeuwarden, the Netherlands, is included even though it may be slightly later than the others (5th–8th century CE, according to Küchelmann's database). When publications did not list numbers, I assumed the minimum number necessary, i.e., if a publication referred to "horse radius skates," I only counted one.

42. The sites in each country follow. Austria: Gerasdorf am Steinfeld, Mautern, Prellenkirchen, and Zwingendorf; Belgium: Vlissegem (de Haan); Bosnia: Donja Dolina; Czech Republic: Ivanovice na Hané and Přemýšlení; Denmark: Bornholm; France: Altorf, Annecy, Gepolsheim, Rosheim, Saône-et-Loire (le Tertre, d'Authumes), and Wiwersheim; Germany: Cologne, Feddersen Wierde, Gera-Tinz, Auf Esch (Groß Gerau); Kliestow bei Frankfurt an der Oder, Mühlhausen, and Oberdorla, Hungary: Endrőd 170, Gyoma 133, Győr, Lébény-Kaszás domb, Törökbálint, and Zamárdi 89; Italy: Bovolone, Castelliere di Pozzuolo del Friuli; the Netherlands: Dongjum, Englum, Ezinge, Leeuwarden, and Wijnaldum-Tjitsma; Poland: Polwica and Wroclaw; Romania: Gara Banca; Slovakia: Štúrovo; Sweden: Köping, Öland, Torslunda, and Vallhagar; Ukraine: Novokievka, Sabatinovka, and Zlatopol.'

43. Choyke, "Worked Animal Bone at the Sarmatian Site of Gyoma 133," 310–317.

44. The neglected skates are two from London whose dates are questionable—they are not shown on the map—from Library Committee of the Corporation of the City of London, *Catalogue of the Collection of London Antiquities in the Guildhall Museum*, 154, and MacGregor, "Bone Skates: A Review of the Evidence," 59–60, fig. 1C, one skate from each of Vlissegem, Belgium; Sorte Muld, Denmark; Auf Esch (Groß Gerau; possibly more than one skate) and Kliestow bei Frankfurt (Oder), Germany; Lébény-Kaszás domb, Hungary; Dongjum and Wijnaldum-Tjitsma, the Netherlands; Štúrovo, Slovakia; and Hässleby, Öland, and Torslunda, Sweden because neither the species nor the bone type is given; a pair of metapodium skates from Győr, Hungary, whose species is not listed in Küchelmann's database; and horse bone skates from the Czech Republic (the Knoviz skate), Donja Dolina, Bosnia, and Wijnaldum-Tjitsma, the Netherlands, whose bone types are not listed.

45. Pankovskiy, *Bone and Antler Industry from the Late Bronze Age Settlement Sabatynivka-I (Ukraine, South Bug)*.

46. Marschalleck, "Die ostgermanische Siedlung von Kliestow bei Frankfurt (Oder)," 302.

47. Putelat, "Les relations homme-animal dans le monde des vivants et des morts: Étude archéozoologique des établissements et des regroupements funéraires ruraux de l'Arc jurassien et de la Plaine d'Alsace de la fin de l'Antiquité Tardive au premier Moyen Âge," 614.

48. Petrucci, Malerba, and Giacobini, "Manufatti in osso dal castelliere di Pozzuolo del Friuli," 146–147.

49. Hubert Berke, "Mit den Wölfen 'Schlitten fahren' … Ein römisches Eisbein aus einem Wolfsknochen," *Archäologie im Rheinland*, 2003, 122–123.

50. Edberg and Karlsson, *Isläggar från Birka och Sigtuna. En undersökning av ett vikingatida och medeltida fyndmaterial*, 31.

51. Neil Christie and Matt Edgeworth, "Living, Working and Trading in Medieval Wallingford," in *Transforming Townscapes: From* burh *to borough: The Archaeology of Wallingford, AD 800–1400*, ed. Neil Christie et al., Monograph 35 (London: Society for Medieval Archaeology, 2013), 313.

52. Choyke, "Worked Animal Bone at the Sarmatian Site of Gyoma 133," 310–317.

53. Putelat, "Les relations homme-animal dans le monde des vivants et des morts: Étude archéozoologique des établissements et des regroupements funéraires ruraux de l'Arc jurassien et de la Plaine d'Alsace de la fin de l'Antiquité Tardive au premier Moyen Âge," 614.

54. Petrucci, Malerba, and Giacobini, "Manufatti in osso dal castelliere di Pozzuolo del Friuli," 142–151.

55. Edberg and Karlsson, *Isläggar från Birka och Sigtuna. En undersökning av ett vikingatida och medeltida fyndmaterial*, 49–50.

56. Choyke, "The Bone Tool Manufacturing Continuum," 67.

57. Jane Eva Baxter, *The Archaeology of Childhood: Children, Gender, and Material Culture*, Gender and Archaeology 10 (Walnut Creek, CA: Altamira Press, 2005), 49–53.

58. Edberg and Karlsson, "Bone Skates and Young People in Birka and Sigtuna," 10, 15.

59. Choyke, "Worked Animal Bone at the Sarmatian Site of Gyoma 133," 310–317.

60. *Ibid.*, 316.

61. Edberg and Karlsson, *Isläggar från Birka och Sigtuna. En undersökning av ett vikingatida och medeltida fyndmaterial*, 15.

62. The features in the tables are the modifications listed in Küchelmann's database. "Upswept end(s)" refers to modifications 2 and 3; "Chopped" refers to 4–7; "Binding apparatus" refers to 7–16; "Pointed toe" is modification 1; "Visibly flattened gliding surface" is modification 18, which refers to "Visible grinding or chopping marks" in addition to the flattening provided by use; "Roughened standing surface" is modification 17. Modification 19, which occurs when the palmar side of the bone is deliberately removed to expose the medullary cavity, is not to be confused with substantial use wear. None of the skates discussed in this chapter exhibit it.

63. Berg, "Skates and Punt Sleds: Some Scandinavian Notes," 6.

64. Edberg and Karlsson, "Bone Skates and Young People in Birka and Sigtuna," 15.

65. Edberg and Karlsson, *Isläggar från Birka och Sigtuna. En undersökning av ett vikingatida och medeltida fyndmaterial*, 42.

66. Choyke and Bartosiewicz, "Skating with Horses: Continuity and Parallelism in Prehistoric Hungary," 321.

67. Alice M. Choyke, email message to author, September 18, 2018.

68. Choyke and Bartosiewicz, "Skating with Horses: Continuity and Parallelism in Prehistoric Hungary," 320; Choyke and Schibler, "Prehistoric Bone Tools and the Archaeozoological Perspective: Research in Central Europe," 58–59, fig. 11.

69. Oberdorla is known to have been a West Germanic site occupied between the second and fourth centuries of the common era, according to Teichert ("Die Rinder aus dem Opfermoor Oberdorla," 74), and an "Opfermoor" (sacrificial bog: a place where sacrifices, including human sacrifices, were made), accord-

ing to Barthel ("Schlittknochen oder Knochengeräte?" 211).
70. Barthel, "Schlittknochen oder Knochengeräte?" 211–212, 217, 222–224, pl. 32.1–2, 34.1–2.
71. Berke, "Mit den Wölfen 'Schlitten fahren' ... Ein römisches Eisbein aus einem Wolfsknochen."
72. *Ibid.*, 122.
73. Küchelmann, *Bone Skates Database*.
74. The description of the Zwingendorf skates makes use of information provided by an anonymous reviewer.
75. Stenberger, "The Finds and the Dating of the Vallhagar Settlement," 1105.
76. Putelat ("Les relations homme-animal dans le monde des vivants et des morts: Étude archéozoologique des établissements et des regroupements funéraires ruraux de l'Arc jurassien et de la Plaine d'Alsace de la fin de l'Antiquité Tardive au premier Moyen Âge," 614) attributes modifications "2, 3, 6, 7 (17)" to this skate.
77. After reviewing Choyke, "Worked Animal Bone at the Sarmatian Site of Gyoma 133," I made the following changes to the listings in the database: Added modification 3 (upswept proximal end) to the skate blank from pit 45 and the second skate from pit 46; added modification 2 (upswept distal end) to the skate from pit 166; removed modification 7 (palmar side flattened at the proximal end) from the skate from pit 319; added the skate from pit 320, a distal fragment of a horse metacarpus that exhibits an upswept distal end and chopping of the proximal side of the distal end (modifications 2 and 6); added modifications 4 and 6 (lateral and palmar sides flattened at the distal end) to the skate from pit 379; removed modification 6 from the skate from square B13; removed modifications 4 and 6 from the skate from square C3; and changed modifications 4 and 5 (sides flattened at both ends) to 7 for the skate from square D6. These changes are primarily because many of the skates are fragmentary; a proximal fragment cannot exhibit any modifications to its distal end, for example, though some could have been made. All modification numbers are those used in Küchelmann's database.
78. Choyke, "Worked Animal Bone at the Sarmatian Site of Gyoma 133," 308–10.
79. Schiffer, *Formation Processes of the Archaeological Record*, 75.
80. Robson Bonnichsen, "Millie's Camp: An Experiment in Archaeology," *World Archaeology* 4, no. 3 (1973): 277–291, doi:10.1080/00438243.1973.9979539.
81. Gawain Hammond and Norman Hammond, "Child's Play: A Distorting Factor in Archaeological Distribution," *American Antiquity* 46 (1981): 634–636.
82. Bonnichsen, "Millie's Camp: An Experiment in Archaeology," 286–287.
83. Andrea Vaday and Katalin Berecz, "Roman Period Barbarian Settlement at the Site of Gyoma 133," in *Cultural and Landscape Changes in South-East Hungary*, ed. Andrea H. Vaday, vol. 2 (Budapest: Archaeological Institute of the Hungarian Academy of Sciences, 1996), 65–82.
84. Brückner, "Über den heutigen Gebrauch von Schlittknochen in Schlesien," 42.
85. Edberg and Karlson, "Bone Skates and Young People in Birka and Sigtuna," 14.
86. Choyke, "Hidden Agendas: Ancient Raw Material Choice for Worked Osseous Objects in Central Europe and Beyond," 8.

87. Bökönyi, "The Earliest Waves of Domestic Horses in East Europe," 57.
88. Eberhard May, "Widerristhöhe und Langknochenmasse bei Pferden—Ein immer noch aktuelles Problem," *Zeitschrift für Säugetierkunde* 50, no. 6 (1985): 375.
89. Edberg and Karlsson, *Isläggar från Birka och Sigtuna. En undersökning av ett vikingatida och medeltida fyndmaterial*.
90. Choyke and Bartosiewicz, "Skating with Horses: Continuity and Parallelism in Prehistoric Hungary," 322.
91. *Ibid.*, 321.
92. Dresbeck, "The Ski: Its History and Historiography."
93. Allen, "The Primitive Bone Skate," 22–35.
94. Choyke and Bartosiewicz, "Skating with Horses: Continuity and Parallelism in Prehistoric Hungary," 318, fig. 2–3.
95. Drzewicz, *Wyroby z kości i poroża z osiedla obronnego ludności kultury łużyckiej w Biskupinie*, 81–82.
96. Lukasiewicz and Rajewski, "Przedmioti rogowe i kościane z grodu kultury 'lużyckiej' w Biskupinie," 46–47, plate XXXVIII.
97. Choyke, "Hidden Agendas: Ancient Raw Material Choice for Worked Osseous Objects in Central Europe and Beyond," 8.
98. Robert Park, "Size Counts: The Miniature Archaeology of Childhood in Inuit Societies," *Antiquity* 72 (1998): 269–281, doi:10.1017/S0003598X00086567.
99. *Ibid.*, 276.
100. *Ibid.*, 274.
101. *Ibid.*, 279.

Chapter 6

1. The Viking Age is generally considered to begin with the raid on Lindisfarne in 793 and to end with the Norman Conquest of England in 1066.
2. Readers unfamiliar with the sagas would do well to start with the collection in Jane Smiley, ed., *The Sagas of Icelanders* (New York: Penguin Classics, 2001). In this chapter, I have kept the Icelandic spellings of names in translations. These include two characters that may be unfamiliar to modern readers: ð (eth) and þ (thorn). These are both pronounced like modern English "th." Details of pronunciation can be found in Jesse L. Byock, *Viking Language I: Learn Old Norse, Runes, and Icelandic Sagas* (Pacific Palisades, CA: Jules William Press, 2013), 330–334.
3. Throughout this chapter, the titles of the sagas are given in Old Norse, as is usual in the field, with translations provided the first time a saga is mentioned.
4. Edberg and Karlsson, *Isläggar från Birka och Sigtuna. En undersökning av ett vikingatida och medeltida fyndmaterial*.
5. Küchelmann, *Bone Skates Database* lists this skate as undated, but E. Levin Nielsen, "Pederstraede i Viborg: Købstadarkaeologiske undersøgelser 1966/67," *Kuml* 18, no. 18 (1968): 38, 78 attributes it to the Viking Age.
6. Data from Küchelmann, *Bone Skates Database*, except for Lödöse, Sweden. Information on the Lödöse skates was graciously provided by Marie Jonasson Schmidt (email message to author, August 31, 2018).

7. Berg, "The Origin and the Development of the Skis throughout the Ages."
8. Steinar Sørensen, "Daterte skifunn fra middelalderen: Et omriss av middelalderens skihistorie," *Collegium Medievale* 1–2 (1996): 7–55.
9. Clark, *Prehistoric Europe: The Economic Basis*, 295.
10. Weinstock, *Skis and Skiing: From the Stone Age to the Birth of the Sport*, 9–10.
11. Björn Bjarnason, *Íþróttir fornmanna á norðurlöndum*, 2nd ed. (Reykjavík: Sigurður Kristjansson, 1950), 122–137.
12. Snorri Sturluson, *Heimskringla*, ed. Bjarni Aðalbjarnarson, Íslenzk fornrit, 26–28 (Reykjavík: Hið íslenzka fornritafélag, 1941–1951), II.259.
13. *Ibid.*, II.259–260.
14. Geir T. Zoëga, *A Concise Dictionary of Old Icelandic* (1910; repr., Mineola, NY: Dover Publications, 2004), s.v. "kefja."
15. *Ibid.*, s.v. "svima," s.v. "skammr."
16. *Ibid.*, s.v. "kafsund."
17. It is not clear why Cleasby and Vigfusson (*An Icelandic-English Dictionary*, s.v. "ísleggir") specify sheep; the bones of horses and cattle were much more common.
18. Jón Jóhánnesson, ed., *Austfirðinga sǫgur*, Íslenzk fornrit 11 (Reykjavík: Hið íslenzka fornritafélag, 1950), 250.
19. *Ibid.*
20. John Porter, "The Saga of the People of Fljotsdal," in *The Complete Sagas of Icelanders*, ed. Viðar Hreinsson, vol. 4 (Reykjavík: Leifur Eiríksson Publishing, 1997), 402.
21. Eleanor Haworth and Jean Young, *The Fljotsdale Saga and the Droplaugarsons* (London: J. M. Dent and Sons, 1990), 31.
22. Halldór Halldórsson, "Leggir og skautar," 76–80.
23. Helle Degnbol et al., *Ordbog over det norrøne prosasprog / A Dictionary of Old Norse Prose*, University of Copenhagen, 2010, s.v. "beinspýta," http://onp.hum.ku.dk/.
24. Thorsteinn Einarsson, "Winter Sport in Iceland," 59.
25. Gunnar Olof Hyltén-Cavallius, *Wärend och wirdarne. Ett försök i svensk ethnologi* (Stockholm: P. A. Norstedt och söner, 1863–1868), II.464.
26. Guðni Jónsson, *Edda Snorra Sturlusonar, nafnaþulur og skáldatal*, 45.
27. Snorri Sturluson, *Heimskringla*, II.260.
28. Degnbol et al., *Ordbog over det norrøne prosasprog / A Dictionary of Old Norse Prose*, s.v. "skíð."
29. Zoëga, *A Concise Dictionary of Old Icelandic*, s.v. "skíð."
30. Cleasby and Vigfusson, *An Icelandic-English Dictionary*, s.v. "skíð."
31. Haraldr harðráði is generally known as Harald Hardrada or Harald Hardruler today; *harðráði* translates roughly to "hard counsel." The story of his life and his time as king of Norway in the eleventh century is told in *Haralds saga Sigurðarsonar (The Saga of Harald Sigurðarson)* in *Heimskringla*.
32. A good starting point for information on *Morkinskinna* is Eleanor Rosamund Barraclough, *Beyond the Northlands: Viking Voyages and the Old Norse Sagas* (Oxford: Oxford University Press, 2016), 186–187.
33. Andersson and Gade, *Morkinskinna: The Earliest Icelandic Chronicle of the Norwegian Kings (1030–1157)*, 149.
34. *Ibid.*
35. Finnbogi Guðmundsson, *Orkneyinga saga. Legenda de Sancto Magno. Magnúss saga skemmri. Magnúss saga lengri. Helga þáttr ok Úlfs*, 130.
36. Hermann Pálsson and Paul Edwards, *Orkneyinga Saga: The History of the Earls of Orkney* (New York: Penguin, 1978), 108.
37. R. Keyser, P. A. Munch, and C. R. Unger, eds., *Speculum regale. Konungs-Skuggsjá. Konge-Speilet. Et philosophisk-didaktisk Skrift, Forfattet i Norge mod Slutningen af det 12. Aarhundrede. Tilligemed et samtidigt Skrift om den norske Kirkes Stilling til Staten* (Christiania: Carl C. Werner, 1848), 20.
38. Guðni Jónsson, *Edda Snorra Sturlusonar, nafnaþulur og skáldatal*, 41.
39. Hollander translates Atti's epithet as "the Fool" (Snorri Sturluson, *Heimskringla: History of the Kings of Norway*, trans. Lee M. Hollander (Austin: University of Texas Press, 1964), 344); Bjarni Aðalbjarnarson suggests "from the dales" in a footnote (Snorri Sturluson, *Heimskringla*, II.149).
40. Snorri Sturluson, *Heimskringla*, II.149.
41. *Ibid.*, I.259.
42. Olav Bø, *Skiing Traditions in Norway* (Oslo: J. Petlitz Boktrykkeri, 1968), 92.
43. Jóhannes Halldórsson, ed., *Kjalnesinga saga. Jökuls þáttr Búasonar. Víglundar saga. Króka-Refs saga. Þórðar saga hreðu. Finnboga saga. Gunnars saga Keldugnúpsfífls*, Íslenzk fornrit 14 (Reykjavík: Hið íslenzka fornritafélag, 1959), 209–210.
44. Bø, *Skiing Traditions in Norway*, 62–63.
45. Säve, *Svenska lekar*, 77; Katajisto, "Turun Kaupunkialuen luuluistimet: tarkasteltuna osteologiselta ja historialliselta kannalta," 4, quoted in Küchelmann and Zidarov, "Let's Skate Together! Skating on Bones in the Past and Today," 440. Both are reprinted in the Appendix.
46. Roland Huntford, *Two Planks and a Passion: The Dramatic History of Skiing* (New York: Continuum, 2008), 51.
47. Zoëga, *A Concise Dictionary of Old Icelandic*, s.v. "broddstöng."
48. Cleasby and Vigfusson, *An Icelandic-English Dictionary*, s.v. "broddstöng."
49. Degnbol et al., *Ordbog over det norrøne prosasprog / A Dictionary of Old Norse Prose*, s.v. "broddstǫng."
50. Hjalmar Falk, *Altnordische Waffenkunde*, Videnskabsselskabets skrifter 2, Historisk-filosofiske klasse 6 (Kristiania: J. Dybwad, 1914), 69.
51. Björn Bjarnason, *Íþróttir fornmanna á norðurlöndum*, 121.
52. Thorsteinn Einarsson, "Winter Sport in Iceland," 59.
53. Allen, *The Culture and Sport of Skiing: From Antiquity to World War II*, 17.
54. Weinstock, *Skis and Skiing: From the Stone Age to the Birth of the Sport*, 73–74.
55. Dresbeck, "The Ski: Its History and Historiography," 477; Berg, "The Origin and the Development of the Skis throughout the Ages," 41.
56. Sørensen, "Daterte skifunn fra middelalderen: Et omriss av middelalderens skihistorie," 54.
57. Berg, "The Origin and the Development of the Skis throughout the Ages," 41.
58. This refers to Eurasian elk, which are called moose in North America.
59. R. Keyser and P. A. Munch, eds., *Norges gamle love indtil 1387*, vol. 2: Lovgivningen under Kong

Mangus Haakonssöns Regjeringstid fra 1263 til 1280, tilligemed et Supplement til förste Bind (Christiania: Chr. Gröndahl, 1848), 143.

60. Finnur Jónsson, ed., *Morkinskinna* (Copenhagen: J. Jørgensen, 1932), 309.

61. Degnbol et al., *Ordbog over det norrøne prosasprog / A Dictionary of Old Norse Prose*, s.v. "andr."

62. Guðni Jónsson, *Edda Snorra Sturlusonar, nafnaþulur og skáldatal*, 41.

63. *Ibid.*, 126.

64. Cleasby and Vigfusson, *An Icelandic-English Dictionary*, s.v. "andrar."

65. Einar Haugen, ed., *Norwegian-English Dictionary* (Madison: University of Wisconsin Press, 1974), 498.

66. Cleasby and Vigfusson, *An Icelandic-English Dictionary*, s.v. "andrar."

67. Weinstock, *Skis and Skiing: From the Stone Age to the Birth of the Sport*, 77.

68. Edberg and Karlsson, "Bone Skates and Young People in Birka and Sigtuna," 15.

69. Zoëga, *A Concise Dictionary of Old Icelandic*, s.v. "skríða."

70. Jan de Vries, *Altnordisches etymologisches Wörterbuch* (Leiden: Brill, 1962), s.v. "skríða."

71. Sveinbjörn Egilsson and Finnur Jónsson, *Lexicon poeticum antiquae linguae septentrionalis*, 2nd ed. (Copenhagen: Atlas Bogtryk, 1966), s.v. "skríða."

72. Ruben Nöjd, Astrid Tornberg, and Margareta Ångström, *McKay's Modern English-Swedish and Swedish-English Dictionary* (New York: David McKay, 1965), s.v. "skrida," "skridsko."

73. de Vries, *Altnordisches etymologisches Wörterbuch*.

74. Haugen, *Norwegian-English Dictionary*, s.v. "skri."

75. Björn Sigfússon, ed., *Ljósvetninga saga*, Íslenzk fornrit 10 (Reykjavík: Hið íslenzka fornritafélag, 1940), 200–201.

76. Jónas Kristjánsson, ed., *Eyfirðinga sögur*, Íslenzk fornrit 9 (Reykjavík: Hið íslenzka fornritafélag, 1956), 253.

77. Einar Ól. Sveinsson, ed., *Brennu-Njáls saga*, Íslenzk fornrit 12 (Reykjavík: Hið íslenzka fornritafélag, 1954), 233.

78. Robertson, *Materials for the History of Thomas Becket, Archbishop of Canterbury*, 11.

79. Douglas and Greenaway, *English Historical Documents*, 961.

80. Björn Sigfússon, *Ljósvetninga saga*, 200–201.

81. Jónas Kristjánsson, *Eyfirðinga sögur*, 253.

82. Einar Ól. Sveinsson, *Brennu-Njáls saga*, 233.

83. Zoëga, *A Concise Dictionary of Old Icelandic*, s.v. "skríða."

84. Guðni Jónsson, ed., *Grettis saga Ásmundarsonar, Bandamanna saga. Odds Þáttr Ófeigssonar*, Íslenzk fornrit 7 (Reykjavík: Hið íslenzka fornritafélag, 1936), 232.

85. Barraclough, *Beyond the Northlands: Viking Voyages and the Old Norse Sagas*, 77.

86. Weinstock, *Skis and Skiing: From the Stone Age to the Birth of the Sport*, 66–69.

87. Einar Ól. Sveinsson, *Brennu-Njáls saga*, 479.

88. Ernst A. Kock, *Den norsk-isländska skaldediktningen* (Lund: C. W. K. Gleerup, 1946–1949), I.71.

89. Weinstock, *Skis and Skiing: From the Stone Age to the Birth of the Sport*, 68.

90. R. Quirk, *The Saga of Gunnlaug Serpent-Tongue* (London, NY: T. Nelson, 1957), 25.

91. Snorri Sturluson, *Heimskringla*, I.202–203.

92. Snorri Sturluson, *Heimskringla or the Chronicle of the Kings of Norway*, Release #15b, 1996, trans. Samuel Liang (Online Medieval and Classical Library, 1844), §1, http://omacl.org/Heimskringla/.

93. Rasmus B. Anderson, "Introduction," in *The Heimskringla or the Sagas of the Norse Kings*, 2nd ed., by Snorri Sturluson, trans. Samuel Liang (London: John C. Nimmo, 1889), xiii.

94. Snorri Sturluson, *Heimskringla: History of the Kings of Norway*, 131.

95. Grammaticus, *Historia Danica*, 131.

96. Grammaticus, *The History of the Danes, Books I–IX*, I.79.

97. The carved skate's accession number at the Swedish History Museum is 1197905.

98. Nielsen, "Pedersstraede i Viborg: Købstadarkaeologiske undersøgelser 1966/67," 38, 73, fig. 9.

99. Einar Ól. Sveinsson and Matthías Þórðarson, eds., *Eyrbyggja saga*, Íslenzk fornrit 4 (Reykjavik: Hið íslenzka fornritafélag, 1965), 127–28.

100. Magnus, *Historia de gentibus septentrionalibus*, 41–42.

101. Magnus, *Description of the Northern Peoples*, 57–58.

102. Ursula Dronke, ed., *The Poetic Edda* (Oxford: Clarendon Press, 1969–2011), III.18.

103. Halldór Halldórsson, "Leggir og skautar," 75–76.

104. Björn Bjarnason, *Íþróttir fornmanna á norðurlöndum*, 121–122.

105. Lee M. Hollander, trans., *The Poetic Edda* (Austin: University of Texas Press, 1962), 26.

106. Dronke, *The Poetic Edda*, III.18.

107. Jackson Crawford, trans., *The Poetic Edda: Stories of the Norse Gods and Heroes*, Hackett Classics (Indianapolis: Hackett Publishing Company, 2015), 31.

108. Carolyne Larrington, trans., *The Poetic Edda*, Oxford World's Classics (Oxford: Oxford University Press, 2009), 25.

109. Andy Orchard, trans., *The Elder Edda: A Book of Viking Lore* (New York: Penguin, 2011), 26.

110. Dronke, *The Poetic Edda*, II.243.

111. Hollander, *The Poetic Edda*, 160; Orchard, *The Elder Edda: A Book of Viking Lore*, 101–102; Larrington, *The Poetic Edda*, 102.

112. Zoëga, *A Concise Dictionary of Old Icelandic*, s.v. "skíð."

113. Cleasby and Vigfusson, *An Icelandic-English Dictionary*, s.v. "skíð."

114. Weinstock, *Skis and Skiing: From the Stone Age to the Birth of the Sport*, 59. Although snowshoes were uncommon in medieval Scandinavia, they were not unknown. Olaus Magnus, *Historia de gentibus septentrionalibus*, 147–148 describes people attaching baskets to their feet and the feet of their horses to cross the snowy mountains in northern Sweden and Norway.

115. Dronke, *The Poetic Edda*, II.243.

116. Zoëga, *A Concise Dictionary of Old Icelandic*, s.v. "hlaupa."

117. Adam Lewenhaupt, ed., *Calendaria Caroli IX* (Stockholm: P. A. Norstedt och söner, 1903), 104.

118. Cleasby and Vigfusson, *An Icelandic-English Dictionary*, s.v. "hlaupa."

Chapter 7

1. "Middle Ages" is a very broad term that refers to different dates in different areas. The Middle Ages began as early as the sixth century in central Europe and continued until as late as the sixteenth century. In this chapter, I follow Küchelmann's lead (in his *Bone Skates Database*) in identifying medieval skates. As of this writing, the database includes over 1800 skates that may date to the Middle Ages (some dates are questionable, and the number of skates found at some sites is not given).

2. Jane Kershaw's recent papers, "Culture and Gender in the Danelaw: Scandinavian and Anglo-Scandinavian Brooches," *Viking and Medieval Scandinavia* 5 (2009): 295–325, doi:10.1484/J.VMS.1.100682; "An Early Medieval Dual-Currency Economy: Bullion and Coin in the Danelaw," *Antiquity* 91, no. 355 (2017): 173–190, doi:10.15184/aqy.2016.249; and (with Ellen C. Røyrvik) "The 'People of the British Isles' Project and Viking Settlement in England," *Antiquity* 90, no. 354 (2016): 1670–1680, doi:10.15184/aqy.2016.193, are also of interest.

3. Thomas Himstedt, "How the Vikings Got to the Rhine: A Historical-Geographical Survey over the Rhinelands in the Early Middle Ages," in *Vikings on the Rhine: Recent Research on Early Medieval Relations between the Rhinelands and Scandinavia*, ed. Rudolf Simek and Ulrike Engel, Studia Medievalia Septentrionalia 11 (Vienna: Verlag Fassbaender, 2004), 24.

4. Malcolm K. Hughes and Henry F. Diaz, "Was There a 'Medieval Warm Period,' and If So, Where and When?" *Climate Change* 26 (1994): 109–142.

5. Lamb, *Climate, History, and the Modern World*, 163–164.

6. W. Dansgaard et al., "Climate Changes, Norsemen, and Modern Man," *Nature* 255 (1975): figure 3, doi:10.1038/255024a0.

7. Oxygen-18 (an isotope of oxygen with atomic mass 18) is slightly heaver than regular oxygen (atomic mass 16) because of it has two extra neutrons in each atom.

8. Dansgaard et al., "Climate Changes, Norsemen, and Modern Man," 27.

9. Moinuddin Ahmed et al., "Continental-Scale Temperature Variability during the Past Two Millennia," *Nature Geoscience* 6, no. 5 (2013): 339–346, doi:10.1038/NGEO1797.

10. James W. Hurrell et al., eds., *The North Atlantic Oscillation: Climatic Significance and Environmental Impact*, Geophysical Monographs (Washington, DC: American Geophysical Union, 2003), 17.

11. Brian Fagan, *The Little Ice Age: How Climate Made History 1300–1850* (New York: Basic Books, 2000), 27.

12. Jean Manco, *Ancestral Journeys: The Peopling of Europe from the First Venturers to the Vikings*, Revised ed. (New York: Thames and Hudson, 2015), 224–231.

13. The 28 May 2018 update of Küchelmann's *Bone Skates Database* is used. When the number of skates found is not listed, I assume it is one. I neglect the Migration Period/Early Middle Age skate from Leeuwarden in the Netherlands because it may be too early and the skate from Grimmersum, Germany because its date is questionable.

14. The additional skates are from Edberg and Karlsson, *Isläggar från Birka och Sigtuna. En undersökning av ett vikingatida och medeltida fyndmaterial* (which provides details of the 112 skates from Birka reported in Stenberger, "The Finds and the Dating of the Vallhagar Settlement," 1106); J. E. Mann, *Early Medieval Finds from Flaxengate I: Objects of Antler, Bone, Stone, Horn, Ivory, Amber, and Jet* (Lincoln: Lincoln Archaeological Trust, 1982); A. Götze, "Der Schlossberg bei Burg im Spreewald," *Praehistorische Zeitschrift* 4 (1912): 264–350; Christie and Edgeworth, "Living, Working and Trading in Medieval Wallingford"; John Thomas, "Evidence for the Dissolution of Thorney Abbey: Recent Excavations and Landscape Analysis at Thorney, Cambridgeshire," *Medieval Archaeology* 50, no. 1 (2006): 179–241, doi:10.1179/174581706x124257; Jennifer Browning, *Archaeological Excavations on Land off Stapleford Road, Whissendine, Rutland (NGR SK 553 086)*, University of Leicester Archaeological Services Report Number 2007–066, 2007; David Baker et al., "Excavations in Bedford 1967–1977," *Bedfordshire Archaeological Journal* 13 (1979): 1–309; Arthur MacGregor, "Skates," in *Artifacts from Medieval Winchester*, ed. Martin Biddle, vol. 7.2: Object and Economy in Medieval Winchester (Oxford: Clarendon Press, 1990), 708–709. The details of the skates described in MacGregor, "Bone Skates: A Review of the Evidence," some of the skates described in Becker, "Bemerkungen über Schlittknochen, Knochenkufen und ähnliche Artefakte, unter besonderer Berücksichtigung der Funde aus Berlin Spandau," and the skates found at Starigard, a Slavic site at Oldenburg in Holstein, Germany, described in Wietske Prummel, *Starigard/Oldenburg: Hauptburg der Slawen in Wagrien*, vol. 4: *Die Tierknochenfunde unter besonderer Berücksichtigung der Beizjagd* (Neumünster: Karl Wachholtz Verlag, 1993), are added.

15. Edberg and Karlsson, *Isläggar från Birka och Sigtuna. En undersökning av ett vikingatida och medeltida fyndmaterial*.

16. Locations are counted at the city level. For example, "General Post Office," "Lloyds," "Pudding Lane," "Swan Lane," and "Watling Court," which are all in London, are grouped together under "London" instead of being counted as separate sites. Some skates have no definite provenance and are simply listed as having been found in, for example, "England"; these are accorded one location per general entry, i.e., all the "England" skates are grouped together.

17. The sites in Great Britain are Bedford, Durham, Empingham, Huntington, Ipswich, Lincoln, London, Northampton, Norwich, Oxford, Reigate, Thetford, Thorney, Torksey, Wallingford, Waltham Abbey, Whissendine, Winchester, and York.

18. The Central skates and their locations follow. Austria: Baumgarten an der March, Podersdorf am See, Stockerau, and Tulln an der Donau. Bulgaria: Durankulak and Garvan Village. Czech Republic: Kadan, Krnov, Mikulčice, and Pohansko, plus some unlocalized skates from Moravia. Germany: Spreewald, Oldenburg in Holstein (Starigard), Berlin, Bernshausen, Böhl-Iggelheim, Halberstadt, Harste, Igersheim, Constance, Mühlhausen, Niederdorla, and Obertraubling, plus some unlocalized skates from Hessens and Saxony, plus, along the northern coast, Groß Raden, Groß Strömkendorf, Hamburg, Lübeck, Mecklenburg, Menzlin, Teterow, and Usedom. Poland: Gniezno, Kalisz, Opole, Poznań, and Wrocław, plus unlocalized skates. Serbia: Sapaja, Botra, Crna Bara, and Crvenka. Slovakia: Bajč. The numbers of skates from Gniezno and of unlocalized skates from Saxony and Poland are not included in Küchelmann's data-

base; therefore, one skate from each site is assumed in the statistics.

19. The sites in the Northeast follow. Estonia: Kuusalu, Lihula, Otepää, Pada, Pärnu, Suure-Jaani, Tallin, and Viljandi. Finland: Kuusisto and Turku. Latvia: Laukskola and Märtinsala in Salaspils, the Aizkraukle and Daugmale Hill-Forts, Riga, Turaida Castle, and Ventspils. Lithuania: Palanga. Russia: Mokhovoye and Novgorod.

20. The sites in the Northwest follow. Belgium: Antwerp, Ghent, and Grimbergen (in Brabant), plus unspecified numbers of skates from Brussels, Nuy, Namur, and Flanders; these are included in the statistics as one per site. France: Saint-Denis and Wandignies-Hamage. Germany: Borken, Bremen, Cologne (including Blumenberg), Emden, Esens, Feddersen Wierde (now Cuxhaven), Jever, Oldorferwarf, and Wirdum. The Netherlands: Aalsum, Anjum, Dongjum, Dorestad, Kastell Lent, Leeuwarden, Medemblik, Oldeboorn, Oost-Souburg, Schagen, Wijk bij Duurested, and Wijnaldum-Tjitsma.

21. The Scandinavian locations follow. Denmark: Århus, Ribe, Viborg, and Trelleborg (Zealand);Germany: Eiderstadt (Elisenhof), Haithabu, Olsborg, and Wellinghusen, plus various sites in Schleswig, including Plessenstrasse and Schild, and some unlocalized skates from Schleswig-Holstein. Norway: Bergen, Oslo, and Trondheim. Sweden: Birka, Lund, Lödöse, and Sigtuna, plus Hellvi, Lau, and Burs Church in Stäng on Gotland and some unlocalized skates from Öland.

22. MacGregor, "Problems in the Interpretation of Microscopic Wear Patterns: The Evidence from Bone Skates"; MacGregor, "Bone Skates: A Review of the Evidence"; MacGregor, *Bone, Antler, Ivory and Horn: The Technology of Skeletal Materials since the Roman Period.*

23. Küchelmann and Zidarov developed the list of features for their project, "Let's Skate Together! Skating on Bones in the Past and Today." Edberg and Karlsson (*Isläggar från Birka och Sigtuna. En undersökning av ett vikingatida och medeltida fyndmaterial*, 18, 22) extended it to include some features of Swedish skates.

24. For example, some of the 50 skates from Novgorod, Russia, feature binding holes, but how many is unknown. They are made from horse and cattle metapodia and radii, but how many of each is not specified. I have left them all as featureless skates of unknown bone type due to the lack of clear documentation. This is another area where better and broadly disseminated documentation would help immensely.

25. Edberg and Karlsson, *Isläggar från Birka och Sigtuna. En undersökning av ett vikingatida och medeltida fyndmaterial*; Edberg and Karlsson, "Bone Skates and Young People in Birka and Sigtuna."

26. W. Eekhoff, "Oude beenen schaatsen," *Nieuwe friesche volks-almanak* 11 (1863): 108.

27. Richard Hall, "Markets of the Danelaw," in *The Vikings in England and in Their Danish Homeland*, ed. Else Roesdahl et al. (London: The Anglo-Danish Viking Project, 1981), 116.

28. Matthew Townend, *Language and History in Viking Age England: Linguistic Relationships between Speakers of Old English and Old Norse*, Studies in the Early Middle Ages 6 (Turnhout, Belgium: Brepols, 2002).

29. Bryan Sykes, *Saxons, Vikings, and Celts: The Genetic Roots of Britain and Ireland* (New York: W. W. Norton, 2006); Stephen Oppenheimer, *The Origins of the British: A Genetic Detective Story* (New York: Carroll and Graf, 2006).

30. D. M. Hadley, *The Vikings in England: Settlement, Society and Culture*, Manchester Medieval Studies (Manchester: Manchester University Press, 2006).

31. Kershaw, "Culture and Gender in the Danelaw: Scandinavian and Anglo-Scandinavian Brooches."

32. Bertram Colgrave, Judith McClure, and Roger Collins, eds., *Bede. The Ecclesiastical History of the English People*, Oxford World's Classics (Oxford: Oxford University Press, 1999), 27.

33. Annet Nieuwhof, "Anglo-Saxon Immigration or Continuity? Ezinge and the Coastal Area of the Northern Netherlands in the Migration Period," *Journal of Archaeology in the Low Countries* 5, no. 1 (2013): 73.

34. D. A. Gerrets, "Conclusions," chap. 20 in *The Excavations at Wijnaldum*, ed. J. C. Besteman et al., Reports on Frisia in Roman and Medieval Times 1 (Rotterdam: A. A. Balkema, 1999), n.p. (§6).

35. Nieuwhof, "Anglo-Saxon Immigration or Continuity? Ezinge and the Coastal Area of the Northern Netherlands in the Migration Period," 77.

36. Wietske Prummel, Hülya Halici, and Annemieke Verbaas, "The Bone and Antler Tools from the Wijnaldum-Tjitsma *Terp*," *Journal of Archaeology in the Low Countries* 3, nos. 1/2 (2011): 68, 97.

37. Annet Nieuwhof, "The Excavation at Englum, a Wierde in the Province of Groningen, the Netherlands: Summary of the Excavation Results," chap. 13 in *De Leege Wier van Englum: Archeologisch onderzoek in het Reitdiepgebied*, ed. Annet Nieuwhof, Jaarverslagen van de Vereniging voor Terpenonderzoek 91 (Groningen: Vereniging voor Terpenonderzoek / Tienkamp en Verheij, 2008), 258.

38. Nieuwhof, "Anglo-Saxon Immigration or Continuity? Ezinge and the Coastal Area of the Northern Netherlands in the Migration Period," 56.

39. Malcolm Todd, "Feddersen Wierde," in *The Oxford Companion to Archaeology*, ed. Brian Fagan (Oxford: Oxford University Press, 1996), 236.

40. Todd, *The Early Germans*, 222.

41. MacGregor, "Bone Skates: A Review of the Evidence," 59–60.

42. Library Committee of the Corporation of the City of London, *Catalogue of the Collection of London Antiquities in the Guildhall Museum*, 154.

43. Lamb, *Climate, History, and the Modern World*, 151.

44. Else Roesdahl, *The Vikings* (New York: Penguin, 1987), 235–235.

45. Hadley, *The Vikings in England: Settlement, Society and Culture*, 29–31.

46. MacGregor, "Bone Skates: A Review of the Evidence."

47. For details, see Arthur MacGregor, Ailsa J. Mainman, and Nicola S. H. Rogers, *Craft, Industry and Everyday Life: Bone, Antler, Ivory and Horn from Anglo-Scandinavian and Medieval York*, The Archaeology of York 17: The Small Finds, fasc. 12 (London: Council for British Archaeology, 1999), 1984–1989.

48. Jeffrey Radley, "Economic Aspects of Anglo-Danish York," *Medieval Archaeology* 15 (1971): 57, 55, doi:10.1080/00766097.1971.11735336.

49. MacGregor, "Bone Skates: A Review of the Evidence," 66–67.

50. Roesdahl, *The Vikings*, 243.

51. MacGregor, "Bone Skates: A Review of the Evidence," 71.

52. MacGregor, Mainman, and Rogers, *Craft, Industry and Everyday Life: Bone, Antler, Ivory and Horn from Anglo-Scandinavian and Medieval York, 1984–1989*, 2024.

53. MacGregor, "Bone Skates: A Review of the Evidence"; Library Committee of the Corporation of the City of London, *Catalogue of the Collection of London Antiquities in the Guildhall Museum* (it is not clear whether these skates are included in MacGregor, "Bone Skates: A Review of the Evidence"); Herman, "Knochenschlittschuh, Knochenkufe, Knochenkeitel: Ein Beitrag zur näheren Kenntnis der řrähistorischen Langknochenfunde"; MacGregor, *Bone, Antler, Ivory and Horn: The Technology of Skeletal Materials since the Roman Period*; Julian Ayre and Robin Wroe-Brown, "The Post-Roman Foreshore and the Origins of the Late Anglo-Saxon Waterfront and Dock of Aethelred's Hithe: Excavations at Bull Wharf, City of London," *Archaeological Journal* 172, no. 1 (2015): 138, doi:10.1080/00665983.2014.984534; Barbara West, "A Note on Bone Skates from London," *Transactions of the London and Middlesex Archaeological Society* 32 (1982): 303; Smith, "Ancient Bone Skates."

54. For the Thetford skates, see MacGregor, "Bone Skates: A Review of the Evidence," App. 1, and Andrew Rogerson and Carolyn Dallas, *Excavations in Thetford 1948–59 and 1973–80*, East Anglian Archaeology 22 (Norfolk: The Norfolk Archaeological Unit, 1984).

55. MacGregor, "Bone Skates: A Review of the Evidence," App. 1.

56. MacGregor, *Bone, Antler, Ivory and Horn: The Technology of Skeletal Materials since the Roman Period*, 144.

57. Some, or perhaps all, of the Bedford skates are more fully documented in Baker et al., "Excavations in Bedford 1967–1977," nos. 1512, 1513, 1515. MacGregor, *Bone, Antler, Ivory and Horn: The Technology of Skeletal Materials since the Roman Period*, 144, gives an eighth-century date for the skates from Bedford, which would also be problematic, but Baker et al., "Excavations in Bedford 1967–1977," 288, lists two as being from the eleventh or twelfth century and the third, a fragment, as dating to the fourteenth century.

58. Thomas, "Evidence for the Dissolution of Thorney Abbey: Recent Excavations and Landscape Analysis at Thorney, Cambridgeshire," 199.

59. Mann, *Early Medieval Finds from Flaxengate I: Objects of Antler, Bone, Stone, Horn, Ivory, Amber, and Jet*, 16–18.

60. Thomas, "Evidence for the Dissolution of Thorney Abbey: Recent Excavations and Landscape Analysis at Thorney, Cambridgeshire," 199.

61. MacGregor, "Skates," 708–709.

62. Christie and Edgeworth, "Living, Working and Trading in Medieval Wallingford," 313.

63. Browning, *Archaeological Excavations on Land off Stapleford Road, Whissendine, Rutland (NGR SK 553 086)*, 70.

64. MacGregor, "Bone Skates: A Review of the Evidence," 65.

65. Ryan Lavelle and Simon Roffey, "Introduction: Danes in Wessex," in *Danes in Wessex: The Scandinavian Impact on Southern England, c. 800–c. 1100*, ed. Ryan Lavelle and Simon Roffey (Oxford: Oxbow Books, 2016), 2.

66. Anders Winroth, *The Age of the Vikings* (Princeton, NJ: Princeton University Press, 2014), 54.

67. John Blair, *Anglo-Saxon Oxfordshire* (Phoenix Mill, UK: Sutton Publishing, 1998), 167.

68. The skates from Dublin are excluded from this part of the analysis because no information beyond their date is available.

69. Edberg and Karlsson, *Isläggar från Birka och Sigtuna. En undersökning av ett vikingatida och medeltida fyndmaterial*, 44–45.

70. Edberg and Karlsson, "Bone Skates and Young People in Birka and Sigtuna," 15.

71. Douglas and Greenaway, *English Historical Documents*, 961.

72. R. E. Latham, D. R. Howlett, and R. K. Ashdowne, eds., *Dictionary of Medieval Latin from British Sources* (Oxford: British Academy, 1975–2013), s.v. "juvenis."

73. Lawrence M. Larson, *The Earliest Norwegian Laws: Being the Gulathing Law and the Frostathing Law* (New York: Columbia University Press, 1935), 272.

74. Edberg and Karlsson, *Isläggar från Birka och Sigtuna. En undersökning av ett vikingatida och medeltida fyndmaterial*, 45.

75. MacGregor, Mainman, and Rogers, *Craft, Industry and Everyday Life: Bone, Antler, Ivory and Horn from Anglo-Scandinavian and Medieval York*, 1989.

76. Ibid., 1987.

77. Kershaw, "Culture and Gender in the Danelaw: Scandinavian and Anglo-Scandinavian Brooches," 296.

78. Kershaw, "An Early Medieval Dual-Currency Economy: Bullion and Coin in the Danelaw."

79. Kershaw, "Culture and Gender in the Danelaw: Scandinavian and Anglo-Scandinavian Brooches," 302–303.

80. Ibid., 303.

81. Kershaw, "An Early Medieval Dual-Currency Economy: Bullion and Coin in the Danelaw," 184.

82. Kershaw, "Culture and Gender in the Danelaw: Scandinavian and Anglo-Scandinavian Brooches," 302.

83. Townend, *Language and History in Viking Age England: Linguistic Relationships between Speakers of Old English and Old Norse*, 189.

84. Stephen Leslie et al., "The Fine-Scale Genetic Structure of the British Population," *Nature* 519 (2015): 309–314, doi:10.1038/nature14230.

85. Kershaw and Røyrvik, "The 'People of the British Isles' Project and Viking Settlement in England," 1679.

86. Sykes, *Saxons, Vikings, and Celts: The Genetic Roots of Britain and Ireland*, 282–283, 286.

87. Dirk Meier, *Seafarers, Merchants and Pirates in the Middle Ages* (Woodbridge: Boydell Press, 2006), 57.

88. Annemarieke Willemsen, "Scattered across the Waterside: Viking Finds from the Netherlands," in *Vikings on the Rhine. Recent Research on Early Medieval Relations between the Rhinelands and Scandinavia*, ed. Rudolf Simek and Ulrike Engel, Studia Medievalia Septentrionalia 11 (Vienna: Verlag Fassbaender, 2004), 77–78.

89. D. P. Block, "De Wikingen in Friesland," *Naamkunde* 10 (1978): 25–47, cited in Willemsen, "Scattered across the Waterside: Viking Finds from the Netherlands," 77–78.

90. Annemarieke Willemsen, *Vikings! Raids in the Rhine/Meuse Region 800–1000* (Utrecht: Centrall Museum, 2004), 35.

91. Edberg and Karlsson, *Isläggar från Birka och*

Sigtuna. En undersökning av ett vikingatida och medeltida fyndmaterial, 31.

92. Willemsen, "Scattered across the Waterside: Viking Finds from the Netherlands," 73–74.

93. Meier, *Seafarers, Merchants and Pirates in the Middle Ages*, 58.

94. Winroth, *The Age of the Vikings*, 47–48.

95. Willemsen, "Scattered across the Waterside: Viking Finds from the Netherlands," 71.

96. Lauwerier and Van Heeringen, "Skates and Prickers from the Circular Fortress of Oost-Souburg, the Netherlands (AD 900–975)," 121.

97. Meier, *Seafarers, Merchants and Pirates in the Middle Ages*, 139.

98. Winroth, *The Age of the Vikings*, 125.

99. Christiane Zimmermann and Hauke Jöns, "Cultural Contacts between the Western Baltic, the North Sea Region and Scandinavia," in *Frisians and their North Sea Neighbours: From the Fifth Century to the Viking Age*, ed. John Hines and Nelleke IJssennagger (Woodbridge: Boydell Press, 2017), 254.

100. P. M. Barford, *The Early Slavs* (London: British Museum Press, 2001), 117–118.

101. Zimmermann and Jöns, "Cultural Contacts between the Western Baltic, the North Sea Region and Scandinavia," 252.

102. Z. Kratochvíl and O. Štěrba, "Osteologische Analyse der heimischen Knochenindustrie aus Mikulčice und Pohansko," *Archeologické rozhledy* 22 (1970): 452, Tab. 2.

103. Barford, *The Early Slavs*, 258.

Chapter 8

1. Munro, "Notes on Ancient Bone Skates," 194.
2. Hagberg, "Fundort und Fundgebiet der Modeln aus Torslunda," 330.
3. Küchelmann, *Bone Skates Database*.
4. Thorsteinn Einarsson, "Winter Sport in Iceland," 59.
5. János Makkay, personal communication to Alice M. Choyke, n.d., cited in Choyke, "Bone Skates: Raw Material, Manufacturing and Use," 152.
6. Blauw, *Van glis tot klapschaats: Schaatsen en schaatsenmakers in Nederland, 1200 tot heden*, 57.
7. Niko Mulder, "Ten IJse (1)," *Kouwe Drukte* 12, no. 33 (October 2008): 27.
8. Olaf Goubitz, "Patten (type 110)," in *Stepping through Time: Archaeological Footwear from Prehistoric Times until 1800* (Zwolle, The Netherlands: Stichting Promotie Archeologie, 2001), 249–252.
9. The laws are quoted in English translation by Goubitz, "Patten (type 110)," 252. Pattens and their use on ice are described in more detail in W. P. Dezutter, "Slijkschoenen: Twee aanwinsten voor het Stedelijk Museum voor Volkskunde te Brugge," *Biekorf* 75 (1974): 289–304.
10. Goubitz, "Patten (type 110)," 252.
11. Niko Mulder and Jos Pronk, *Acht eeuwen schatsen in en om Amsterdam* (Bergen, Netherlands: De Poolster, 2011), 10.
12. Mulder, "Ten IJse (4)—Met scaetzen en souerding op de hofgracht," 37; B. R. de Melker, *Oorkondenboek van Amsterdam tot 1400: Supplement*, Apparaat voor de geschiedenis van Holland 12 (Amsterdam: Historische Vereniging Holland, 1995), 54–55.
13. Mulder, "Ten IJse (4)—Met scaetzen en souerding op de hofgracht," 37.
14. Jan de Vries, *Nederlands etymologisch woordenboeck* (Leiden: Brill, 1971), s.v. "schaats."
15. Goubitz, "Patten (type 110)," 249.
16. Fowler, *On the Outside Edge: Being Diversions in the History of Skating*, 43.
17. Roes, *Bone and Antler Objects from the Frisian Terp-Mounds*, 58.
18. J. H. van Dale and C. Kruyskamp, *Groot Woordenboek der Nederlandse Taal*, 8th ed. (The Hauge: Nijhoff, 1961), 1757.
19. Herman, "Knochenschlittschuh, Knochenkufe, Knochenkeitel: Ein Beitrag zur näheren Kenntnis der prähistorischen Langknochenfunde," 226.
20. Rudolf Virchow, "Einige Ueberlebsel in pommerschen Gebräuchen." In Verhandlungen der Berliner Gesellschaft für Anthropologie, Ethnologie und Urgeschichte, *Zeitschrift für Ethnologie* 19 (1887): 361–362.
21. Balfour, "Sledges with Bone Runners in Modern Use," 250.
22. Thunig, "Über Schlittknochen und Gräbeurnen," 280.
23. Olaf Goubitz, "Nederland's oudste schaats?" *Kouwe Drukte* 3, no. 9 (September 2000): 5.
24. Mulder, "Ten IJse (1)," 28.
25. Mulder, "Ten IJse (4)—Met scaetzen en souerding op de hofgracht," 39.
26. Edberg and Karlsson, *Isläggar från Birka och Sigtuna. En undersökning av ett vikingatida och medeltida fyndmaterial*, 23, 31.
27. Goubitz, "Nederland's oudste schaats?" 249.
28. Magnus, *Historia de gentibus septentrionalibus*, 41–42.
29. Magnus, *Description of the Northern Peoples*, 58.
30. Formenti and Minetti, "Human Locomotion on Ice: The Evolution of Ice-Skating Energetics through History," 1826.
31. Mulder, "Ten IJse (1)," 28.
32. *Ibid.*, 26.
33. *Ibid.*, 28.
34. Fowler, *On the Outside Edge: Being Diversions in the History of Skating*, 39.
35. Marie Jonasson Schmidt, email message to author, September 14, 2018.
36. Formenti and Minetti, "Human Locomotion on Ice: The Evolution of Ice-Skating Energetics through History," 1831.
37. This manuscript image could not be reproduced here, but it has been published in Lilian M. C. Randall, *Images in the Margins of Gothic Manuscripts*, California Studies in the History of Art 4 (Berkeley: University of California Press, 1966), 31, no. 471 and mentioned in Blauw, *Van glis tot klapschaats: Schaatsen en schaatsenmakers in Nederland, 1200 tot heden*, 11 and Küchelmann and Zidarov, "Let's Skate Together! Skating on Bones in the Past and Today," 432. It can be found online in the Oxford Digital Library, http://www.odl.ox.ac.uk/.
38. Mulder, "Ten IJse (4)—Met scaetzen en souerding op de hofgracht," 39.
39. Blauw, *Van glis tot klapschaats: Schaatsen en schaatsenmakers in Nederland, 1200 tot heden*, 57.
40. Niko Mulder, "Ten IJse (2)—Schaatsles voor graaf Floris," *Kouwe Drukte* 12, no. 34 (December 2008): 18–23.
41. Niko Mulder, "Ten IJse (6)—De revolutionaire puntschaats," *Kouwe Drukte* 14, no. 38 (April 2010): 19.

42. Willie Soon and Sallie Baliunas, "Proxy Climate and Environmental Changes of the Past 1000 Years," *Climate Research* 23 (2003): 95.
43. Fagan, *The Little Ice Age: How Climate Made History 1300–1850*, 107.
44. Morgan Kelly and Cormac Ó Gráda, "The Waning of the Little Ice Age: Climate Change in Early-Modern Europe," *Journal of Interdisciplinary History* 44, no. 3 (2014): 301–325, doi:10.1162/JINH_a_00573.
45. See Dennis L. Hartmann, *Global Physical Climatology* (San Diego: Academic Press, 1994), 225; Jared Diamond, *Collapse: How Societies Choose to Fail or Succeed* (New York: Penguin, 2005), 267; and Gwyn Jones, *The Norse Atlantic Saga: Being the Norse Voyages of Discovery and Settlement to Iceland, Greenland, America* (Oxford: Oxford University Press, 1964), 55–59, for examples.
46. Sam White, "The Real Little Ice Age," *Journal of Interdisciplinary History* 44, no. 3 (2014): 327–352, doi:10.1162/JINH_a_00574.
47. Ulf Büntgen and Lena Hellmann, "The Little Ice Age in Scientific Perspective: Cold Spells and Caveats," *Journal of Interdisciplinary History* 44, no. 3 (2014): 354, doi:10.1162/JINH_a_00575.
48. Morgan Kelly and Cormac Ó Gráda, "Debating the Little Ice Age," *Journal of Interdisciplinary History* 45, no. 1 (2014): 57–68, doi:10.1162/JINH_a_00650.
49. Alfred Thomas, *A Blessed Shore: England and Bohemia from Chaucer to Shakespeare* (Ithaca: Cornell University Press, 2007), 151, 155–156.
50. Johann Andreas Schmeller, ed., *Des böhmischen Herrn Leo's von Rožmital Ritter-, Hof-, und Pilger-Reise durch die Abendlande 1465–1467* (Stuttgart: Literarischer Verein, 1944), 29.
51. Malcolm Letts, trans., *The Travels of Leo of Rozmital through Germany, Flanders, England, France, Spain, Portugal and Italy: 1465–1467*, Hakluyt Society, Second Series 108 (Cambridge: Cambridge University Press, 1957), 39.
52. *Ibid.*
53. Márton Szepsi Csombor, *Europica Varietas*, ed. Péter Kulcsár (Budapest: Szépirodalmi Könyvkiadó, 1979), 165, quoted in Petényi, *Games and Toys in Medieval and Early Modern Hungary*, 114.
54. *Ibid.*, 114–115.
55. Balfour, "Sledges with Bone Runners in Modern Use," 253–254.
56. Johan Ekeblad, *Johan Ekeblads bref*, ed. N. Sjöberg (Stockholm: P. A. Norstedt och Söner, 1911–1915), 1.66, cited in *Svenska Akademiens ordbok* (Lund: Svenska Akademien, 2017), s.v. "skridsko."
57. Johan Ekeblad, *Brev till brodern Claes Ekeblad, 1639–1655*, ed. Sture Allén (Göteborg: Acta Universitates Gothoburgensis, 1965), 86, cited in *Svenska Akademien, Svenska Akademiens ordbok*, s.v. "skridsko."
58. *Ibid.*
59. Abraham Sahlstedt, *Svensk Ordbok* (Stockholm: Carl Stolpe, 1773), 495.
60. *Ibid.*, 251.
61. Lödöse Museum, *Sportlov*, 2018, http://www.lodosemuseum.se/arkiv-program/2018/februari/sportlov/.
62. A parallel has occasionally been drawn for the German words *Schlittschuh* (ice skate) and *Eisbein* (pork knuckles); the latter looks as if it ought to descend from *Eis* (ice) plus *Bein* ("leg" in Modern German, earlier "bone"). See Hans Sperber, "Beiträge zur germanischen Wortkunde," *Wörter und Sachen* 6 (1914): 14–57, and Friedrich Thiele, "Deutscher und englischer Sprachgebrauch in gegenseitiger Erhellung," *The German Quarterly* 11, no. 1 (January 1938): 42–50, for examples of papers linking *Eisbein* with bone skates. This etymology has not been fully accepted by recent German etymological dictionaries; Friedrich Kluge and Elmar Seebold, eds., *Etymologisches Wörterbuch der deutschen Sprache*, 24th ed. (Berlin: Walter de Gruyter, 2002), s.v. "Eisbein" derives *Eisbein* from the Latin word *ischia* (hip joint) but also cites Sperber's paper.
63. This story was collected from W. H. Barrett in Porter, "Fen Skating," 44–45.
64. Foster, *A Bibliography of Skating*, 24.
65. Oxford University Press, *OED Online*, s.v. "skate, n.2."
66. E. S. de Beer, ed., *The Diary of John Evelyn* (Oxford: Oxford University Press, 1959), 448–449.
67. Mynors Bright, ed., *Diary and Correspondence of Samuel Pepys, Esp., F.R.S.: From His MS. Cypher in the Pepysian Library* (New York: Dodd, Mead, 1904), 41.
68. *Ibid.*, 45.
69. *Ibid.*, 48–49.
70. Brückner, "Über den heutigen Gebrauch von Schlittknochen in Schlesien," 42.
71. Berg, "Skates and Punt Sleds: Some Scandinavian Notes," 5.
72. Balfour, "Sledges with Bone Runners in Modern Use," 252–253.
73. Herman, "Knochenschlittschuh, Knochenkufe, Knochenkeitel: Ein Beitrag zur näheren Kenntnis der prähistorischen Langknochenfunde," 225, fig. V.2–3, 5–13.
74. A. Treichel, "Vom Schlittknochen, sogenannten Hund und Bock." In Verhandlungen der Berliner Gesellschaft für Anthropologie, Ethnologie und Urgeschichte," *Zeitschrift für Ethnologie* 17 (1885): 398.
75. Thorsteinn Einarsson, "Winter Sport in Iceland," 59.
76. Säve, *Hafvets och fiskarens sagor samt spridda drag ur Gotlands odlingssaga och strandallmogens lif*, 87, quoted in Edberg and Karlsson, *Isläggar från Birka och Sigtuna. En undersökning av ett vikingatida och medeltida fyndmaterial*, 42.
77. Allen, "The Primitive Bone Skate," 33–35.
78. Brückner, "Über den heutigen Gebrauch von Schlittknochen in Schlesien," 42; von Luschan, "Mitteilungen aus dem Museum der Gesellschaft," 145–147; Säve, *Svenska lekar*, 76–77.
79. An anonymous Romanian child quoted in Petényi, *Games and Toys in Medieval and Early Modern Hungary*, 111–112; Ernst Friedel Friedel, "Mitteilungen über altertümliche Geräte. d. Schlittknochen," quoted in Herman, "Knochenschlittschuh, Knochenkufe, Knochenkeitel: Ein Beitrag zur näheren Kenntnis der prähistorischen Langknochenfunde," 221; and Thorsteinn Einarsson, "Winter Sport in Iceland," 59.
80. Kuckuck ("Über den Gebrauch von Schlittknochen in neuester Zeit," 104) is the exception; he does not recall how skaters attached bone skates to their feet in his youth.
81. Kuckuck, "Über den Gebrauch von Schlittknochen in neuester Zeit," 104.
82. MacGregor, "Bone Skates: A Review of the Evidence."
83. Edberg and Karlsson, *Isläggar från Birka och*

Sigtuna. En undersökning av ett vikingatida och medeltida fyndmaterial.
84. Herman, "Knochenschlittschuh, Knochenkufe, Knochenkeitel: Ein Beitrag zur näheren Kenntnis der prähistorischen Langknochenfunde," 226.
85. Edberg and Karlsson, Isläggar från Birka och Sigtuna. En undersökning av ett vikingatida och medeltida fyndmaterial.
86. This skate is briefly described in V. Schmidt, "Archaeologický výzkum "Údoli Svatojiřského" a okolí. Slánská hora a její předhistoričtí obyvatelé," *Památky archeologické a místopisné* 16, no. 10 (1895): 619, 629, which includes a drawing (figure 37.32).
87. Stemp, Watson, and Evans, "Surface Analysis of Stone and Bone Tools."

Appendix

1. Kuckuck, "Über den Gebrauch von Schlittknochen in neuester Zeit," 104.
2. Herm. Grimm, "Schlittknochen von Wiepersdorf bei Jüterbogk." In Verhandlungen der Berliner Gesellschaft für Anthropologie, Ethnologie und Urgeschichte," *Zeitschrift für Ethnologie* 4 (1872): 3.
3. Küchelmann and Zidarov, "Let's Skate Together! Skating on Bones in the Past and Today," 439.
4. Staatsbibliothek zu Berlin, *Projekt dur Erschliessung historisch wertvoller Altkartenbestände: Anleitung und Sonderregeln mit Beispielen für die Aufnahme*, §4.3.3.
5. Brückner, "Über den heutigen Gebrauch von Schlittknochen in Schlesien," 42.
6. Ibid., 43.
7. Thunig, "Über Schlittknochen und Gräbeurnen," 280.
8. Treichel, "Vom Schlittknochen, sogenannten Hund und Bock," 397–398.
9. For a list of Ernst Friedel's contributions to the Berliner Gesellschaft für Anthropologie, Ethnologie und Urgeschichte, see Rudolf Virchow, ed., *General-Register zu Band I–XX (1869–1888) der Zeitschrift für Ethnologie und der Verhandlungen der Berliner Gesellschaft für Anthropologie, Ethnologie und Urgeschichte* (Berlin: A. Asher und Co., 1894), 2.
10. Friedel, "Mitteilungen über altertümliche Geräte. d. Schlittknochen," quoted in Herman, "Knochenschlittschuh, Knochenkufe, Knochenkeitel: Ein Beitrag zur näheren Kenntnis der prähistorischen Langknochenfunde," 221.
11. A. Treichel, "Vorkommen von Schlittknochen und Rundmarken." In Verhandlungen der Berliner Gesellschaft für Anthropologie, Ethnologie und Urgeschichte," *Zeitschrift für Ethnologie* 19 (1887): 83.
12. Von Luschan, "Mitteilungen aus dem Museum der Gesellschaft," 145–147.
13. O. Herman, "Ósi elemek a magyar népies halászeszközökben (Ancient Elements in Hungarian Popular Fishing Equipment)," *ArchÉrt* 5 (1885): 164–165, quoted in Choyke, "Bone Skates: Raw Material, Manufacturing and Use," 152.
14. Kovacs, "Hogyan gilicseznek Gyergyóban," 17–18, quoted in Petényi, *Games and Toys in Medieval and Early Modern Hungary*, 111, 113.
15. Petényi, *Games and Toys in Medieval and Early Modern Hungary*, 111–113.
16. Makkay, cited in Choyke, "Bone Skates: Raw Material, Manufacturing and Use," 152.
17. Balfour, "Notes on the Modern Use of Bone Skates," 35.
18. P. C. Buckland, personal communication to Arthur Macgregor, n.d., cited in MacGregor, "Bone Skates: A Review of the Evidence," 58.
19. W. H. Barrett, quoted in Enid Porter, "Fen Skating," *Folk Life* 7 (1969): 43.
20. Balfour, "Notes on the Modern Use of Bone Skates," 35–36.
21. Katajisto, "Turun Kaupunkialuen luuluistimet: tarkasteltuna osteologiselta ja historialliseltakannalta," 4, quoted in Küchelmann and Zidarov, "Let's Skate Together! Skating on Bones in the Past and Today," 440n23.
22. Translation by Auli Touronen, Turku University, Finland, quoted in Küchelmann and Zidarov, "Let's Skate Together! Skating on Bones in the Past and Today," 440.
23. Söderbäck, *Rågöborna*, 256–257, quoted in Edberg and Karlsson, *Isläggar från Birka och Sigtuna. En undersökning av ett vikingatida och medeltida fyndmaterial*, 12.
24. Middendorff, *Reise in den äussersten Norden und Osten Sibiriens während der Jahre 1843 und 1844*, 1350.
25. Marcusson, unpublished record, quoted in Edberg and Karlsson, *Isläggar från Birka och Sigtuna. En undersökning av ett vikingatida och medeltida fyndmaterial*, 11.
26. Petter Andersson, unpublished interview, Folkslivarkivet, Lund, No. 3084–16, 1931, quoted in Edberg and Karlsson, *Isläggar från Birka och Sigtuna. En undersökning av ett vikingatida och medeltida fyndmaterial*, 11.
27. Atle Pettersson, unpublished record, Folksminnesarkivet, Göteborg, No. 1932–69, 1932, quoted in Edberg and Karlsson, *Isläggar från Birka och Sigtuna. En undersökning av ett vikingatida och medeltida fyndmaterial*, 11.
28. Klindt-Jensen, "Economic and Daily Life at Vallhagar," 857.
29. F. Nordin, "Gotlands s.k. kämpagrafvar," *KHVAA:s Månadsblad* 5, no. 15 (1886): 163–164, quoted in Edberg and Karlsson, *Isläggar från Birka och Sigtuna. En undersökning av ett vikingatida och medeltida fyndmaterial*, 11.
30. Säve, *Svenska lekar*, 76–77.
31. Munro, "Notes on Ancient Bone Skates," 191.
32. A. Bastian, "Aus einem Briefe des Dr. Hans Hildebrand-Hildebrand zu Stockholm." In Verhandlungen der Berliner Gesellschaft für Anthropologie, Ethnologie und Urgeschichte, *Zeitschrift für Ethnologie* 3 (1871): 103–104.
33. Hyltén-Cavallius, *Wärend och wirdarne. Ett försök i svensk ethnologi*, II.464, quoted in Edberg and Karlsson, *Isläggar från Birka och Sigtuna. En undersökning av ett vikingatida och medeltida fyndmaterial*, 11.
34. Allen, "The Primitive Bone Skate," 33–35.
35. Balfour, "Notes on the Modern Use of Bone Skates," 29–31.
36. Thorsteinn Einarsson, "Winter Sport in Iceland," 59.
37. Hagberg, "Fundort und Fundgebiet der Modeln aus Torslunda," 330.

Bibliography

Addyman, P. V., J. S. R. Hood, H. K. Kenward, A. MacGregor, and D. Williams. "Palaeoclimate in Urban Environmental Archaeology at York, England: Problems and Potential." *World Archaeology* 8, no. 2 (1976): 220–233. doi:10.1080/00438243.1976.9979666.

Ahmed, Moinuddin, Kevin J. Anchukaitis, Asfawossen Asrat, Hemant P. Borgaonkar, Martina Braida, Brendan M. Buckley, Ulf Büntgen, et al. "Continental-Scale Temperature Variability during the Past Two Millennia." *Nature Geoscience* 6, no. 5 (2013): 339–346. doi:10.1038/NGEO1797.

Alekseeva, T., A. Alekseev, B. A. Maher, and V. Demkin. "Late Holocene Climate Reconstructions for the Russian Steppe, Based on Meralogical and Magnetic Properties of Buried Palaeosols." *Palaeogeography, Palaeoclimatology, Palaeoecology* 249 (2007): 103–127. doi:10.1016/j.palaeo.2007.01.006.

Allen, E. John B. *The Culture and Sport of Skiing: From Antiquity to World War II.* Amherst: University of Massachusetts Press, 2007.

Allen, J. Romilly. "The Primitive Bone Skate." *The Reliquary and Illustrated Archaeologist* 2 (1896): 33–36.

Andersen, H. H., P. J. Crabb, and H. J. Madsen. *Århus Søndervold: En byarkaeologisk undersøgelse.* Jyst Arkeologisk Selskabs skrifter 9. Copenhagen: Nordisk Forlag, 1971.

Anderson, Rasmus B. "Introduction." In *The Heimskringla or the Sagas of the Norse Kings*, 2nd ed., by Snorri Sturluson, translated by Samuel Liang, vii-xviii. London: John C. Nimmo, 1889.

Andersson, Petter. *Unpublished interview.* Folkslivarkivet, Lund. No. 3084–16, 1931.

Andersson, Theodore M., and Kari Ellen Gade. *Morkinskinna: The Earliest Icelandic Chronicle of the Norwegian Kings (1030–1157).* Islandica 51. Ithaca: Cornell University Press, 2000.

Anonymous. *Deutscher Eis-Sport*, 31 March 1898.

Anthony, David W. "Bridling Horse Power: The Domestication of the Horse." Chap. 4 in *Horses through Time*, edited by Sandra L. Olsen, 57–82. Boulder, CO: Roberts Rinehart, 1996.

_____. *The Horse, the Wheel, and Language: How Bronze-Age Riders from the Eurasian Steppes Shaped the Modern World.* Princeton: Princeton University Press, 2007.

_____. "Two IE Phylogenies, Three PIE Migrations, and Four Kinds of Steppe Pastoralism." *Journal of Language Relationship (Moscow)* 9 (2013): 1–21.

Arnesen, Liv, Ann Bancroft, and Cheryl Dahle. *No Horizon Is So Far: Two Women and Their Extraordinary Journey across Antarctica.* Cambridge, MA: Da Capo Press, 2003.

Ayre, Julian, and Robin Wroe-Brown. "The Post-Roman Foreshore and the Origins of the Late Anglo-Saxon Waterfront and Dock of Aethelred's Hithe: Excavations at Bull Wharf, City of London." *Archaeological Journal* 172, no. 1 (2015): 121–194. doi:10.1080/00665983.2014.984534.

Baker, David, Evelyn Baker, Jane Hassall, and Angela Simco. "Excavations in Bedford 1967–1977." *Bedfordshire Archaeological Journal* 13 (1979): 1–309.

Balfour, Henry. "Notes on the Modern Use of Bone Skates." *The Reliquary and Illustrated Archaeologist* 4 (1898): 29–37.

_____. "Sledges with Bone Runners in Modern Use." *The Reliquary and Illustrated Archaeologist* 4 (1898): 242–254.

Barford, P. M. *The Early Slavs.* London: British Museum Press, 2001.

Barraclough, Eleanor Rosamund. *Beyond the Northlands: Viking Voyages and the Old Norse Sagas.* Oxford: Oxford University Press, 2016.

Barthel, Hans-Joachim. "Schlittknochen oder Knochengeräte?" *Alt-Thüringen* 10 (1968–1969): 205–226.

Bartosiewicz, László. "Animal Exploitation at the Sarmatian Site of Gyoma 133." In *Cultural and Landscape Changes in South-East Hungary*, edited by Andrea H. Vaday, 2:365–446. Budapest: Archaeological Institute of the Hungarian Academy of Sciences, 1996.

_____. "The Archaeology of Domestic Animals."

In *Hungarian Archaeology at the Turn of the Millennium*, edited by Zsolt Visy, 60–64. Budapest: Ministry of National Cultural Heritage, Teleki László Foundation, 2003.

Bastian, A. "Aus einem Briefe des Dr. Hans Hildebrand-Hildebrand zu Stockholm." In Verhandlungen der Berliner Gesellschaft für Anthropologie, Ethnologie und Urgeschichte. *Zeitschrift für Ethnologie* 3 (1871): 103–104.

Baxter, Jane Eva. *The Archaeology of Childhood: Children, Gender, and Material Culture*. Gender and Archaeology 10. Walnut Creek, CA: Altamira Press, 2005.

Becker, Cornelia. "Bemerkungen über Schlittknochen, Knochenkufen und ähnliche Artefakte, unter besonderer Berücksichtigung der Funde aus Berlin Spandau." In *Festschrift für Hans R. Stampfli*, edited by Jörg Schibler, J. Sedlmeier, and H. Spycher, 19–30. Beiträge zur Archäologie, Anthropologie, Geologie und Paläontologie. Basel: Helbing and Lichtenhahn, 1990.

———. "Bone Points: No Longer a Mystery? Evidence from the Slavic Urban Fortification of Berlin-Spandau." In *Crafting Bone: Skeletal Technologies Through Space and Time, Proceedings of the Second Meeting of the International Council for Archaeozoology Worked Bone Research Group in Budapest, 31 August–5 September 1999*, edited by Alice M. Choyke and László Bartosiewicz, 129–148. British Archaeological Reports International Series. Oxford: Archaeopress, 2001.

Benecke, Norbert, and Angela von den Driesch. "Horse Exploitation in the Kazakh Steppes during the Eneolithic and Bronze Age." In *Prehistoric Steppe Adaptation and the Horse*, edited by Marsha Levine, Colin Renfrew, and Katie Boyle, 69–82. Cambridge: McDonald Institute for Archaeological Research, 2003.

Berg, Gösta. "Isläggar och skridskor." *Fataburen*, 1943, 79–90.

———. "The Origin and the Development of the Skis throughout the Ages." In *Finds of Skis from Prehistoric Time in Swedish Bogs and Marshes*, 9–64. Stockholm: Generalstabens litografiska anstalts förlag, 1950.

———. "Skates and Punt Sleds: Some Scandinavian Notes." In *Vriendenboek voor A. J. Bernet Kempers*, edited by P. J. Meertens and Hermanna W. M. Plettenburg, 4–13. Arnhem: Nederlands Openluchtmuseum, 1971.

———. "Skier und Schlittschuhe: Zwei nordische Fortbewegungsmittel." *Tribus: Jahrbuch des Linden-Museums Stuttgart* 2 (1952): 188–195.

Berke, Hubert. "Mit den Wölfen 'Schlitten fahren' … Ein römisches Eisbein aus einem Wolfsknochen." *Archäologie im Rheinland*, 2003, 122–123.

Bertolini, Marco, and Ursula Thun Hohenstein. "Evidence of Butchery Marks and Anthropic Modifications on Horse Remains in a Late Bronze Age Site of Northern Italy: The Case of Bovolone." *Journal of Archaeological Science: Reports* 9 (2016): 468–480. doi:10.1016/j.jasrep.2016.08.031.

Bharatdwaj, K. *Physical Geography: Atmosphere*. New Delhi: Discovery Publishing House, 2006.

Björn Bjarnason. *Íþróttir fornmanna á norðurlöndum*. 2nd ed. Reykjavík: Sigurður Kristjansson, 1950.

Björn Sigfússon, ed. *Ljósvetninga saga*. Íslenzk fornrit 10. Reykjavík: Hið íslenzka fornritafélag, 1940.

Blair, John. *Anglo-Saxon Oxfordshire*. Phoenix Mill, UK: Sutton Publishing, 1998.

Blauw, Wiebe. *Van glis tot klapschaats: Schaatsen en schaatsenmakers in Nederland, 1200 tot heden*. Franeker: Van Wijnen, 2001.

Block, D. P. "De Wikingen in Friesland." *Naamkunde* 10 (1978): 25–47.

Bø, Olav. *Skiing Traditions in Norway*. Oslo: J. Petlitz Boktrykkeri, 1968.

Bökönyi, Sándor. "The Earliest Waves of Domestic Horses in East Europe." *Journal of Indo-European Studies* 6, nos. 1–2 (1978): 17–76.

Bonnichsen, Robson. "Millie's Camp: An Experiment in Archaeology." *World Archaeology* 4, no. 3 (1973): 277–291. doi:10.1080/00438243.1973.9979539.

Bright, Mynors, ed. *Diary and Correspondence of Samuel Pepys, Esp., F.R.S.: From His MS. Cypher in the Pepysian Library*. New York: Dodd, Mead, 1904.

British Archaeological Association. "Proceedings of the Association." *Journal of the British Archaeological Association* 28 (1872): 71–84.

Brown, Nigel. *Ice-Skating: A History*. London: Nicholas Kaye Limited, 1959.

Browning, Jennifer. *Archaeological Excavations on Land off Stapleford Road, Whissendine, Rutland (NGR SK 553 086)*. University of Leicester Archaeological Services Report Number 2007-066, 2007.

Brückner. "Über den heutigen Gebrauch von Schlittknochen in Schlesien." In Verhandlungen der Berliner Gesellschaft für Anthropologie, Ethnologie und Urgeschichte. *Zeitschrift für Ethnologie* 4 (1872): 42–43.

Brugman, Johannes. *Vita alme virginis Lijdwine*. Schiedam: Otgier Nachtegael, 1498.

Buck, Henry A. "Ice-Sailing." In *Skating*, 407–428. Badminton Library of Sports and Pastimes. London: Longmans, Green, 1892.

Buckland, P. C. Personal communication to Arthur Macgregor, n.d.

Büntgen, Ulf, and Lena Hellmann. "The Little Ice Age in Scientific Perspective: Cold Spells and Caveats." *Journal of Interdisciplinary History* 44, no. 3 (2014): 353–368. doi:10.1162/JINH_a_00575.

Büntgen, Ulf, Willy Tegel, Kurt Nicolussi, Michael McCormick, David Frank, Valerie Trouet, Jed O. Kaplan, et al. "2500 Years of European

Climate Variability and Human Susceptibility." *Science* 331 (February 2011): 578–582. doi:10.1126/science.1197175.

Burov, Grigoriy M. "Some Mesolithic Wooden Artifacts from the Site of Vis I in the European North East of the U.S.S.R." In *The Mesolithic in Europe,* edited by Clive Bonsall, 391–401. Edinburgh: John Donald Publishers, 1985.

Byock, Jesse L. *Viking Language I: Learn Old Norse, Runes, and Icelandic Sagas.* Pacific Palisades, CA: Jules William Press, 2013.

Cargiolli, Carlo, ed. *Il Viaggio Settentrionale di Francesco Negri.* Bologna: Nicola Zanichelli, 1883.

Choyke, Alice M. "Bone Skates: Raw Material, Manufacturing and Use." Special issue *Pannonia and Beyond: Studies in Honor of L. Barkoczi,* ed. A. Vaday. *Antaeus: Communicationes ex Instituto Archaeologico Academiae Scientiarum Hungaricae* 24 (1997/1998): 148–156, 651–654.

———. "The Bone Tool Manufacturing Continuum." *Anthropozoologica* 25–26 (1997): 65–72.

———. "Hidden Agendas: Ancient Raw Material Choice for Worked Osseous Objects in Central Europe and Beyond." In *From These Bare Bones: Raw Materials and the Study of Worked Osseous Objects, Proceedings of the Raw Materials Session at the 11th ICAZ Conference, Paris, 2010,* edited by Alice M. Choyke and Sonia O'Connor, 1–11. Oxford: Oxbow Books, 2013.

———. "Worked Animal Bone at the Sarmatian Site of Gyoma 133." In *Cultural and Landscape Changes in South-East Hungary,* edited by Andrea H. Vaday, 2:307–322. Budapest: Archaeological Institute of the Hungarian Academy of Sciences, 1996.

Choyke, Alice M., and Jörg Schibler. "Prehistoric Bone Tools and the Archaeozoological Perspective: Research in Central Europe." In *Bones as Tools: Current Methods and Interpretations in Worked Bone Studies,* edited by Christian Gates St-Pierre and Renee B. Walker, 1622:51–65. British Archaeological Reports International Series. Oxford: Archaeopress, 2007.

Choyke, Alice M., and László Bartosiewicz. "Skating with Horses: Continuity and Parallelism in Prehistoric Hungary." *Revue de Paléobiologie* spéc. 10 (2005): 317–326.

Christie, Neil, and Matt Edgeworth. "Living, Working and Trading in Medieval Wallingford." In *Transforming Townscapes: From* burh *to* borough*: The Archaeology of Wallingford, AD 800–1400,* edited by Neil Christie, Oliver Creighton, Matt Edgeworth, and Helena Hamerow, 293–342. Monograph 35. London: Society for Medieval Archaeology, 2013.

Clark, J. G. D. *Prehistoric Europe: The Economic Basis.* Stanford: Stanford University Press, 1952.

Cleasby, Richard, and Gudbrand Vigfusson. *An Icelandic-English Dictionary.* Oxford: Clarendon Press, 1874.

Clunies Ross, Margaret. *The Old Norse Poetic Translations of Thomas Percy.* Making the Middle Ages 4. Turnhout, Belgium: Brepols, 2001.

Colgrave, Bertram, Judith McClure, and Roger Collins, eds. *Bede. The Ecclesiastical History of the English People.* Oxford World's Classics. Oxford: Oxford University Press, 1999.

Cornwall, I. W. *Bones for the Archaeologist.* London: Phoenix House, 1956.

Cowell, Cressida. *How to Steal a Dragon's Sword: The Heroic Misadventures of Hiccup the Viking.* How to Train Your Dragon 9. New York: Little, Brown, 2012.

Crawford, Jackson, trans. *The Poetic Edda: Stories of the Norse Gods and Heroes.* Hackett Classics. Indianapolis: Hackett Publishing Company, 2015.

Crossley-Holland, Kevin. *Bracelet of Bones.* Viking Sagas 1. New York: Quercus, 2014.

Culin, Stewart. *Games of the North American Indians.* New York: AMS Press, 1907.

Dale, J. H. van, and C. Kruyskamp. *Groot Woordenboek der Nederlandse Taal.* 8th ed. The Hauge: Nijhoff, 1961.

Dansgaard, W., S. J. Johnsen, N. Reeh, N. Gundestrup, H. B. Clausen, and C. U. Hammer. "Climate Changes, Norsemen, and Modern Man." *Nature* 255 (1975): 24–28. doi:10.1038/255024a0.

Davenport, Charles B. "The Growth of the Human Foot." *American Journal of Physical Anthropology* 17 (1932): 167–211.

Davidson, Daniel Sutherland. *Snowshoes.* Memoirs of the American Philosophical Society 6. Philadelphia: American Philosophical Society, 1937.

Davis, Graeme. *Vikings in America.* Edinburgh: Birlinn Limited, 2009.

de Beer, E. S., ed. *The Diary of John Evelyn.* Oxford: Oxford University Press, 1959.

de Vries, Jan. *Altnordisches etymologisches Wörterbuch.* Leiden: Brill, 1962.

———. *Nederlands etymologisch woordenboeck.* Leiden: Brill, 1971.

Degnbol, Helle, Bent Chr. Jacobsen, Eva Rode, Christopher Sanders, and Þorbjörg Helgadóttir. *Ordbog over det norrøne prosasprog / A Dictionary of Old Norse Prose.* University of Copenhagen, 2010. http://onp.hum.ku.dk/.

Desideri, Jocelyne, and Marie Besse. "Swiss Bell Beaker Population Dynamics: Eastern or Southern Influences?" *Archaeological and Anthropological Science* 2 (2010): 157–173. doi:10.1007/s12520-010-0037-9.

Dezutter, W. P. "Slijkschoenen: Twee aanwinsten voor het Stedelijk Museum voor Volkskunde te Brugge." *Biekorf* 75 (1974): 289–304.

Diamantidi, Demeter, C. von Korper, and Max

Wirth. *Spuren auf dem Eise: Die Entwicklung des Eislaufes auf der Bahn des wiener Eislauf-Vereines.* 2nd ed. Vienna: Alfred Hölder, 1892.

Diamond, Jared. *Collapse: How Societies Choose to Fail or Succeed.* New York: Penguin, 2005.

Dietrich, Laura. "Projectile Weapons of the Late Bronze Age in Eastern Europe. The Case of the Noua-Sabatinovka-Coslogeni Cultural Complex." In *Archäologische Wanderungen zwischen Ost- und Westeuropa: Studien für Tiberius Bader zum 75. Geburtstag,* 29/1:- 181–197. Studii și comunicări Satu Mare. Satu Mare, Romania: Editura Muzeului Sătmărean, 2013.

Dodge, Mary Mapes. *Hans Brinker or the Silver Skates: A Story of Life in Holland.* New York: James O'Kane, 1865.

Douglas, David C., and George W. Greenaway, eds. *English Historical Documents.* Vol. 2: 1042–1189. Oxford: Oxford University Press, 1968.

Dresbeck, LeRoy J. "The Ski: Its History and Historiography." *Technology and Culture* 8, no. 4 (October 1967): 467–479.

Dronke, Ursula, ed. *The Poetic Edda.* Oxford: Clarendon Press, 1969–2011.

Drzewicz, Anna. *Wyroby z kości i poroża z osiedla obronnego ludności kultury łużyckiej w Biskupinie.* Warsaw: Wydawnictwo Naukowe 'Semper', 2004.

Duncan-Jones, R. P. "Economic Change and the Transition to Late Antiquity." Chap. 2 in *Approaching Late Antiquity: The Transformation from Early to Late Empire,* edited by Simon Swain and Mark Edwards, 20–52. Oxford: Oxford University Press, 2004.

Durczewski, Dobromir. *Gród ludności kultury łużyckiej z okresu halsztackiego w Smuszewie, woj. pilskie.* Biblioteka Fontes archaeologici Posnanienses 6. Poznań: Muzeum Archeologiczne w Poznaniu, 1985.

Edberg, Rune, and Johnny Karlsson. "Bone Skates and Young People in Birka and Sigtuna." *Fornvännen* 111 (2016): 7–16.

―――. *Isläggar från Birka och Sigtuna. En undersökning av ett vikingatida och medeltida fyndmaterial.* Stockholm Archaeological Reports 43. Stockholm: Institutionen för arkeologi och antikens kultur, Stockholms universitet, 2015.

Eekhoff, W. "Oude beenen schaatsen." *Nieuwe friesche volks-almanak* 11 (1863): 103–109.

Einar Ól. Sveinsson and Matthías Þórðarson, eds. *Eyrbyggja saga.* Íslenzk fornrit 4. Reykjavik: Hið íslenzka fornritafélag, 1965.

Einar Ól. Sveinsson, ed. *Brennu-Njáls saga.* Íslenzk fornrit 12. Reykjavík: Hið íslenzka fornritafélag, 1954.

Ekeblad, Johan. *Brev till brodern Claes Ekeblad, 1639–1655.* Edited by Sture Allén. Göteborg: Acta Universitates Gothoburgensis, 1965.

―――. *Johan Ekeblads bref.* Edited by N. Sjöberg. Stockholm: P. A. Norstedt och Söner, 1911–1915.

Emonds, Joseph Embley, and Jan Terje Faarlund. *English: The Language of the Vikings.* Olomouc Modern Language Monographs 3. Olomouc: Palacký University, 2014.

Englert, Anton. "Trial Voyages as a Method of Experimental Archaeology: The Aspect of Speed." Chap. 6 in *Connected by the Sea: Proceedings of the Tenth International Symposium on Boat and Ship Archaeology, Denmark 2003,* edited by Lucy Blue, Frederick M. Hocker, and Anton Englert. Oxford: Oxbow Books, 2017.

Fagan, Brian. *The Little Ice Age: How Climate Made History 1300–1850.* New York: Basic Books, 2000.

Falk, Hjalmar. *Altnordische Waffenkunde.* Videnskabsselskabets skrifter 2, Historisk-filosofiske klasse 6. Kristiania: J. Dybwad, 1914.

Fantner, Georg E., Henrik Birkedal, Johannes H. Kindt, Tue Hassenkam, James C. Weaver, Jacquelin A. Cutroni, Bonnie L. Bosma, et al. "Influence of the Degradation of the Organic Matrix on the Microscopic Fracture Behavior of Trabecular Bone." *Bone* 35 (2004): 1013–1022. doi:10.1016/j.bone.2004.05.027.

Ferris, Dave G., Jihong Cole-Dai, Angelica R. Reyes, and Drew M. Budner. "South Pole Ice Core Record of Explosive Volcanic Eruptions in the First and Second Millennia A.D. and Evidence of a Large Eruption in the Tropics around 535 A.D." *Journal of Geophysical Research* 116 (2011): D17308. doi:10.1029/2011JD015916.

Finnbogi Guðmundsson, ed. *Orkneyinga saga. Legenda de Sancto Magno. Magnúss saga skemmri. Magnúss saga lengri. Helga þáttr ok Úlfs.* Íslenzk fornrit 34. Reykjavík: Hið íslenzka fornritafélag, 1965.

Finnur Jónsson, ed. *Morkinskinna.* Copenhagen: J. Jørgensen, 1932.

Formenti, Federico, and Alberto E Minetti. "The First Humans Travelling on Ice: An Energy-Saving Strategy?" *Biological Journal of the Linnean Society* 93 (2008): 1–7. doi:10.1111/j.1095–8312.2007.00991.x.

―――. "Human Locomotion on Ice: The Evolution of Ice-Skating Energetics through History." *Journal of Experimental Biology* 210 (2007): 1825–1833. doi:10.1242/jeb.002162.

Foster, Fred. W. *A Bibliography of Skating.* London: B. W. Warhurst, 1898.

Fowler, G. Herbert. *On the Outside Edge: Being Diversions in the History of Skating.* London: Horace Cox, 1897.

―――. *On the Outside Edge: Being Diversions in the History of Skating.* Edited by B. A. Thurber. Evanston, IL: Skating History Press, 2018.

Friedel, E. "Mitteilungen über altertümliche Geräte. d. Schlittknochen." *Brandenburgia,*

Monatsblatt der Gesellschaft für Heimatskunde u. s. w. zu Berlin 6 (1898): 318–327.

Furmánek, Václav, Ladislav Veliácik, and Jozef Vladář. *Die Bronzezeit in Slowakischen Raum.* Prähistorische Archäologie in Südosteuropa 15. Rahden, Westfalia: Marie Leidorf, 1999.

Gál, E. "Animal Remains from the Multi-Period Site of Hajdúnánás-Fürjhalom-Dűlő." *Acta Archaeologica Academiae Scientarum Hungaricae* 61 (2010): 207–260. doi:10.1556/AArch.61.2010.1.7.

Genito, Bruno. "The Archaeological Cultures of the Sarmatians with a Preliminary Note on the Trial Trenches at Gyoma 133: A Sarmatian Settlement in South-Eastern Hungary (Campaign 1985)." *Annali, Istituto Orientale di Napoli* 48, no. 2 (1988): 81–126.

Gerrets, D. A. "Conclusions." Chap. 20 in *The Excavations at Wijnaldum*, edited by J. C. Besteman, J. M. Bos, D. A. Gerrets, H. A. Heidinga, and J. de Koning. Reports on Frisia in Roman and Medieval Times 1. Rotterdam: A. A. Balkema, 1999.

Gershkovich, Yakov P. "Farmers and Pastoralists of the Pontic Lowland during the Late Bronze Age." In *Prehistoric Steppe Adaptation and the Horse*, edited by Marsha Levine, Colin Renfrew, and Katie Boyle, 307–317. Cambridge: McDonald Institute for Archaeological Research, 2003.

Gerškovič, Jakov Pertrovič. *Studien zur spätbronzezeitlichen Sabatinovka-Kultur am unteren Dnepr und an der Westküste des Azov'schen Meeres.* Archäologie in Eurasien 7. Rahden, Westphalia: Marie Leidorf, 1999.

Götze, A. "Der Schlossberg bei Burg im Spreewald." *Praehistorische Zeitschrift* 4 (1912): 264–350.

Goubitz, Olaf. "Nederland's oudste schaats?" *Kouwe Drukte* 3, no. 9 (September 2000): 4–5.

———. "Patten (type 110)." In *Stepping through Time: Archaeological Footwear from Prehistoric Times until 1800*, 249–266. Zwolle, the Netherlands: Stichting Promotie Archeologie, 2001.

Grammaticus, Saxo. *Historia Danica.* Edited by Petrus Erasmus Müller. Havniae: Librariæ Gyldendalianæ, 1839.

———. *The History of the Danes, Books I-IX.* Translated by Hilda Ellis Davidson and Peter Fisher. Cambridge: D. S. Brewer, 1980.

Gräslund, Bo, and Neil Price. "Twilight of the Gods? The 'Dust Veil Event' of AD 536 in Critical Perspective." *Antiquity* 86, no. 332 (January 2012): 428–443. doi:10.1017/S0003598X00062852.

Green, D. H. *Language and History in the Early Germanic World.* Cambridge: Cambridge University Press, 1998.

Grimm, Herm. "Schlittknochen von Wiepersdorf bei Jüterbogk." In Verhandlungen der Berliner Gesellschaft für Anthropologie, Ethnologie und Urgeschichte. *Zeitschrift für Ethnologie* 4 (1872): 3.

Guagnini, Alessandro. "Arma Magni Ducis Moschouiæ." In *Sarmatiae Europeæ descriptio, quae Regnum Poloniæ, Lituaniam, Samogitiam, Russiam, Masouiam, Prussiam, Pomeraniam, Liuoniam, & Moschouiæ, Tartariaeque partem complectitur.* Krakow: Matthia Wirzbietae, 1578.

Guðni Jónsson, ed. *Edda Snorra Sturlusonar, nafnapulur og skáldatal.* Íslendingasagnaútgáfan. Akureyri: Prentverk Odds Björnssonar, 1954.

———, ed. *Grettis saga Ásmundarsonar, Bandamanna saga. Odds Þáttr Ófeigssonar.* Íslenzk fornrit 7. Reykjavík: Hið íslenzka fornritafélag, 1936.

Gustafsson, Anders, Lars Werdelin, Birgitta S. Tullberg, and Patrick Lindenfors. "Stature and Sexual Stature Dimorphism in Sweden, from the 10th to the End of the 20th Century." *American Journal of Human Biology* 19 (2007): 861–870.

Haak, Wolfgang, Iosif Lazaridis, Nick Patterson, Nadin Rohland, Swapan Mallick, Bastien Llamas, Guido Brandt, et al. "Massive Migration from the Steppe Was a Source for Indo-European Languages in Europe." *Nature* 522 (2015): 207–211. doi:10.1038/nature14317.

Hadley, D. M. *The Vikings in England: Settlement, Society and Culture.* Manchester Medieval Studies. Manchester: Manchester University Press, 2006.

Hagberg, Ulf Erik. "Fundort und Fundgebiet der Modeln aus Torslunda." *Frühmittelalterliche Studien* 10 (1976): 323–349.

———. "Isläggar och pälsskinn—en frusen kulturhistorisk rapsodi." In *Vi får tacka Lamm*, edited by Bente Magnus, Carin Orrling, Monika Rasch, and Göran Tegnér, 79–81. Studies 10. Stockholm: Statens Historiska Museet, 2001.

Hall, Richard. *Book of Viking Age York.* English Heritage. London: B. T. Batsford, 1994.

———. "Markets of the Danelaw." In *The Vikings in England and in Their Danish Homeland*, edited by Else Roesdahl, James Graham-Campbell, Patricia Connor, and Kenneth Pearson, 95–140. London: The Anglo-Danish Viking Project, 1981.

Halldór Halldórsson. "Leggir og skautar." In *Nordæla: Afæmliskveðja til Sigurðar Nordals*, edited by Halldór Halldórsson, Steingrímur J. Þorsteinsson, Jón Jóhannesson, and Þorkell Jóhannesson, 75–89. Reykjavík: Helgafell, 1956.

Hammond, Gawain, and Norman Hammond. "Child's Play: A Distorting Factor in Archaeological Distribution." *American Antiquity* 46 (1981): 634–636.

Hartmann, Dennis L. *Global Physical Climatology.* San Diego: Academic Press, 1994.

Haugen, Einar, ed. *Norwegian-English Dictionary.* Madison: University of Wisconsin Press, 1974.

Haworth, Eleanor, and Jean Young. *The Fljotsdale Saga and the Droplaugarsons.* London: J. M. Dent and Sons, 1990.

Heathcote, J. M. "Skating as a Recreation." In *Skating*, 198–220. Badminton Library of Sports and Pastimes. London: Longmans, Green, 1892.

Herman, O. "Ósi elemek a magyar népies halászeszközökben (Ancient Elements in Hungarian Popular Fishing Equipment)." *ArchÉrt* 5 (1885): 153–167.

Herman, Otto. "Knochenschlittschuh, Knochenkufe, Knochenkeitel: Ein Beitrag zur näheren Kenntnis der prähistorischen Langknochenfunde." *Mittheilungen der anthropologischen Gesellschaft in Wien* 32 (1902): 217–238.

Hermann Pálsson and Paul Edwards. *Orkneyinga Saga: The History of the Earls of Orkney.* New York: Penguin, 1978.

Herodotus. *The History.* Translated by David Grene. London: University of Chicago Press, 1987.

Herteig, Asbjørn E. *Kongers havn og handels sete: Fra de arkeologiske undersøkelser på Bryggen i Bergen 1955–68.* Oslo: H. Aschehoug, 1969.

Heyd, Volker. "Überregionale Verbindungen der süddeutschen Glockenbecherkultur anhand der Siedlungen." In *Siedlungen der Glockenbecherkultur in Süddeutschland und Mitteleuropa,* edited by Volker Heyd, Ludwig Husty, and Ludwig Kreiner, 181–202. Arbeiten zur Archäologie Süddeutschlands 17. Büchenbach: Verlag Dr. Faustus, 2004.

Himstedt, Thomas. "How the Vikings Got to the Rhine: A Historical-Geographical Survey over the Rhinelands in the Early Middle Ages." In *Vikings on the Rhine: Recent Research on Early Medieval Relations Between the Rhinelands and Scandinavia,* edited by Rudolf Simek and Ulrike Engel, 23–38. Studia Medievalia Septentrionalia 11. Vienna: Verlag Fassbaender, 2004.

Hollander, Lee M., trans. *The Poetic Edda.* Austin: University of Texas Press, 1962.

Hucke, Karl. "Frühgeschichtliche Geweih- und Knochengeräte von der Insel Olsborg im Großen Plöner See in Holstein." *Zeitschrift für Morphologie und Anthropologie* 44, nos. 1–2 (1952): 108–114.

Hughes, Malcolm K., and Henry F. Diaz. "Was There a 'Medieval Warm Period,' and If So, Where and When?" *Climate Change* 26 (1994): 109–142.

Huntford, Roland. *Two Planks and a Passion: The Dramatic History of Skiing.* New York: Continuum, 2008.

Hurrell, James W., Yochanan Kushnir, Geir Ottersen, and Martin Visbeck, eds. *The North Atlantic Oscillation: Climatic Significance and Environmental Impact.* Geophysical Monographs. Washington, D.C.: American Geophysical Union, 2003.

Hyltén-Cavallius, Gunnar Olof. *Wärend och wirdarne. Ett försök i svensk ethnologi.* Stockholm: P. A. Norstedt och söner, 1863–1868.

Ingen Schenau, G. J. van. "The Influence of Air Friction in Speed Skating." *Journal of Biomechanics* 15, no. 6 (1982): 449–458. doi:10.1016/0021-9290(82)90081-1.

Isaac, G. R. "The Origins of the Celtic Languages: Language Spread from East to West." Chap. 1 in *Celtic from the West: Alternative Perspectives from Archaeology, Genetics, Language and Literature,* edited by Barry Cunliffe and John T. Koch, 153–166. Celtic Studies Publications 15. Oxford: Oxbow Books, 2012.

Jacobi, H. W. *De Nederlandse Glissen.* student project. University of Amsterdam, June 1976.

Janse, Antheun. *Ridderschap in Holland: Portret van een adellijke elite in de late Middeleeuwen.* Adelgeschiedenis 1. Hilversum, Netherlands: Verloren BV, 2001.

Jóhannes Halldórsson, ed. *Kjalnesinga saga. Jökuls þáttr Búasonar. Víglundar saga. Króka-Refs saga. Þórðar saga hreðu. Finnboga saga. Gunnars saga Keldugnúpsfífls.* Íslenzk fornrit 14. Reykjavík: Hið íslenzka fornritafélag, 1959.

Jón Jóhannesson, ed. *Austfirðinga sǫgur.* Íslenzk fornrit 11. Reykjavík: Hið íslenzka fornritafélag, 1950.

Jónas Kristjánsson, ed. *Eyfirðinga sögur.* Íslenzk fornrit 9. Reykjavík: Hið íslenzka fornritafélag, 1956.

Jones, Gwyn. *The Norse Atlantic Saga: Being the Norse Voyages of Discovery and Settlement to Iceland, Greenland, America.* Oxford: Oxford University Press, 1964.

Jones, R., and W. E. Cormack. *A Treatise on Skating.* Edited by B. A. Thurber. Evanston, IL: Skating History Press, 2017.

Katajisto, J. "Turun Kaupunkialuen luuluistimet: tarkasteltuna osteologiselta ja historialliselta kannalta." Unpublished manuscript, University of Turku, Turku, 2002.

Kelaidis, Panayoti. "Introduction: Principal Steppe Regions." In *Steppes: The Plants and Ecology of the World's Semi-arid Regions,* 8–31. Portland, OR: Timber Press, 2015.

Kelly, Morgan, and Cormac Ó Gráda. "Debating the Little Ice Age." *Journal of Interdisciplinary History* 45, no. 1 (2014): 57–68. doi:10.1162/JINH_a_00650.

———. "The Waning of the Little Ice Age: Climate Change in Early Modern Europe." *Journal of Interdisciplinary History* 44, no. 3 (2014): 301–325. doi:10.1162/JINH_a_00573.

Kennedy, John. *Translating the Sagas: Two Hundred Years of Challenge and Response.* Making the Middle Ages 5. Turnhout, Belgium: Brepols, 2007.

Kershaw, Jane. "An Early Medieval Dual-Currency Economy: Bullion and Coin in the Danelaw."

Antiquity 91, no. 355 (2017): 173–190. doi:10.15184/aqy.2016.249.

Kershaw, Jane, and Ellen C. Røyrvik. "The 'People of the British Isles' Project and Viking Settlement in England." *Antiquity* 90, no. 354 (2016): 1670–1680. doi:10.15184/aqy.2016.193.

Kershaw, Jane F. "Culture and Gender in the Danelaw: Scandinavian and Anglo-Scandinavian Brooches." *Viking and Medieval Scandinavia* 5 (2009): 295–325. doi:10.1484/J.VMS.1.100682.

———. *Viking Identities: Scandinavian Jewellery in England*. Medieval History and Archaeology. Oxford: Oxford University Press, 2013.

Keyser, R., and P. A. Munch, eds. *Norges gamle love indtil 1387*. Vol. 2: Lovgivningen under Kong Mangus Haakonssöns Regjeringstid fra 1263 til 1280, tilligemed et Supplement til förste Bind. Christiania: Chr. Gröndahl, 1848.

Keyser, R., P. A. Munch, and C. R. Unger, eds. *Speculum regale. Konungs-Skuggsjá. Konge-Speilet. Et philosophisk-didaktisk Skrift, Forfattet i Norge mod Slutningen af det 12 Aarhundrede. Tilligemed et samtidigt Skrift om den norske Kirkes Stilling til Staten*. Christiania: Carl C. Werner, 1848.

Kiekebusch, Albert. *Die Ausgrabung des bronzezeitlichen Dorfes Buch bei Berlin*. Vol. 1. Berlin: Dietrich Reimer, 1923.

Kjellberg, S. T. "Gnida, mangla och stryka." *Kulturen*, 1940, 68–91.

Klindt-Jensen, Ole. "Economic and Daily Life at Vallhagar." In *Vallhagar: A Migration Period Settlement on Gotland, Sweden*, edited by Mårten Stenberger and Ole Klindt-Jensen, 2:832–862. Copenhagen: Einar Munksgaards Forlag, 1955.

Kluge, Friedrich, and Elmar Seebold, eds. *Etymologisches Wörterbuch der deutschen Sprache*. 24th ed. Berlin: Walter de Gruyter, 2002.

Kock, Ernst A. *Den norsk-isländska skaldediktningen*. Lund: C. W. K. Gleerup, 1946–1949.

Kohl, Philip L. *The Making of Bronze Age Eurasia*. Cambridge World Archaeology. Cambridge: Cambridge University Press, 2007.

Kolodny, Annette. *In Search of First Contact: The Vikings of Vinland, the Peoples of the Dawnland, and the Anglo-American Anxiety of Discovery*. Durham, NC: Duke University Press, 2012.

Kovacs, J. S. "Hogyan gilicseznek Gyergyóban." *Erdély*, 1908, 17–20.

Kratochvíl, Z., and O. Šterba. "Osteologische Analyse der heimischen Knochenindustrie aus Mikulčice und Pohansko." *Archeologické rozhledy* 22 (1970): 447–470.

Kremenetski, Constantin V., Olga A. Chichagova, and Nathalia I. Shishlina. "Palaeoecological Evidence for Holocene Vegetation, Climate and Land-Use Change in the Low Don Basin and Kalmuk Area, Southern Russia." *Vegetation History and Archaeobotany* 8 (1999): 233–246.

Kremenetski, Konstantin V. "Steppe and Forest-Steppe Belt of Eurasia: Holocene Environmental History." In *Prehistoric Steppe Adaptation and the Horse*, edited by Marsha Levine, Colin Renfrew, and Katie Boyle, 11–27. Cambridge: McDonald Institute for Archaeological Research, 2003.

Küchelmann, Hans Christian. *Bone Skates Database*. Updated May 28, 2018. https://www.knochenarbeit.de/bone-skates-database/?lang=en.

Küchelmann, Hans Christian, and Petar Zidarov. "Let's Skate Together! Skating on Bones in the Past and Today." In *From Hooves to Horns, from Mollusc to Mammoth: Manufacture and Use of Bone Artefacts from Prehistoric Times to the Present, Proceedings of the 4th Meeting of the ICAZ Worked Bone Research Group at Tallinn, 26th-31st of August 2003*, edited by Heidi Luik, Alice M. Choyke, Colleen Batey, and Lembi Lõugas, 425–445. Muinasaja Teadus 15. Tallinn: Ajaloo Instituut, 2005.

Kuckuck. "Über den Gebrauch von Schlittknochen in neuester Zeit." In Verhandlungen der Berliner Gesellschaft für Anthropologie, Ethnologie und Urgeschichte. *Zeitschrift für Ethnologie* 3 (1871): 104.

Kunst, Günther Karl. "Tierreste aus der frühmittelalterlichen Siedlung von Baumgarten an der March, Niederösterreich." *Mitteilungen der Prähistorischen Kommission* 68 (2009): 197–204.

Kustár, Rozália, and Beáta Tugya. "'Csontkorcsolyák' a késő bronzkorban (Bone 'Skates' in the Late Bronze Age)." In *Csont és bőr: Az állati eredetű nyersanyagok feldolgozásának törtnéte, régészete és néprajza (Bone and Leather: History, Archaeology and Ethnography of Crafts Utilizing Raw Materials from Animals)*, edited by János Gömöri and Andrea Korösi, 79–84. Az anyagi kultúra a Kárpát-medencében. Budapest: Magyar Tudományos Akadémia Veszprémi Akadémiai Bizottsága (MTA VEAB), 2010.

Lamb, H. H. *Climate, History, and the Modern World*. First. London: Methuen, 1982.

Länsstyrelsen Gotlands Län. *Bästeträsk*, accessed December 15, 2018. https://www.lansstyrelsen.se/gotland/besok-ochupptack/naturreservat/bastetrask.html.

Larrington, Carolyne, trans. *The Poetic Edda*. Oxford World's Classics. Oxford: Oxford University Press, 2009.

Larsen, L. B., B. M. Vinther, K. R. Briffa, T. M. Melvin, H. B. Clausen, P. D. Jones, M.-L. Siggaard-Andersen, et al. "New Ice Core Evidence for a Volcanic Cause of the A.D. 536 Dust Veil." *Geophysical Research Letters* 35 (2008): L04708. doi:10.1029/2007GL032450.

Larsen, Nils. "The Timeless Skiers of the Altay." In *The Original Place of Skiing—Altai Prefecture of Xingjiang, China*, edited by Shan Zhaojian and Wang Bo, 292–304. Beijing Shi: Ren min ti yu chu ban she : Xinjiang ren min chu ban she, 2011.

Larson, Lawrence M. *The Earliest Norwegian Laws: Being the Gulathing Law and the Frostathing Law.* New York: Columbia University Press, 1935.

Latham, R. E., D. R. Howlett, and R. K. Ashdowne, eds. *Dictionary of Medieval Latin from British Sources.* Oxford: British Academy, 1975–2013.

Lauwerier, Roel C. G. M., and Robert M. Van Heeringen. "Skates and Prickers from the Circular Fortress of Oost-Souburg, the Netherlands (AD 900–975)." *Environmental Archaeology* 3 (1998): 121–126. doi:10.1179/env.1998.3.1.121.

Lavelle, Ryan, and Simon Roffey. "Introduction: Danes in Wessex." In *Danes in Wessex: The Scandinavian Impact on Southern England, c. 800-c. 1100*, edited by Ryan Lavelle and Simon Roffey, 1–6. Oxford: Oxbow Books, 2016.

Layard, Nina F. "Bone Skates and Skating Stakes." *East Anglian Miscellany* 2 (1908): 74.

LeMoine, Genevieve M. "Use Wear on Bone and Antler Tools from the Mackenzie Delta, Northwest Territories." *American Antiquity* 59, no. 2 (April 1994): 316–334. doi:10.2307/281935.

Lennon, Joan. *Ice Road.* The Wickit Chronicles 3. La Jolla, CA: Kane/Miller Book Publishers, 2008.

Leslie, Stephen, Bruce Winney, Garrett Hellenthal, Dan Davison, Abdelhamid Boumertit, Tammy Day, Katarzyna Hutnik, et al. "The Fine-Scale Genetic Structure of the British Population." *Nature* 519 (2015): 309–314. doi:10.1038/nature14230.

Letts, Malcolm, trans. *The Travels of Leo of Rozmital through Germany, Flanders, England, France, Spain, Portugal and Italy: 1465–1467.* Hakluyt Society, Second Series 108. Cambridge: Cambridge University Press, 1957.

Levine, Marsha A. "Eating Horses: The Evolutionary Significance of Hippophagy." *Antiquity* 72, no. 275 (1998): 90–100. doi:10.1017/S0003598X00086300.

Lewenhaupt, Adam, ed. *Calendaria Caroli IX.* Stockholm: P. A. Norstedt och söner, 1903.

Library Committee of the Corporation of the City of London, ed. *Catalogue of the Collection of London Antiquities in the Guildhall Museum.* London: Blades, East and Blades, 1908.

Lödöse Museum. *Sportlov.* 2018. http://www.lodosemuseum.se/arkiv-program/2018/februari/sportlov/.

Long, Clifford D. "Excavations in the Medieval City of Trondheim, Norway." *Medieval Archaeology* 19 (1975): 1–32.

Lovgren, Stefan. "Bone Ice Skates Invented by Ancient Finns, Study Says." *National Geographic News*, 2008.

Luik, Heidi. "Luust Uisud Eesti Arheoloogilises Leiumaterjalis." *Eesti Arheoloogia Ajakiri* 4, no. 2 (2000): 129–150.

Lukasiewicz, Kazimierz, and Zdzislaw Rajewski. "Przedmioti rogowe i kościane z grodu kultury 'łużyckiej' w Biskupinie." In *Gród prasłowiański w Biskupinie w powiecie żnińskim. Sprawozdanie z prac wykopaliskowych w latach 1937 i 1937 z uwględnieniem lat 1934–1935*, edited by Józefa Kostrzewskiego, 46–54+plates. Poznań: Nakladem Instytutu Prehistorycznego Uniwersytetu Poznańskiego, 1938.

Lunde, Paul, and Caroline Stone, eds. *Ibn Fadlān and the Land of Darkness.* New York: Penguin, 2012.

Luschan, F. von. "Mitteilungen aus dem Museum der Gesellschaft." *Mitteilungen der anthropologischen Gesellschaft in Wien* 6 (1876): 137–160.

———. "Mitteilungen aus dem Museum der Gesellschaft." *Mitteilungen der anthropologischen Gesellschaft in Wien* 10 (1881): 298–336.

MacGregor, Arthur. "Bone Skates: A Review of the Evidence." *Archaeological Journal* 133 (1976): 57–74.

———. "Bone, Antler and Horn: An Archaeological Perspective." *Journal of Museum Ethnography* 2 (1991): 29–38.

———. *Bone, Antler, Ivory and Horn: The Technology of Skeletal Materials since the Roman Period.* 1985. Reprint, London: Routledge, 2015.

———. "Problems in the Interpretation of Microscopic Wear Patterns: The Evidence from Bone Skates." *Journal of Archaeological Science* 2 (1975): 385–390.

———. "Skates." In *Artifacts from Medieval Winchester*, edited by Martin Biddle, vol. 7.2: *Object and Economy in Medieval Winchester*, 708–709. Oxford: Clarendon Press, 1990.

MacGregor, Arthur, Ailsa J. Mainman, and Nicola S. H. Rogers. *Craft, Industry and Everyday Life: Bone, Antler, Ivory and Horn from Anglo-Scandinavian and Medieval York.* The Archaeology of York 17: The Small Finds, fasc. 12. London: Council for British Archaeology, 1999.

Magnus, Olaus. *Description of the Northern Peoples.* Edited by P. G. Foote. Translated by Peter Fisher and Humphrey Higgens. London: Hakluyt Society, 1996.

———. *Historia de gentibus septentrionalibus.* Rome: J. M. de Viottis, 1555.

Magnusson, Magnus. *Vikings!* New York: Elsevier-Dutton, 1980.

Makkay, János. Personal communication to Alice M. Choyke, n.d.

Malinowski, Tadeusz. "Osadnictwo kultury łużyckiej wczesnej epoki żelaza w Słupcy." *Fontes Archaeologici Posnanienses: Annales Musei Archaeologici Posnaniensis* 8–9 (1957–1958): 98–114.

Mallory, J. P. *In Search of the Indo-Europeans.* New York: Thames and Hudson, 1991.

———. "The Indo-Europeanization of Atlantic Europe." Chap. 1 in *Celtic from the West 2: Rethinking the Bronze Age and the Arrival of Indo-European in Atlantic Europe,* edited by John T. Koch and Barry Cunliffe, 17–39. Oxford: Oxbow Books, 2013.

Manco, Jean. *Ancestral Journeys: The Peopling of Europe from the First Venturers to the Vikings.* Revised ed. New York: Thames and Hudson, 2015.

———. *Blood of the Celts.* New York: Thames and Hudson, 2015.

Mann, J. E. *Early Medieval Finds from Flaxengate I: Objects of Antler, Bone, Stone, Horn, Ivory, Amber, and Jet.* Lincoln: Lincoln Archaeological Trust, 1982.

Marcusson, Christina. Unpublished record. Folkslivarkivet, Lund. No. 2898–69, 1929.

Markwart, Josef. "Ein arabischer Bericht über die arktischen (uralischen) Länder aus dem 10. Jahrhundert." *Hungarische Jahrbücher* 4 (1924): 261–334.

Marriott, Alice. "Indians on Horseback." In *Saynday's People*, 91–226. New York: University of Nebraska Press, 1963.

Marschalleck, Karl H. "Die ostgermanische Siedlung von Kliestow bei Frankfurt (Oder)." *Praehistorische Zeitschrift* 30–31, nos. 1–2 (1940): 253–307.

Martin, Toby F. *The Cruciform Brooch and Anglo-Saxon England.* Woodbridge: Boydell Press, 2015.

May, Eberhard. "Widerristhöhe und Langknochenmasse bei Pferden—Ein immer noch aktuelles Problem." *Zeitschrift für Säugetierkunde* 50, no. 6 (1985): 368–381.

Meier, Dirk. *Seafarers, Merchants and Pirates in the Middle Ages.* Woodbridge: Boydell Press, 2006.

Melker, B. R. de. *Oorkondenboek van Amsterdam tot 1400: Supplement.* Apparaat voor de geschiedenis van Holland 12. Amsterdam: Historische Vereniging Holland, 1995.

Meredith, Howard V. "Human foot length from embryo to adult." *Human Biology* 16, no. 4 (December 1944): 207–282.

Middendorff, Alexander Theodor von. *Reise in den äussersten Norden und Osten Sibiriens während der Jahre 1843 und 1844.* Vol. 4.2. St. Petersburg: Kaiserlichen Akademie der Wissenschaften, 1875.

Mierow, Charles Christopher, trans. *The Gothic History of Jordanes.* Princeton: Princeton University Press, 1915.

Millet, Guillaume Y., Martin D. Hoffman, Robin B. Candau, and Philip S. Clifford. "Poling Forces during Roller Skiing: Effects of Technique and Speed." *Medicine and Science in Sports and Exercise* 30, no. 11 (November 1998): 1645–1653. doi:10.1097/00005768-199811000-00014.

Minorsky, V. *Sharaf Al-Zamān Ṭāhir Marvazī on China, the Turks, and India.* London: The Royal Asiatic Society, 1942.

Mogielnicka-Urban, Małgorzata. *Warsztat ceramiczny w kulturze łużyckiej.* Wrocław: Zakład Narodowy imienia Ossolinskich, 1984.

Mongyat, A. L. "XII vek. Puteschestvie v Rossiyu." *Nauka i zhizn'* 1 (1965): 34–38.

Mulder, Niko. "Ten IJse (1)." *Kouwe Drukte* 12, no. 33 (October 2008): 25–30.

———. "Ten IJse (2)—Schaatsles voor graaf Floris." *Kouwe Drukte* 12, no. 34 (December 2008): 18–23.

———. "Ten IJse (4)—Met scaetzen en souerding op de hofgracht." *Kouwe Drukte* 13, no. 36 (October 2009): 37–40.

———. "Ten IJse (6)—De revolutionaire puntschaats." *Kouwe Drukte* 14, no. 38 (April 2010): 19–21.

Mulder, Niko, and Jos Pronk. *Acht eeuwen schatsen in en om Amsterdam.* Bergen, Netherlands: De Poolster, 2011.

Munro, Robert. "Notes on Ancient Bone Skates." *Proceedings of the Society of Antiquaries of Scotland* 28 (1894): 185–197.

Nielsen, E. Levin. "Pedersstraede i Viborg: Købstadarkaeologiske undersøgelser 1966/67." *Kuml* 18, no. 18 (1968): 23–81.

Nieuwhof, Annet. "Anglo-Saxon Immigration or Continuity? Ezinge and the Coastal Area of the Northern Netherlands in the Migration Period." *Journal of Archaeology in the Low Countries* 5, no. 1 (2013): 53–84.

———. "The Excavation at Englum, a Wierde in the Province of Groningen, the Netherlands: Summary of the Excavation Results." Chap. 13 in *De Leege Wier van Englum: Archeologisch onderzoek in het Reitdiepgebied,* edited by Annet Nieuwhof, 255–259. Jaarverslagen van de Vereniging voor Terpenonderzoek 91. Groningen: Vereniging voor Terpenonderzoek / Tienkamp en Verheij, 2008.

Nöjd, Ruben, Astrid Tornberg, and Margareta Ångström. *McKay's Modern English-Swedish and Swedish-English Dictionary.* New York: David McKay, 1965.

Nordin, F. "Gotlands s.k. kämpagrafvar." *KHVAA:s Månadsblad* 5, no. 15 (1886): 145–174.

Oostrom, Fritz Pieter van. *Court and Culture: Dutch Literature, 1350–1450.* Berkeley: University of California Press, 1992.

Oppenheimer, Stephen. *The Origins of the British: A Genetic Detective Story.* New York: Carroll and Graf, 2006.

Orchard, Andy, trans. *The Elder Edda: A Book of Viking Lore.* Penguin Classics. New York: Penguin, 2011.

Ostoja-Zagórski, Janusz. *Gród halsztacki w Jankowie nad Jeziorem Pakoskim.* Wrocław: Zakład Narodowy imienia Ossolińskich, 1978.

Oxford University Press. *OED Online,* 2011. http://www.oed.com/.

Pankovskiy, Valentin. *Bone and Antler Industry from the Late Bronze Age Settlement Sabatynivka-I (Ukraine, South Bug).* Abstract #31 of the Sixth Meeting of the Worked Bone Research Group of the International Council for Archaeozoology, August 27th—31th [sic] 2007, Paris. August 2007.

Park, Robert. "Size Counts: The Miniature Archaeology of Childhood in Inuit Societies." *Antiquity* 72 (1998): 269–281. doi:10.1017/S0003598X00086567.

Parma, David, Jiří Kala, Miriam Nývltová Fišáková, and Michaela Rašková Zelinková. "Netradični materiál, neobvyklý předmět Opomíjený segment kostěné industrie mladší doby bronzové (Non-Traditional Material and a Non-Traditional Object: A Neglected Sort of the Late Bronze Age Bone Industry)." *Archeologické rozhledy* 43, no. 1 (2011): 136–150.

Percy, Thomas, trans. *Five Pieces of Runic Poetry, Translated from the Islandic Language.* London: R. and J. Dodsley, 1763.

———. *Northern Antiquities: Or a Description of the Manners, Customs, Religion and Laws of the Ancient Danes, Including Those of Our Own Saxon Ancestors.* 2nd ed. Edinburgh: C. Stewart, 1809.

Petényi, Sándor. *Games and Toys in Medieval and Early Modern Hungary.* Studia archaeologica mediae recentisque aevorum Universitatis Scientiarum de Rolando Eötvös nominatae 1. Krems: Medium Aevum Quotidianum, 1994.

Petrucci, Gabriella, Giancarla Malerba, and Giacomo Giacobini. "Manufatti in osso dal castelliere di Pozzuolo del Friuli." Chap. 2 in *La prima età del ferro nel settore meridionale del castelliere: Le attività produttive e i resti faunistici,* edited by Paolo Càssola Guida, Silvia Pettarin, Gabriella Petrucci, and Alessandra Giumlia-Mair, 2.2:139–179. Pozzuolo del Friuli. Rome: Edizioni Quasar di Severino Tognon s.r.l., 1998.

Pettersson, Atle. Unpublished record. Folksminnesarkivet, Göteborg. No. 1932–69, 1932.

Porter, Enid. "Fen Skating." *Folk Life* 7 (1969): 43–59.

Porter, John. "The Saga of the People of Fljotsdal." In *The Complete Sagas of Icelanders,* edited by Viðar Hreinsson, 4:379–433. Reykjavík: Leifur Eiríksson Publishing, 1997.

Prummel, Wietske. *Starigard/Oldenburg: Hauptburg der Slawen in Wagrien.* Vol. 4: *Die Tierknochenfunde unter besonderer Berücksichtigung der Beizjagd.* Neumünster: Karl Wachholtz Verlag, 1993.

Prummel, Wietske, Hülya Halici, and Annemieke Verbaas. "The Bone and Antler Tools from the Wijnaldum-Tjitsma *Terp.*" *Journal of Archaeology in the Low Countries* 3, nos. 1/2 (2011): 65–106.

Putelat, Olivier. "Les relations homme-animal dans le monde des vivants et des morts: Étude archéozoologique des établissements et des regroupements funéraires ruraux de l'Arc jurassien et de la Plaine d'Alsace de la fin de l'Antiquité Tardive au premier Moyen Âge." PhD diss., University of Paris Panthéon-Sorbonne, 2015.

Quirk, R. *The Saga of Gunnlaug Serpent-Tongue.* London, NY: T. Nelson, 1957.

Radley, Jeffrey. "Economic Aspects of Anglo-Danish York." *Medieval Archaeology* 15 (1971): 37–57. doi:10.1080/00766097.1971.11735336.

Rajewski, Z. "Przedmioty z kości i rogu i obróbka obu surowców w grodach 'łużyckich' z wczesnego okresu żelaza," *Sprawozdanie z prac wykopaliskowych w grodzie kultury łużyckiej w Biskupinie w powiecie żnińskim* (Poznań) 3 (1950): 171–185.

Randall, Lilian M. C. *Images in the Margins of Gothic Manuscripts.* California Studies in the History of Art 4. Berkeley: University of California Press, 1966.

Robertson, James Craigie, ed. *Materials for the History of Thomas Becket, Archbishop of Canterbury.* Vol. 3. London: Longman, 1877.

Roes, Anna. *Bone and Antler Objects from the Frisian Terp-Mounds.* Haarlem: H. D. Tjeenk Willink and Zoon N.V., 1963.

Roesdahl, Else. *The Vikings.* New York: Penguin, 1987.

Rogers, Nicola S. H. *Anglian and Other Finds from Fishergate.* The Archaeology of York 17: The Small Finds, fasc. 9. York: York Archaeological Trust for Excavation and Research, 1993.

Rogerson, Andrew, and Carolyn Dallas. *Excavations in Thetford 1948–59 and 1973–80.* East Anglian Archaeology 22. Norfolk: The Norfolk Archaeological Unit, 1984.

Rubruquensis, Wilhelmus. "Itinerarium ad partes orientales." In *Sinica Franciscana,* edited by Anastasius van den Wyngaert, vol. 1: Itinera et relationes Fratrum Minorum saeculi XIII et XIV. Florence: Quaracchi-Florence, 1929.

Sahlstedt, Abraham. *Svensk Ordbok.* Stockholm: Carl Stolpe, 1773.

Säve, P. A. *Hafvets och fiskarens sagor samt spridda drag ur Gotlands odlingssaga och strandallmogens lif.* 2nd ed. Visby: Gotlands Allehandas tryckeri, 1892.

———. *Svenska lekar.* Edited by Herbert Gustavson. Vol. 1: Gotländska lekar. Uppsala: Almqvist och Wicksells Boktryckeri AB, 1948.

Schiffer, Michael B. *Formation Processes of the Archaeological Record.* Albuquerque: University of New Mexico Press, 1987.

Schiffer, Michael Brian. *Studying Technological Change: A Behavioral Approach.* Salt Lake City: University of Utah Press, 2011.

Schmeller, Johann Andreas, ed. *Des böhmischen Herrn Leo's von Rožmital Ritter-----, Hof-----, und Pilger-Reise durch die Abend-*

lande 1465–1467. Stuttgart: Literarischer Verein, 1944.
Schmidt, V. "Archaeologický výzkum "Údoli Svatojiřského" a okolí. Slánská hora a její předhistoričtí obyvatelé." *Památky archeologické a místopisné* 16, no. 10 (1895): 539–636.
Schott, W. "Über die ächten Kirgisen." *Abhandlungen der Königlichen Akademie der Wissenschaften zu Berlin*, 1864, 429–474.
Schranil, Josef. *Die Vorgeschichte Böhmens und Mährens*. Berlin: W. de Gruyter, 1928.
Schuldt, Ewald. *Altslawisches Handwerk: Ausstellung zur 800-Jahrfeier der Stadt Schwerin: Sonderausstellung 1960*. Bildkataloge des Museums für Ur- und Frühgeschichte Schwerin 2. Schwerin: Museum für Ur- und Frühgeschichte, 1960.
Sellmann, O. "Latènezeitliche Grab- und Wohngrubenfunde von der Aue bei Mühlhausen i. Th." *Jahresschrift für die Vorgeschichte der sächsisch-thüringischen Länder* 10 (1911): 61–70 + figures 8–9.
Semenov, S. A. *Prehistoric Technology: An Experimental Study of the Oldest Tools and Artefacts from Traces of Manufacture and Wear*. Translated by M. W. Thompson. Bath: Adams and Dart, 1964.
Short, William R. *Viking-Age Ice Skates*. Hurstwic, accessed December 15, 2018. http://www.hurstwic.com/history/articles/daily_living/text/ice_skates.htm.
Smiley, Jane, ed. *The Sagas of Icelanders*. New York: Penguin Classics, 2001.
Smith, Charles Roach. "Ancient Bone Skates." *Collectanea Antiqua* 1 (1848): 167–169.
Smith, Charles Roach, and Alfred Smee. "Bone Skate Found at Moorfield." *Archæologia, or Miscellaneous Tracts Relating to Antiquity* 29 (1842): 397–399.
Snorri Sturluson. *Heimskringla or the Chronicle of the Kings of Norway*. Release #15b, 1996. Translated by Samuel Liang. Online Medieval and Classical Library, 1844. http://omacl.org/Heimskringla/.
———. *Heimskringla: History of the Kings of Norway*. Translated by Lee M. Hollander. Austin: University of Texas Press, 1964.
———. *Heimskringla*. Edited by Bjarni Aðalbjarnarson. Íslenzk fornrit, 26–28. Reykjavík: Hið íslenzka fornritafélag, 1941–1951.
Söderbäck, P. *Rågöborna*. Stockholm: P. A. Norstedt, 1940.
Sofaer, Joanna. "Technology and Craft." In *Organizing Bronze Age Societies: The Mediterranean, Central Europe, and Scandinavia Compared*, edited by Timothy Earle and Kristian Kristiansen, 185–217. Cambridge: Cambridge University Press, 2010.
Sofaer, Joanna, Lise Bender Jørgensen, and Alice M. Choyke. "Craft Production: Ceramics, Textiles, and Bone." Chap. 26 in *The Oxford Handbook of the European Bronze Age*, edited by Harry Fokkens and Anthony Harding, 469–491. Oxford: Oxford University Press, 2013.
Soon, Willie, and Sallie Baliunas. "Proxy Climate and Environmental Changes of the Past 1000 Years." *Climate Research* 23 (2003): 89–110.
Sørensen, Steinar. "Daterte skifunn fra middelalderen: Et omriss av middelalderens skihistorie." *Collegium Medievale* 1–2 (1996): 7–55.
Sperber, Hans. "Beiträge zur germanischen Wortkunde." *Wörter und Sachen* 6 (1914): 14–57.
Staatsbibliothek zu Berlin. *Projekt dur Erschliessung historisch wertvoller Altkartenbestände: Anleitung und Sonderregeln mit Beispielen für die Aufnahme*, 2001. http://ikar.sbb.spk-berlin.de/werkzeugkasten/sonderregeln/index.htm.
Stanc, Simina, and Luminita Bejenaru. *Bone and Antler Manufacturing in the IVth-VIth AD Centuries, on the Territory of Romania*. presented at the 6th Meeting of the International Council for Archaeozoology Worked Bone Research Group, August 2007.
Stemp, W. James, Adam S. Watson, and Adrian A. Evans. "Surface Analysis of Stone and Bone Tools." *Surface Topography: Metrology and Properties* 4 (2016): 013001.
Stenberger, Mårten. "The Finds and the Dating of the Vallhagar Settlement." In *Vallhagar: A Migration Period Settlement on Gotland, Sweden*, edited by Mårten Stenberger and Ole Klindt-Jensen, 2:1065–1160. Copenhagen: Einar Munksgaards Forlag, 1955.
Steve. "Bone Skates or, Vikings on Ice." Dark Ages Re-Creation Company. updated August 26, 2017. http://www.darkcompany.ca/projects/skates/index.php.
Stow, John. *A Survey of London: Contayning the Originall, Antiquity, Increase, Moderne Estate, and Description of that Citie: Written in the Yeare 1598*. London: John Wolfe, 1598.
Straten, Roelof van. *An Introduction to Iconography: Symbols, Allusions and Meaning in the Visual Arts*. Revised ed. Translated by Patricia de Man. Documenting the Image. Yverdon: Gordon and Breach Science Publishers, 1994.
Sutherland, Patricia D., and Robert McGhee. *Vergessene Welten unter Schnee und Eis: Die Vorläufer der Eskimos vor 4000 Jahren. Begleitheft zur Ausstellung: Vergessene Welten unter Schnee und Eis, Übersee-Museum, 29 November 1998 bis 14. März 1999*. Bremen: Bremen Übersee-Museum, 1998.
Sveinbjörn Egilsson and Finnur Jónsson. *Lexicon poeticum antiquae linguae septentrionalis*. 2nd ed. Copenhagen: Atlas Bogtryk, 1966.
Svenska Akademien. *Svenska Akademiens ordbok*. Lund: Svenska Akademien, 2017.
Sykes, Bryan. *Saxons, Vikings, and Celts: The Genetic Roots of Britain and Ireland*. New York: W. W. Norton, 2006.
Szamałek, Krzysztof. *Kruszwicki zespół osadniczy*

w młodszej epoce brązu i w początkach epoki żelaza. Wrocław: Zakład Narodowy imienia Ossolinskich, 1987.

Szepsi Csombor, Márton. *Europica Varietas*. Edited by Péter Kulcsár. Budapest: Szépirodalmi Könyvkiadó, 1979.

Tacitus. *Agricola, Germany*. Translated by A. R. Birley. Oxford World's Classics. Oxford: Oxford University Press, 1999.

Teichert, M. "Die Rinder aus dem Opfermoor Oberdorla." *Zeitschrift für Tierzüchtung und Züchtungsbiologie* 77 (1962): 74–86. doi:0.1111/j.1439–0388.1962.tb01233.x.

Tergast, Petrus. *Die heidnischen Alterthümer Ostfrieslands*. Emden: W. Haynel, 1879.

Thiele, Friedrich. "Deutscher und englischer Sprachgebrauch in gegenseitiger Erhellung." *The German Quarterly* 11, no. 1 (January 1938): 42–50.

Thomas, Alfred. *A Blessed Shore: England and Bohemia from Chaucer to Shakespeare*. Ithaca: Cornell University Press, 2007.

Thomas, John. "Evidence for the Dissolution of Thorney Abbey: Recent Excavations and Landscape Analysis at Thorney, Cambridgeshire." *Medieval Archaeology* 50, no. 1 (2006): 179–241. doi:10.1179/174581706x124257.

Thorsteinn Einarsson. "Winter Sport in Iceland." In *Winter Games Warm Traditions*, edited by Matti Goksøyr, Gerd von der Lippe, and Kristen Mo, 54–62. Oslo: The Norwegian Society of Sports History and The International Society for the History of Physical Education and Sport, 1994.

Thunig. "Über Schlittknochen und Gräbeurnen." In *Verhandlungen der Berliner Gesellschaft für Anthropologie, Ethnologie und Urgeschichte. Zeitschrift für Ethnologie* 4 (1872): 280.

Thurber, B. A. "A New Interpretation of Frithiof's Steel Shoes." *Scandinavica* 50, no. 2 (2011): 6–30.

———. "The Similarity of Bone Skates and Skis." *Viking and Medieval Scandinavia* 9 (2013): 199–217. doi:10.1484/J.VMS.1.103882.

Točík, Anton. "Knochen- und Geweihindustrie der Maďarovce-Kultur in der Südwestslovakei." *Studijné Zvesti Archeologicky Ustav Slovenskej Akadémie Vied* 3 (1959): 42–53.

Todd, Malcolm. *The Early Germans*. The Peoples of Europe Series. Malden, MA: Blackwell Publishers, 1992.

———. "Feddersen Wierde." In *The Oxford Companion to Archaeology*, edited by Brian Fagan, 236. Oxford: Oxford University Press, 1996.

———. "Germans and Germanic Invasions." In *The Oxford Companion to Archaeology*, edited by Brian Fagan, 250–251. Oxford: Oxford University Press, 1996.

Tolkien, J. R. R. *The Lord of the Rings*. 50th anniversary ed. Edited by Wayne G. Hammond and Christina Scull. Boston: Houghton Mifflin Harcourt, 2004.

Tóth, Ágnes B. "The Gepids." In *Hungarian Archaeology at the Turn of the Millennium*, edited by Zsolt Visy, 294–298. Budapest: Ministry of National Cultural Heritage, Teleki László Foundation, 2003.

Townend, Matthew. *Language and History in Viking Age England: Linguistic Relationships between Speakers of Old English and Old Norse*. Studies in the Early Middle Ages 6. Turnhout, Belgium: Brepols, 2002.

Treichel, A. "Vom Schlittknochen, sogenannten Hund und Bock." In *Verhandlungen der Berliner Gesellschaft für Anthropologie, Ethnologie und Urgeschichte. Zeitschrift für Ethnologie* 17 (1885): 397–398.

———. "Vorkommen von Schlittknochen und Rundmarken." In *Verhandlungen der Berliner Gesellschaft für Anthropologie, Ethnologie und Urgeschichte. Zeitschrift für Ethnologie* 19 (1887): 83–84.

Tschumi, Otto. "Die steinzeitlichen Epochen." In *Urgeschichte der Schewiz*, edited by Otto Tschumi, 1:567–727. Frauenfeld: Verlag Huber, 1949.

Vaday, Andrea. "Introduction." In *Cultural and Landscape Changes in South-East Hungary*, edited by Andrea H. Vaday, 2:9–14. Budapest: Archaeological Institute of the Hungarian Academy of Sciences, 1996.

Vaday, Andrea, and Katalin Berecz. "Roman Period Barbarian Settlement at the Site of Gyoma 133." In *Cultural and Landscape Changes in South-East Hungary*, edited by Andrea H. Vaday, 2:51–306. Budapest: Archaeological Institute of the Hungarian Academy of Sciences, 1996.

Vandervell, H. E., and T. Maxwell Witham. *A System of Figure-Skating: Being the Theory and Practice of the Art as Developed in England, with a Glance at Its Origin and History*. London: Horace Cox, 1869.

Vaughan, I., ed. *Strangeways' Veterinary Anatomy*. 4th ed. Edinburgh: Bell and Bradfute, 1892.

Vereeniging ter Beoefening der Geschiedenis van 's-Gravenhage. "'s Gravenhage onder de regering van de Huizen van Holland, Henegouwen en Beijeren." *Mededeelingenvan de Vereeniging ter Beoefening der Geschiedenis van 's-Gravenhage* 1 (1863): 207–342.

Vicze, Magdolna. "The Prehistoric Settlement at the Site of Gyoma 133." In *Cultural and Landscape Changes in South-East Hungary*, edited by Andrea H. Vaday, 2:15–26. Budapest: Archaeological Institute of the Hungarian Academy of Sciences, 1996.

Virchow, Rudolf. "Einige Ueberlebsel in pommerschen Gebräuchen." In *Verhandlungen der Berliner Gesellschaft für Anthropologie, Ethnologie und Urgeschichte. Zeitschrift für Ethnologie* 19 (1887): 361–362.

———. "Geglättete Knochen zum Gebrauche beim

Schlittschuhlaufen und Weben." In Verhandlungen der Berliner Gesellschaft für Anthropologie, Ethnologie und Urgeschichte. *Zeitschrift für Ethnologie* 3 (1870): 19–21.

———, ed. *General-Register zu Band I-XX (1869–1888) der Zeitschrift für Ethnologie und der Verhandlungen der Berliner Gesellschaft für Anthropologie, Ethnologie und Urgeschichte.* Berlin: A. Asher und Co., 1894.

Wang, Bo. "Research on Altay, the Original Place of Skiing." In *The Original Place of Skiing—Altai Prefecture of Xingjiang, China,* edited by Zhaojian Shan and Bo Wang, 227–248. Beijing Shi: Ren min ti yu chu ban she : Xinjiang ren min chu ban she, 2011.

Wawn, Andrew. *The Vikings and the Victorians: Inventing the Old North in Nineteenth-Century Britain.* Woodbridge: Boydell and Brewer, 2000.

Weinstock, John. *Skis and Skiing: From the Stone Age to the Birth of the Sport.* Lewiston, NY: Edwin Mellen Press, 2003.

Weiss, Guido. "Ueber den Eissport im sechzehnten Jahrhundert." In Verhandlungen der Berliner Gesellschaft für Anthropologie, Ethnologie und Urgeschichte, *Zeitschrift für Ethnologie* 3:60. 1871.

West, Barbara. "A Note on Bone Skates from London." *Transactions of the London and Middlesex Archaeological Society* 32 (1982): 303.

White, Sam. "The Real Little Ice Age." *Journal of Interdisciplinary History* 44, no. 3 (2014): 327–352. doi:10.1162/JINH_a_00574.

Wikipedia contributors. *Ice Skating.* Wikipedia, accessed September 14, 2017. https://en.wikipedia.org/wiki/Ice_skating.

Willemsen, Annemarieke. "Scattered across the Waterside: Viking Finds from the Netherlands." In *Vikings on the Rhine. Recent Research on Early Medieval Relations between the Rhinelands and Scandinavia,* edited by Rudolf Simek and Ulrike Engel, 65–82. Studia Medievalia Septentrionalia 11. Vienna: Verlag Fassbaender, 2004.

———*Vikings! Raids in the Rhine/Meuse Region 800–1000.* Utrecht: Centrall Museum, 2004.

Winroth, Anders. *The Age of the Vikings.* Princeton, NJ: Princeton University Press, 2014.

Zeydel, Edwin H. "More Oddities and Novelties for the German Literary Historian." *The German Quarterly* 8, no. 1 (1945): 26–31.

Zimmermann, Christiane, and Hauke Jöns. "Cultural Contacts between the Western Baltic, the North Sea Region and Scandinavia." In *Frisians and their North Sea Neighbours: From the Fifth Century to the Viking Age,* edited by John Hines and Nelleke IJssennagger, 243–271. Woodbridge: Boydell Press, 2017.

Zindel, Christian Siegmund. *Der Eislauf.* Nürnberg: Friedrich Campe, 1825.

Zoëga, Geir T. *A Concise Dictionary of Old Icelandic.* 1910. Reprint, Mineola, NY: Dover Publications, 2004.

Index

Aberdeen 110
Abingdon 110
Abu-Hamid 56
adolescents 4, 9, 36, 66, 69, 72–74, 79–83, 90, 105, 107, 110–111, 115, 131, 146
Afanasievo culture 59
Albertfalva 49–50, 52–53, 56, 59, 61–62, 83–84
Alfred-Guthrum boundary 108–109
Alps 64, 108
Alta skier 60
Altai City 56
Altai Mountains 4, 6, 51, 54, 56–61
Altorf 45
Amsterdam 119–122
Andersson, Petter 143
andrar 91–92
Angles 107
Anglo-Saxon Chronicle 107
Anglo-Saxons 107–108, 113
Anthropologischen Gesellschaft in Wien 36, 138
Arabic 26, 57, 66
Århus 86
Asia 57
ass 116
attachment *see* bindings
Atti in dœlski 90
Austria 11, 62, 63, 71, 101, 104, 119

Baltic languages 62
Baltic Sea 1, 104, 116
Bastarnae 63
Bede 107
Bedford 110, 165n57
beinspýtum 87–88, 97–98
Belgium 63, 71, 101, 104, 115; *see also* Brussels; Ghent; Low Countries; Vlessegem
Bell Beaker culture 62, 83
Bergen 86
Berlin 30, 36; *see also* Buch
bicycle 70
bindings 5, 6, 17–19, 20–24, 28–29, 43–44, 75–77, 81, 132, 135–140, 142–146, 167n80; *see also* holes
Birka 2, 9, 11, 17, 20, 24, 29, 34, 69, 74, 85, 86, 92, 96, 111, 115, 117, 122
Birmingham 112, 141
Bīrūnī 58
Biskupin 48
Black Sea 48, 53
boasting 4, 40, 86–90
boat 33, 67, 84, 93–97, 130
Bohemia 128
Bohuslän 67
Bølamannen 60
Bosch, Hieronymus 126
Bosnia 63, 71
bow 56, 89
breakage 35
Bremen 115–116
Brinker, Hans 7
British Isles 64; *see also* Great Britain
brodd- 91
Bronze Age 1, 4, 6, 9, 34, 35, 48, 49, 52–53, 56, 60, 62–63, 66, 69, 71, 75–76, 78, 82–84, 157n100
Brussels 127–128
Buch 49, 52
Bulgaria 101, 104
Bulghār 57–58
bullion 112–113
Burgundians 64

Calendaria Caroli 98
Canada 84
Carpathian Basin 52, 63; *see also* Csepel-Háros group; Great Hungarian Plain
cascade model 59
cattle 87; bones 1, 5, 6, 8, 11, 12, 13, 14, 15, 17, 18, 19, 23, 34, 43, 47, 52, 58, 62, 70–82, 88, 96, 105, 110–111, 114–115, 117, 119, 123–124, 130–132, 138–141, 143, 145
Celts 63
ceramics 48
chariot 55
children 2, 4, 6, 29, 35, 36, 61, 68–69, 71, 73–74, 80, 82, 83, 90, 117–118, 131–132, 134, 135, 142; *see also* adolescents; fun; games; toys

China 56
chopping 75–80, 119
Christmas 47, 137, 145
church 32, 67, 143–144
climate 54–56, 57, 62–64, 108–109, 112, 133–134; *see also* Little Ice Age; Medieval Warm Period; weather
clothing 112–113
Cnut 113
Cologne 72, 75, 79
"The Complaint of Harold" 40, 89
Coriolis force 103
crampons 67–68, 96, 98
Crowland 110
Csepel-Háros group 49, 62
Czech Republic 11, 53, 63, 71, 101, 104; *see also* Ivanovice na Hané; Mikulčice; Olomouc; Slánská Hora

Dalarna 86
Danelaw 108, 110, 113
danger 5, 20, 27, 28, 33, 120, 142–144
Danube Valley 62, 63
deer 11, 26, 48, 58, 71–73, 90, 92, 111, 116, 123
Denmark 63, 71; *see also* Århus; Ribe; Trelleborg; Viborg
DNA *see* genetic evidence
Dneiper basin 53
dogsleds 57–58
Dongjum 107
donkey 11, 72, 73, 78
Dordrecht 119–120, 122–123
Dorestad 115
Drevja 52, 57
Dublin 110, 112, 165n68
Dundebulake Valley 52 (fig. 29), 56
Dürer, Albrecht 71
Durham 110
Dutch 41–42, 121–122

edges 28
Eisbein 167n62
Ekeblad, Johan 129
Ely Cathedral 130
Empingham 110

183

Index

Endröd 170 52 (fig. 29), 53
England 66, 102, 127, 128, 132–133; *see also* Great Britain
Englum 107
Erik the Red 102
Estonia 101, 104, 119, 131, 133, 142
Europe 85–86, 100; Central 52–53, 127, 134, 138–141; Eastern 55
Europica Varietas 128
Evelyn, John 129–130
experiments 10–31, 44–45
Eyrbyggja saga 96, 98
Eysteinn, King 86–90, 97, 98–99
Ezinge 72–73, 78

Falun 25
features 16–20, 74–82, 96, 104–107, 114–116, 132; *see also* bindings; chopping; gliding surface; pointed toe; split bones; standing surface; upswept ends
Feddersen Wierde 72–73, 107
feet 68–69
femur 133
fens 66, 130, 141
fimbulvetr 64
Finland 51–52, 56, 67, 86, 101, 104, 119, 131, 142; *see also* Turku
fishing 66, 67, 80, 82
fitz Stephen, William 2, 5, 10, 20, 24, 26, 31, 33, 36, 38, 68, 93, 111
Flaxengate 110
Fljótsdæla saga 66, 87, 98–99
foot-pushing 8, 43, 124–126, 132
Fowler, George Herbert 1, 3, 10, 21, 38–39
France 63, 71, 73, 101, 104, 115; *see also* Altorf; Geispolsheim; Saint-Denis; Wiwersheim
Franks 64, 107, 157n101
French 122
friction 15, 29, 32, 35, 71, 123–124, 133
Friðþjófs saga ins frækna 38
Frisia 53
Frithiofs saga 38
Frostathing law code 111
fun 1–2, 5, 68, 70, 82, 111, 131, 138, 142

games 5, 7, 41, 111, 113, 144; *see also* fun
Gara Banca 72–73
geisl 91
Geispolsheim 80–81
genetic evidence 107, 112–113
Gepids 63–64
Germanic 1, 62, 104; *see also* Bastarnae; Gepids; Goths
Germany 51, 62, 63, 71, 101, 103, 104, 114–116, 119, 135–138; *see also* Berlin; Bremen; Buch; Cologne; Feddersen Wierde; Grimmersum; Groß Strömkendorf; Höngeda; Kliestow; Menzlin; Munich; Oberdorla; Schliersee; Starigard; Westphalia
Ghent 124
gliding surface 16, 19, 35, 74–75, 81, 88, 136–137, 139–140, 142, 144
goats 48, 71
Goths 63–64, 103
Gotland 23, 32, 67, 86, 143–144
Grammaticus, Saxo 31, 33, 95, 97
Great Britain 9, 100–114, 117, 119; *see also* Aberdeen; Abingdon; Bedford; Birmingham; Crowland; Durham; Ely Cathedral; Empingham; England; Flaxengate; Huntingdon; Ipswich; Lincoln; London; Mucking; Norfolk; Northampton; Norwich; Oxford; Reigate; Thetford; Thorney; Tickhill; Torksey; Wallingford; Waltham Abbey; Wellingford; Whissendine; Winchester; York
Great Hungarian Plain 63
Greenland 84, 100, 102–103, 127; *see also* Grœnlendinga saga
Grettis saga 94
Grimmersum 163n13
Grœndlinga saga 102–103
Groß Strömkendorf 116
Guagnini, Allesandro 43
Gylfaginning 40–41, 64–65, 88, 90, 91
Gyoma 133 11, 12, 16, 52, 53, 71–73, 74, 81–82, 160n77

the Hague 41, 122, 124
Hallstatt period 154n82
Hälsingland 67
Hamadani, Rashīd al-Dīn 57
Hanseatic League 115
Haraldr harðráði 89, 161n31
Haralds saga Gráfeldar 94
Hávamál see Poetic Edda
head-and-hoof burials 71
Heimskringla 86; *see also* Haralds saga Gráfeldar; Magnússona saga; Óláfs saga helga
Herodotus 53
Hexham, Henry 130
Hippophagy *see* horses, eating
Hlaupa 97–98
hockey 7
holes 47, 49, 77–80, 112, 117, 119, 136–141; *see also* bindings
Höngeda 49, 52
horses 11, 52–55, 72, 103; bones 1, 3, 5, 6, 12, 13, 14, 15, 16, 19, 23, 36, 47, 56–62, 70–82, 88, 105, 110–111, 114–117, 124, 132, 136–145; eating 11, 53, 55, 70–71
Howell, Frederick W. W. 145
Hungary 20, 48, 53, 62, 63, 67, 71, 119, 128, 131, 140–141; *see also* Albertfalva; Csepel-Háros; Endröd 170; Gyoma 133; Lébény-Kaszás domb; Százha-lombatta-Földvár; Törökbálint; Zamárdi
hunting 1, 42, 53, 56, 65, 66–67, 84–86, 90–92, 97–99
Huntingdon 110

ice princesses 142
Iceland 30, 85–99, 103, 119, 131–132; *see also* Káldárhöfði farm
iconography 42
Indo-Iranian 62
Inuit 84
íþrott 90
Ipswich 34, 110
Ireland 101, 104; *see also* Dublin
Irkutsk 58
Iron Age 62–63, 157n101
ísleggr 87–88, 98, 129–131
Isū 57–58
Italic 62
Italo-Celtic 63
Italy 53, 63, 71; *see also* Pozzuolo del Friuli hill-fort
Ivanovice na Hané 52 (fig. 29), 60–61, 79

Jankowie 48, 154n82
jaw 7
jewelry *see* metalwork
Jones, Robert 2, 37–38
Jordanes 63–64, 86, 103
Jutes 107
Juvel, Antii 143
juvenis 111

Kaldárhöfði farm 145–146
Kalevala 92
Kali Kolsson 89–90
Karelia district 60
Kazakhstan 54–55, 57, 62
kennings 94–95
Khan, Genghis 57
Khvalynsk 71
Kliestow 64, 72
Klopstock, Gottfried 41
Knoviz 157n100, 158n41
Konungs skuggsjá 90, 98

Łagniewnikach 48, 154n83
Lake Baikal *see* Irkutsk
Lake Mälaren 133
Lake Onega 52 (fig. 29)
Lake Sindor 56
Latvia 101, 104, 119, 133; *see also* Riga
leatherworking 46, 48–50
Lébény-Kaszás domb 75–77, 78
Leeuwarden 107, 157n101, 159n41, 163n13
Leo of Rožmitál and Blatná 127
Letchworth Village 69
Ligurian 62
Lincoln 110
linguistic evidence 54, 62, 101, 112–113, 122
Lithuania 101, 104
Little Ice Age 127
Lödöse 11, 12, 13, 18, 86
London 2, 5, 36, 110–113, 163n16

Low Countries 132; *see also* Belgium; Netherlands
Lund 86
Lusatian culture 154*n*80–81
Lusitanian 62

MacGregor, Arthur 2, 8, 9, 10, 23, 24, 29, 34, 44, 46, 47, 48, 51, 104, 108–110, 117, 132
Magnus, Olaus 10, 14, 15, 26, 27, 31, 33, 36, 42, 66, 67–68, 90, 95, 97, 123–124, 127, 129
Magnús Lagabœtir 91
Magnússona saga 86–90, 97, 98
mandible-based tools 62
mandibular sled 7–8
manufacturing quality continuum 69–70, 73–74, 77–80, 82, 83
Marcusson, Christina 143
Marvazī, Sharaf al-Zamān Ṭāhir 26, 58
mathematical model 32–33
Maximilian's Triumphal Arch 71
Medieval Warm Period 101–103
Menzlin 116
metalwork 100, 107, 112–114, 116; *see also* bullion
metapodium 6, 11, 12–14, 15, 16, 17–18, 19, 20, 23, 34, 35, 36, 44, 45, 46, 47, 53, 55, 58, 59, 60, 62, 70–82, 88, 96, 105, 110–111, 114–117, 119, 124, 132, 136–137, 139–142, 145
metrology 133
Middendorff, Alexander von 58–59, 61, 142
Middle Ages 1, 4, 8–9, 31, 34, 54, 57, 65, 66, 85–118, 131, 163*n*1
migration 1, 59, 62, 71, 75, 157*n*101
Migration Period 63–64, 66, 69, 71–75, 78, 82, 85, 107–108, 114
Mikulčice 116
mistranslation 39–43, 94–95
modifications *see* features
Morkinskinna 40, 89, 91
Mucking 108
Munich 138

Negri, Francesco 57
Neolithic 1, 56, 71
Netherlands 51, 63, 71, 101, 104, 115, 123, 125–128, 130–131, 133; *see also* Amsterdam; Dongjum; Dordrecht; Dorestad; Englum; Ezinge; the Hague; Leeuwarden; Low Countries; Oost-Souburg; Rotterdam; Wierengen; Wijnaldum-Tjitsma
Nitriansky Hrádock 48
Njáls saga 93–94
nomads 1, 62, 71
Nordin, Fredrik 143–144
Norfolk 112, 141
Norman Conquest 113, 130
North America 3, 7, 85, 100, 101; *see also* Canada; Greenland

North Atlantic Oscillation 103
North Pole 37, 57
North Sea 114
Northampton 110
Norway 6, 17, 51, 64, 67; *see also* Bergen; Drevja; Oslo; Østerdal; Rødøy; Trondheim
Norwich 110
Novgorod 164*n*24
Novokievka 52 (fig. 29)

Oberdorla 75, 79, 159*n*69
Odin 89
Óláfs saga helga 90–91
Öland 86
Olbia 44
Old English 101, 113
Old European 62
Old Norse 3, 7, 36, 38, 39–43, 85–99, 100, 113
Olomouc 139–140
Omsk 52, 56–57
Oost-Souburg 115–116
Orkneyinga saga 40, 89–90
Orphan Asylum of Brooklyn 69
Oslo 85–86
Østerdal 91
Oxford 110
Ozero Sindorskoye 52

Palaeolithic 56
pattens 120–122
Paulsen, Axel 40, 152*n*25
Pécel 48
Pepys, Samuel 129–130
Percy, Thomas 7, 36, 39–41
Petersson, Matilda 143
Petersson, Nils Johan 143
Pettersson, Atle 143
Poetic Edda 85, 97, 99
pointed toe 75–76, 78–79, 112, 117
points 24
Poland 63, 71, 101, 104, 119, 135–138; *see also* Biskupin
pole 5, 11, 24–25, 26–30, 32, 57, 90–93, 98, 122, 124–125, 128, 136–146
Þórðar saga hreðu 91
Pozzuolo del Friuli hillfort 72–73
practical use 61, 65, 71, 74, 79–81, 83; *see also* hunting; tools
Přemýšlení 157*n*100
Procopius 86, 103
projectile points 56
Prose Edda 85, 95; *see also* Gylfaginning; Skáldskaparmál
Proto-Indo-European 1, 3, 55, 62

races 67–68, 143
radius 11, 12, 46, 47, 58, 59, 70–75, 79–80, 105, 111, 115–117, 145–146; mysterious artifacts 48–50, 52–53, 61, 62, 83–84, 116, 133
Ragnarök 64–65
recreation 61, 65, 79–80, 83, 123; *see also* toys

Reigate 110, 111
Reykdæla saga 93
rib 7, 11
Ribe 86
Riga 114
rock art 52, 56–57, 59–60, 86
Rødøy 52, 60, 86
Roman period 53, 71, 79, 110, 120
Romania 16, 18, 21, 63, 71, 119, 140–141; *see also* Gara Banca; Sîntana de Mureş/Černjahov culture
Romans 63–64, 108
Roric 115
Rotterdam 124
runes 95–96, 139–140
runners 1, 3, 7–8, 18, 31, 37, 39, 44, 47, 49, 52, 84, 122, 126, 129, 140–142, 145
Russia 30, 55, 57, 63, 101, 104, 119; *see also* Irkutsk; Karelia district; Lake Baikal; Lake Onega; Lake Sindor; Novgorod; Omsk; Ozero Sindorskoye; Siberia; Sintashta; Vis I; Zalavrouga

Sabatinovka 47, 52, 62, 71, 157*n*100
Sabatinovka culture 47, 52–53, 55; *see also* Novokievka; Sabatinovka; Zlatopol'
sagas 85–99, 160*n*2; *see also* Eyrbyggja saga; Fljótsdæla saga; Grettis saga; Grænlendinga saga; Heimskringla; Njáls saga; Þórðar saga hreðu; Reykdæla saga; Valla-Ljóts saga
Sahlstedt; Abraham 129
Saint-Denis 115
St. Lydwina 125–126
St. Pierre of Blandigny 124
Sarkel 44
Sarmatians 11, 53, 63–64
Šašek, Václav 127–129
Saxons 64, 107–108
Scandinavia 1, 3, 4, 6, 8–9, 26, 38, 54, 56, 57, 66, 69, 85–118, 142–146; expansion 100–101, 114–118, 133–134; *see also* Denmark; Iceland; Norway; Sweden
schaats 122
schenkel 122
Schliersee 138
Schlittschuh 167*n*62
Sea of Azov 47, 57
Semenov, S.A. 44–45, 47
Sephton, John 38
Serbia 101, 104
Serooskerke 36
sheep 8, 48, 66, 71, 87, 137–138, 141
ship *see* boat
shoes 7, 12, 17, 20, 21, 22
Siberia 58–59, 142; *see also* Yakut
Sigtuna 2, 9, 11, 18–19, 29, 69, 72, 74, 82, 85, 86, 88, 92, 110–111, 115, 117, 122

Sigurðr 87–90
Sîntana de Mureş/Černjahov culture 64
Sintashta 52, 55
Skaði 90–92, 97
Skáldskaparmál 91
skate-sailing 30–31, 58, 136–138
skateboarding 30, 58, 92, 98
skates, wood 7, 128–129; bone with wood 8, 122, 137; metal-bladed 1–3, 8, 38–39, 119–131, 143
skating, speed 40; incomparable 139–140; public 119
skíð 40, 89, 91
skiing 51, 54, 57, 89
skis 1, 3–4, 6, 8, 39, 40, 54, 56–61, 62, 84, 85–99, 134, 142; in bogs 57, 86; *see also andar*; *skíð*; skiing
Skrið-Finnar 86
skríða 41, 86, 93–99, 129
skridskor 129–130
Slánská Hora 133
Slavic languages 62, 103
Slavs 100, 103–104, 116–117
Slovakia 49, 52, 53, 63, 71, 101, 104; *see also* Nitriansky Hrádock; Veselé; Vrablé
Słupcy 48, 154*n*81
Småland 67, 86
Smith, Charles Roach 36
Smuszewie 48, 154*n*80
Snorri Sturluson 40–41, 85, 88, 93; *see also Heimskringla*; *Prose Edda*
snowshoes 39, 40, 59, 62, 89, 97, 162*n*114
Spain 103
speed 32–33
spinning 28, 68, 151*n*116
spita 88
split bones 88, 106, 114, 117, 145
standing surface 75–76, 79, 81, 88, 141–142, 144
Staraya Ladoga 44
Starigard 163*n*14
steppes 1, 3, 6, 51, 53, 54–56, 62, 63, 133–134; Pontic-Caspian 59; *see also* Khalvynsk
Stockholm 25
stopping 7–8, 28, 124, 130, 142
Stow, John 38

Sweden 2, 6, 34, 51, 63, 66–68, 71, 74, 81, 98, 119, 127, 128, 131; *see also* Birka; Bohuslän; Dalarna; Gotland; Hälsingland; Lake Mälaren; Lödöse; Lund; Öland; Sigtuna; Småland; Swedish; Tjörn; Uppland; Uppsala; Vallhagar; Västerbotten
Swedish 129
swimming 87, 89, 96, 98
Switzerland 49, 52
Százhalombatta-Földvár 49, 56, 83
Szepsi Csombor, Márton 128

Tacitus 63
Tegnér, Esaias *see Frithiofs saga*
textile smoothing 44–46, 47, 48, 83, 135
Thetford 110–113
Thorney 110
Thrymheim 90
Thule 84
tibia 5, 20, 48, 58, 64, 72, 110, 111, 116, 123
Tickhill 141
Tjörn 67
tools 66–70, 85; *see also* ceramics; leatherworking; mandible-based tools; textile smoothing
Torksey 110
Törökbálint 35, 52, 60, 75, 77, 78
toys 66–70, 71, 74, 77–78, 82, 117; *see also* fun; games; recreation
Transylvania 128–129, 139–140
travel 66–67, 82, 83–84, 85, 99
Trelleborg 86
tricks *see* spinning
Trondheim 47, 86
Tungus 59
Turku 114
turning 7–8, 28, 91–92, 124, 128, 142, 144–145

Ukraine 53, 55, 63, 71; *see also* Novokievka; Sabatinovka; Zlatopol'
Ullr 88, 91, 95–97, 99
ulna 49
Uppland 67, 86
Uppsala 86

upswept ends 19, 60, 75, 78–79, 81, 117
Ural Mountains 55, 56; *see also* Yugra
Ural River 59
Urnfield culture 60
use wear 6–8, 15–16, 24, 33–35, 44–50, 69–70, 79, 83, 109, 133, 146, 159*n*62

Valla-Ljóts saga 91, 93
Vallhagar 75, 80
Västerbotten 52, 57
Verhandlungen der Berliner Gesellschaft für Anthropologie, Ethnologie und Urgeschichte 10, 36, 131, 135–138, 144–145
Veselé 48
Viborg 86, 96
Vienna 36
Vikings 2, 6, 10, 37, 100, 108, 115
Vis I 56–57, 59–60
Vlissegem 107
Vrablé 52

Wallingford 110
Waltham Abbey 110
wear *see* use wear
weather 58
Wellingford 72
Westphalia 130
Whissendine 110
Wieringen 114
Wijnaldum-Tjitsma 107
William of Rubruck 31, 58, 61, 67
Winchester 110
Wiwersheim 72
wolf 72, 75, 79–81
women 32, 34, 53, 67, 128, 143, 144, 151*n*142
wooden horses 56

Yakut 60
Yamnaya culture 55, 59, 62
York 9, 24, 101, 109, 111–113
Yugra 57

Zalavrouga 60
Zamárdi 76–77
Zlatopol' 47, 52, 60

www.ingramcontent.com/pod-product-compliance
Ingram Content Group UK Ltd.
Pitfield, Milton Keynes, MK11 3LW, UK
UKHW050524150426
5217IPUK00026B/1780